Fred Espen Benth • Dan Crisan
Paolo Guasoni • Konstantinos Manolarakis
Johannes Muhle-Karbe • Colm Nee
Philip Protter

Paris-Princeton Lectures on Mathematical Finance 2013

Editors:
Vicky Henderson
Ronnie Sircar

Springer

Fred Espen Benth
Department of Mathematics
University of Oslo
Oslo, Norway

Dan Crisan
Department of Mathematics
Imperial College London
London, United Kingdom

Paolo Guasoni
School of Mathematical Sciences
Dublin City University
Dublin, Ireland

Konstantinos Manolarakis
Department of Mathematics
Imperial College London
London, United Kingdom

Johannes Muhle-Karbe
Department of Mathematics
ETH Zurich
Zurich, Switzerland

Colm Nee
Department of Mathematics
Imperial College London
London, United Kingdom

Philip Protter
Department of Statistics
Columbia University
New York, NY, USA

[The addresses of the volume editors appear on page ix]

ISBN 978-3-319-00412-9　　ISBN 978-3-319-00413-6 (eBook)
DOI 10.1007/978-3-319-00413-6
Springer Cham Heidelberg New York Dordrecht London

Lecture Notes in Mathematics ISSN print edition: 0075-8434
ISSN electronic edition: 1617-9692

Library of Congress Control Number: 2013941989

Mathematics Subject Classification (2010): 91B28, 91B70, 60G49, 49J55, 60H07, 90C46

Printed on acid-free paper

Springer is part of Springer Science+Business Media (www.springer.com)

Lecture Notes in Mathematics 2081

For further volumes:
http://www.springer.com/series/304

Preface

It has been our pleasure to have been invited by René Carmona and Nizar Touzi to edit this 5th volume of the Paris-Princeton Lectures on Mathematical Finance. The present volume contains four chapters touching on some of the most important and modern areas of research in mathematical finance: asset price bubbles; energy markets; investment under transaction costs; and numerical methods for solving stochastic equations.

In the first chapter, Philip Protter presents a comprehensive survey of the Mathematical Theory of Financial Bubbles. Understanding, defining, and detecting a bubble from observed prices is a long-standing challenge spanning ideas from economics, stochastic analysis, mathematical finance, and statistics. This chapter presents a broad review of the history and literature of the problem as well as the mathematics and an empirical analysis of some recent data.

The second chapter, by Fred Espen Benth, concerns analysis and models for energy markets, particularly electricity (or power) prices. Here, short-term spikes are of paramount importance and the author describes using Lévy processes to try and capture this. In particular, he focuses on introducing stochastic volatility effects and on valuation of energy derivatives such as forwards and cross-commodity spread options. This covers part of a growing area as energy markets become more financialized and a greater part of the financial economy.

In the third chapter, by Paolo Guasoni and Johannes Muhle-Karbe, the problem of investment choice in the presence of transaction costs is surveyed in the form of a User's Guide. The problem has a long history and brings interesting problems in singular stochastic control when analyzed in the typical framework of continuous time models. Recently there have been some important breakthroughs, involving the so-called shadow prices, that have led to a burst of activity in understanding how to optimally invest under this friction. The authors bring us up to date on this progress and list some open problems.

The final chapter, by Crisan, Manolarakis and Nee, discusses numerical methods for solving stochastic differential equations (SDEs) based on cubature. The theory and analysis, including tools from Malliavian calculus, are introduced from scratch, and some financial applications are studied. The fourth section also develops cubature methods for backward SDEs, including numerical examples demonstrating the impressive performance.

We thank the authors for their outstanding contributions, as well as those we enlisted as anonymous referees for their hard work and valuable suggestions that improved the chapters.

Oxford, UK Vicky Henderson
Princeton, NJ Ronnie Sircar

Contents

Editors

Vicky Henderson
Oxford-Man Institute of Quantitative Finance
Oxford University
Walton Well Road
OX2 6ED Oxford, UK
Vicky.Henderson@oxford-man.ox.ac.uk

Ronnie Sircar
ORFE Department
Princeton University
Charlton Street
Princeton, NJ 08544, USA
sircar@princeton.edu

A Mathematical Theory of Financial Bubbles

Philip Protter

Recurrent speculative insanity and the associated financial deprivation and larger devastation are, I am persuaded, inherent in the system. Perhaps it is better that this be recognized and accepted.
–John Kenneth Galbraith, *A Short History of Financial Euphoria*, Forward to the 1993 Edition, p. viii.

Abstract Over the last 10 years or so a mathematical theory of bubbles has emerged, in the spirit of a martingale theory based on an absence of arbitrage, as opposed to an equilibrium theory. This paper attempts to explain the major developments of the theory as it currently stands, including equities, options, forwards and futures, and foreign exchange. It also presents the recent development of a theory of bubble detection. Critiques of the theory are presented, and a defense is offered. Alternative theories, especially for bubble detection, are sketched.

1 Introduction

The economic phenomenon that the popular media refers to as a financial bubble has been with us for a long time. A short adumbration of some economy wide bubbles would include the following major events (see [55] for a comprehensive history of bubbles through the ages):

Supported in part by NSF grant DMS-0906995.
The author wishes to thank the hospitality of the Courant Institute of Mathematical Sciences, of NYU, as well as INRIA at Sophia Antipolis, for their hospitality during the writing of this paper.

P. Protter (✉)
Statistics Department, Columbia University, New York, NY 10027, USA
e-mail: pep2117@columbia.edu

F.E. Benth et al., *Paris-Princeton Lectures on Mathematical Finance 2013*,
Lecture Notes in Mathematics 2081, DOI 10.1007/978-3-319-00413-6_1,

- The bubble known as Tulipmania which occurred in Amsterdam in the seventeenth century (circa 1630s) is the first documented bubble of the modern era. Some merchants had excessive wealth due to Holland's role in shipping and world commerce, and as tulips became a fad, some rare and complicated bulbs obtained through hybrid techniques led to massive speculation in the prices of bulbs. One bulb in particular came to be worth the price of two buggies with horses, the then equivalent of two automobiles. As often happens with economy wide bubbles, when the bubble burst the economy of Holland went into a tailspin.
- In the eighteenth century, John Law advised the Banque Royale (Paris, 1716–1720) to finance the crown's war debts by selling off notes giving rights to the gold yet to be discovered in the Louisiana territories. When no gold was found, the bubble collapsed, leading to an economic catastrophe, and helped to create the French distrust of banks which lasted almost 100 years.
- Not to be outdone by the French, the South Sea Company of London (1711–1720) sold the rights to the gold pillaged from the Inca and Aztec civilizations in South America, neglecting the detail that the Spanish controlled such trade and had command of the high seas at the time. As this was realized by the British public, the bubble collapsed.
- The real king of bubbles, however, is the United States. A list of nineteenth, twentieth and now even twenty-first century bubbles would include the following, detailing only the crashes:

  - The 1816 crash due to real estate speculation.
  - With the construction of the spectacular Erie Canal connecting New York to Chicago through inland waterways, "irrational exuberance" (in the words of Alan Greenspan) led to the Crash of 1837.
  - Not having the learned its lesson in 1837, irrational exuberance due to the construction of the railroad system within the U.S. led to The Panic of 1873.
  - The Wall Street panic of October, 1907, where the market fell by 50 %, helped to solidify the fame of J.P. Morgan, who (as legend has it) stepped into the fray[1] and ended the panic by announcing he would buy everything. It also had some good effects, as its aftermath created the atmosphere that led to the creation and development of the Federal Reserve in 1913, via the Glass–Owen bill.[2]
  - And of course the mother of all bubbles began with Florida land speculation as people would buy swamp land that was touted as beautiful waterfront

[1]More precisely, J.P. Morgan's role was to organize and pressure a group of important bankers to themselves add liquidity to the system and help to stem the panic. Ron Chernow describes the scene dramatically, as a crucible in which every minute counted [28, pp. 124–125] as the 70 year old J.P. Morgan's prestige and personality prevailed to save the day.

[2]Even in 1907, in his December 30 speech in Boston, President Taft pointed out that an impediment to resolving the crisis was the government's inability to increase rapidly and temporarily the money supply; one can infer from his remarks that he was already thinking along the lines of creating a Federal Reserve system [149].

property; this then segued into massive stock market speculation, ending with The Great Crash of 1929.

- There was no runaway speculation in the US markets, nor major panics, in the 1940s and 1950s. But it began again with minor stock market crashes in the 1960s and 1980s.
- The marvel of "junk bond financing" led to the fame of Michael Milken, the movie *Wall Street* and the stock market crash of 1987.
- While it did not occur in the U.S., we need to mention the Japanese housing bubble, circa 1970–1989, which upon bursting led to Japan's "lost decade," one of a stagnant economy and "zombie" banks.
- Back to the U.S. next, where speculation due to the commercial promise of the internet led to the "dot com" crash, from March 11th, 2000 to October 9th, 2002. Many of the internet dot-coms were listed on the Nasdaq Composite index, and it lost 78 % of its value as it fell from 5,046.86 to 1,114.11; a truly dramatic crash.
- Finally, we are all familiar with the recent US housing bubble tied to subprime mortgages, and the creation of many three letter acronym financial products, such as ABS, CDO, CDS, and even CDO^2. It is worth noting that the crash of 2007/2008, along with the one of 1929, escaped the economic borders of North America and thrust much of the world into economic depression.

It is of intrinsic interest to investigate the causes of financial bubbles, and there is a wealth of often insightful economic literature on the subject. This is not the purpose of this paper, which is rather to analyze prices and to try to determine if or if not a bubble is occurring, regardless of how it came about. For those with an understandable interest in the causes of bubbles, the author can recommend the little book of J.K. Galbraith [55], where Galbraith makes the case that speculation on a grand scale occurs when there is a new, or perceived as new, technological breakthrough (such as trade with the new world, the building of canals, the advent of railroads, junk bond financing, the internet, etc.) and that this can result in over enthusiasm and uncontrolled speculation. The more modern analysis of economists suggest that varied opinions among investors and short sales constraints can create financial bubbles (see for example [26, 41, 118, 138], just as a sampling). And recently, the interesting paper of Hong, Scheinkman, and Xiong [67] agrees with the conclusions of Galbraith, but takes the analysis further, beyond an explanation of simple overreaction on the part of investors to news. Hong et al. focus on the relations between investors and their advisers, the latter being classified into two types, "tech savvies" and "old fogies." They discuss how reputation incentives create an upward bias among the recommendations of the tech savvy investors, which are taken at face value by those investors who are naïve. For an interpretation of how the recent housing bubble arose, one can consult [129]. Other interesting references are [19,49,139,153,154,158].

To mathematically model a bubble, we start small, and consider an individual stock, rather than a sector (such as the technology sector), or an entire economy. If there is a bubble in the price of the stock, then the price is too high, relative to what one should pay for the stock. This seems intuitively obvious. But what is not obvious is: What then is the correct price of the stock? We assume such a stock is traded on an established exchange, and the theory of rational markets tells us that the price of the stock reflects exactly what the stock is worth, since if it were overvalued, people would sell it, and if it were undervalued, people would buy it. Such a theory eliminates the possibility of bubbles, and if we believe bubbles do in fact occur, we are forced not to accept this idea wholesale. Therefore we need a fair value for the stock.

This raises the question: Why does one buy a stock in the first place? Your brother-in-law might have a start-up and want you to participate by buying some stock in his company. This may be a bad investment, but good for your marriage. We will simplify life by assuming one buys stocks based only on their perceived investment potential. Moreover we will further simplify by assuming when one buys a stock, one is not speculating, and tries to pay a fair price for a long term investment, to the point where whether or not the stock goes up or down in the short run is irrelevant, and the only issue that matters is the future cash flow of the company. Nevertheless there is more of a risk in buying stocks than there is in banking money (especially when the deposits are insured by the government), so one can expect a rate of return with stocks that is higher than that of bank deposits, at least in the long run.[3] This return premium for taking an extra risk to buy stocks is known as "the market price of risk."

So the compelling question we must first answer is: How do we determine what we call a *fundamental price* for a stock?

Organization

After the introduction, we first explain in Sect. 2 how to model the *fundamental price* of a risky asset. Since the fundamental price is expressed as a conditional expectation of future cash flows, with the conditional expectation being taken under the risk neutral measure, it is more easily explained in a complete market, since then the risk neutral measure is unique. We can then define a bubble as the difference between the market price of the risky asset in question, and the fundamental price. When the risky asset is simply a stock price, then the bubbles are always nonnegative. In Sect. 3 we establish the relationship between strict local martingales[4] and bubbles, and give a theorem classifying bubbles into three types. In Sect. 4 we give examples of mathematical models of financial bubbles by

[3]Classic economic theory tells us that it makes no difference *in the short run* whether or not a company pays out dividends or reinvests its returns in the company in order to grow, in terms of wealth produced for the stockholders. However eventually investors are going to want a cash flow, as even Apple has recently discovered [160], and dividends will be issued.

[4]A strict local martingale is a local martingale which is not a martingale. More precisely, a process M with $M_0 = 1$, is a local martingale if there exists a sequence of stopping times $(\tau_n)_{n\geq 1}$ increasing to ∞ a.s. such that for each n one has that the process $(M_{t\wedge\tau_n})_{t\geq 0}$ is a martingale.

exhibiting a method of generating strict local martingales as solutions of a certain kind of stochastic differential equation. This is based on a theorem of Mijatovic and Urysov [116], and we provide a detailed proof of the theorem (Corollary 5 of Sect. 4 in this paper). Special attention is given to the inverse Bessel process. We also present results on strict local martingales in Heston type models with stochastic volatility that go beyond the framework of Corollary 5, and we discuss the multidimensional case. We end by giving a criterion to determine whether or not the system is a strict local martingale through the use of Hellinger Processes.

In Sects. 5 and 6 we consider incomplete markets arising from a risky asset price process $S = (S_t)_{t \geq 0}$. Since incomplete markets have an infinite number of risk neutral measures, and since the fundamental price is defined using "the" risk neutral measure within the framework of complete markets, this is a bit of a thorny issue. Hence we review the method of letting the market choose the risk neutral measure originally proposed in [73] (see also [141,142]), which works essentially by artificially completing the market through the use of call option prices. Once the risk neutral measure is chosen and temporarily fixed, the analysis proceeds analogously to the complete market case, with one important exception. The exception is that we allow the market choice of the risk neutral measure to change at random times, in a type of regime shift. This basically assumes the market is fickle, and while it always prices options in internally consistent ways (since otherwise there would be arbitrage), it can change this pricing from time to time, which actually represents a change in the selection of the risk neutral measure, from the infinite number of them compatible with the underlying risky asset price S. This method keeps the coefficients of the underlying stochastic differential equation unchanged, but we could equally and instead introduce a regime change where we change the underlying SDEs; this too may alter the structure of risk neutral measures, or it may not, depending on how dramatic is the change.

In Sect. 7 we consider what happens with calls and puts in the presence of bubbles. There are some surprising results, such as the loss of put-call parity (!) when bubbles are present, and that Merton's "No Early Exercise" theorem for American calls no longer need hold, a fact first observed (to our knowledge) by Heston et al. [63] and by Cox and Hobson [30]. An analysis of the behavior of options in the presence of bubbles can be found in [122]. We then introduce the concept, originally due to Merton in 1973 [114] but refined mathematically successively in [88, 89], and finally in [131], and known as *No Dominance*. This extra assumption restores put call parity. Section 8 is devoted to a study of bubbles in foreign exchange, which is related to inflation. Here negative bubbles can occur, and Sect. 9 covers forwards and futures. Section 10 covers the controversial topic of trying to identify (in real time) when a given risky asset (such as a stock) is undergoing bubble pricing. This seems to be a question of great current interest, as the quotes given in this paragraph seem to indicate. Indeed, the quotations are from none other than Ben Bernanke (Chairman of the U.S. Federal Reserve system), William Dudley (President of the New York Federal Reserve), Charles Evans (President of the Chicago Federal Reserve), and Donald Kohn, Federal Reserve Board Vice Chairman.

Finally, in Sect. 11 we attempt to defend the local martingale approach to the study of bubbles from its critics. These criticisms seem to revolve around the use of strict local martingales, and the (technically mistaken) belief that they exist only in continuous time. Jacod and Shiryaev, in a 1998 paper [75], clarify the relationship between local martingales and *generalized martingales* in discrete time, and give necessary and sufficient conditions for a local martingale to be a martingale in discrete time. It is true that when a finite horizon price process in discrete time is nonnegative (such as a stock price) then as a consequence of the results of Jacod and Shiryaev, a nonnegative discrete time local martingale is indeed a martingale. So in this sense, when modeling stock prices (as we often are doing in this paper), it is indeed true that strict local martingale models do not exist for discrete time. But we argue in Sect. 11, as we have in [86], that this is just another reason of several that discrete time models are in fact inadequate to understand the full range of ideas required for a profound understanding of financial models.

We also discuss in Sect. 11 two of the leading alternative approaches to the study of bubbles, the first associated with P.C.B. Phillips and his co-authors, and the second associated with Didier Sornette and his co-authors. The key difference between these alternative approaches (of Phillips et al. and Sornette et al.) with the one presented here, is that both alternative approaches make assumptions (albeit very different ones) on the drift that leads to bubbles (under their understanding of what constitutes a bubble), whereas in our presentation the key assumptions related to bubbles revolve around the diffusive part of the model.

2 The Fundamental Price in a Complete Market

We begin with a complete probability space $(\Omega, \mathcal{F}, P)$ and a filtration $\mathbb{F} = (\mathcal{F}_t)_{t\geq 0}$ satisfying the "usual hypotheses."[5] We let $r = (r_t)_{t\geq 0}$ be at least progressively measurable, and it denotes the instantaneous default-free spot interest rate, and

$$B_t = \exp\left(\int_0^t r_u du\right) \tag{1}$$

is then the time t value of a money market account. We work on a time interval $[0, T^\star]$ where $T^\star$ can be a finite fixed time T, or it can be ∞. We find that it is more interesting to consider a compact time interval (the *finite horizon* case, where $T^\star = T < \infty$), but for now let us consider the general case. Next we let τ be the lifetime of the risky asset (or stock, to be specific), where τ is a stopping time,

[5]The "usual hypotheses" are defined in [128]. For convenience, what they are is that on the underlying space $(\Omega, \mathcal{F}, P)$ with filtration $\mathbb{F} = (\mathcal{F}_t)_{t\geq 0}$, the filtration $\mathbb{F}$ is right continuous in the sense that $\mathcal{F}_t = \bigcap_{u>t}\mathcal{F}_u$, and also $\mathcal{F}_0$ contains all the P null sets of $\mathcal{F}$. For all other unexplained stochastic calculus terms and notation, please see [128].

and $\tau \leq T^{\star}$. τ can occur due to bankruptcy, to a buyout of the company by another company, to a merger, to being broken up by antitrust laws, etc.[6]

Next we let $D = (D_t)_{0 \leq t < \tau} \geq 0$ be the dividend process, and we assume it is a semimartingale. We let $S = (S_t)_{0 \leq t < \tau}$ be nonnegative and denote the price process of the risky asset (again, we are thinking of a stock price here), and again, we are assuming it is a semimartingale. Since S has càdlàg paths,[7] it represents the price process *ex cash flow*. By ex cash flow we mean that the price at time t is after all dividends have been paid, including the time t dividend. But now we have to be a little more careful, since while the assumption that S is a semimartingale on the stochastic interval $[0.\tau)$ is necessary to exclude arbitrage opportunities, it is not sufficient. (See for example [83, 98, 127, 130]). That is, only a subclass of semimartingales exclude arbitrage opportunities. Let $\Delta \in \mathcal{F}_\tau$ be the time τ terminal payoff or liquidation value of the asset. We assume that $\Delta \geq 0$.

Finally, we let W be the *wealth process* associated with the market price of the risky asset plus accumulated cash flows:

$$W_t = \mathbf{1}_{\{t<\tau\}} S_t + B_t \int_0^{t\wedge\tau} \frac{1}{B_u} dD_u + \frac{B_t}{B_\tau} \Delta \mathbf{1}_{\{\tau \leq t\}}. \tag{2}$$

Note that all cash flows are invested in the money market account.

It is standard (and desirable) to have a market which excludes arbitrage opportunities. There are different mathematical formulations of an arbitrage opportunity, but if one formulates them the right way then one has the validity of the first fundamental theorem of asset pricing: namely that the absence of arbitrage is mathematically equivalent to the existence of another probability measure Q, with the same null sets, that turns the price process into a martingale, or more generally a local martingale.[8] The correct formulation for the absence of arbitrage to hold and for the first fundamental theorem to hold in full generality was established by Delbaen and Schachermayer [34, 35]. (See alternatively [127].) If is called *No Free Lunch with Vanishing Risk* and if often referred to by its acronym *NFLVR*. Note that it need not be applied directly to the price process S but can be assumed as a hypothesis relative to any risky asset in question. In our case we want to assume that NFLVR holds for the wealth process defined in (2).

Henceforth, we assume NFLVR holds (and hence there are no arbitrage opportunities) which implies there exists at least one probability measure Q, with the same null sets as P (we write $Q \sim P$), such that under Q we have that W is a

---

[6]No company, government, or economic system can last forever. Of the original 12 companies from the 1896 Dow Jones Industrial Average, only General Electric and Laclede Gas still exist under the same name with remarkable continuity. National Lead is now NL Industries, and Laclede Gas (a utility in St. Louis) was removed from the DJIA in 1899. See [51] for more details.

[7]Càdlàg paths refers to paths that are right continuous and have left limits, a.s.

[8]In the general case one must also consider sigma martingales, but if the price process is assumed to be nonnegative, we use the fortuitous fact that sigma martingales bounded from below are local martingales (see for example [76] or [128]).

local martingale. We make two more assumptions, both of which will be weakened later:

1. The equivalent probability measure Q is unique (and hence the market is complete; see e.g. [83] and Sect. 5).
2. The random variable $W_t = \mathbf{1}_{\{t<\tau\}} S_t + B_t \int_0^{t\wedge\tau} \frac{1}{B_u} dD_u + \frac{B_t}{B_\tau} \Delta \mathbf{1}_{\{\tau \le t\}}$ is assumed to be in $L^1(dQ)$ for each t, $0 \le t \le T^\star$.

The (now assumed to be) unique equivalent probability measure Q is often called a *risk neutral measure*, or the Equivalent Local Martingale Measure, sometimes abbreviated with the acronym ELMM. The term "risk neutral" comes from equilibrium theory. While individual people are risk averse when trading with their own money (and this is often mathematically modeled using utility functions), and perhaps people trading large sums with other people's money are much less risk adverse, nevertheless the market in the whole is assumed to have risk aversion. By changing from the underlying probability P to an ELMM Q, we have an artificial transformation that generates risk neutral pricing in the market.[9] We use this risk neutral measure to give the market's *fundamental value* for the risky asset; this should be the best guess for the future discounted cash flows, given one's knowledge at the present time. If we take conditional expectations in (2) and rearrange the terms, this translates into:

$$S_t^\star = E_Q\left(\int_t^{\tau\wedge T^\star} \frac{1}{B_u} dD_u + \frac{\Delta}{B_\tau} \mathbf{1}_{\{\tau \le T^\star\}} \middle| \mathcal{F}_t\right) B_t. \tag{3}$$

The superscript $\star$ will be used systematically to denote fundamental values.

Definition 1. We define β_t by

$$\beta_t = S_t - S_t^\star,$$

the difference between the market price and the fundamental price. (In a well functioning market, this difference is 0.) The process β is called a *bubble*.

[9]One way to think of risk aversion is to consider the following game one time, and one time only: I toss a fair coin, and you pay me \$2 if it comes up heads, and I pay you \$5 if it comes up tails. Most people would gladly play such a game. But if the stakes were raised to \$20,000 and \$50,000, most people short of the 2012 US Presidential candidate and über rich Mitt Romney would not play the game, unwilling to risk losing \$20,000 in one toss of a coin. (In a 2012 presidential race debate Romney offered to bet \$10,000 about something an opponent said; he did it casually, as if this were a frequent type of bet for him.) Exceptions it is easy to imagine are Wall Street and Connecticut Hedge Fund traders, who deal with large sums of other people's money; they might well take advantage of such an opportunity for a quick profit (or loss) since the game is a good bet, irrespective of the high stakes. The hedge fund traders are still risk averse of course, but in ways quite different from the small "retail" investor.

3 Characterization of Bubbles

Our first observation is that we always must have $S_t \geq S_t^\star$, $t \geq 0$. This is of course equivalent to saying that the bubble β has the property $\beta_t \geq 0$ for all t, i.e., bubbles are always nonnegative.[10] This is an important point, so even though it is quite simple, we formalize it as a theorem. For simplicity we consider only the case where the stock pays no dividends, and the spot interest rate is 0. Note that it is only for simplicity, and an analogous result holds if the spot rate is not 0, and also if dividends are paid. If the spot rate is not 0, one needs to discount the final term. In the case of futures however, it matters whether or not the interest rates are deterministic, or random. We treat this is Sect. 9. For dividends, there are details to keep track of (for example when the stock is ex dividend, etc.), but the ideas are the same.

Theorem 1. *Let S be the nonnegative price process of a stock and assume S pays no dividends. Moreover assume the spot interest rate is constant and equal to 0. Let Q be a risk neutral measure under which S is a local martingale (and hence a supermartingale). Let $S^\star$ be the fundamental value of the stock calculated under Q, and let $\beta_t = S_t - S_t^\star$. Then $\beta \geq 0$.*

Proof. Under these simplifying hypotheses of no dividends and 0 interest, the fundamental value of the stock is nothing more than

$$S_t^\star = E_Q(\Delta 1_{\{\tau \leq T^\star\}} | \mathcal{F}_t). \tag{4}$$

Since under Q the process S is a supermartingale, we have

$$E_Q(S_\tau | \mathcal{F}_t) \leq S_t \tag{5}$$

and since $S_\tau = \Delta 1_{\{\tau \leq T^\star\}}$, combining (4) and (5) gives the result. □

We can classify bubbles into three types, as shown in the following theorem, which was originally proved in [88]. For this theorem, we assume fixed a risk neutral measure Q under which both S and W are local martingales.

Theorem 2. *If in an asset's price there exists a bubble $\beta = (\beta_t)_{t \geq 0}$ that is not identically zero, then we have three and only three possibilities:*

1. β_t is a local martingale (which could be a uniformly integrable martingale) if $\mathbb{P}(\tau = \infty) > 0$.

---

[10]One can ask if it is not possible to have bubbles which are negative? In our models, *for stocks*, the answer is no. However for risky assets other than stocks, such as foreign exchange, it is possible to have negative bubbles. For example when the dollar is in a bubble relative to the euro, then the euro would be in a negative bubble relative to the dollar.

2. *β_t is a local martingale but not a uniformly integrable martingale if τ is unbounded, but with $\mathbb{P}(\tau < \infty) = 1$.*
3. *β_t is a strict $\mathbb{Q}$-local martingale, if τ is a bounded stopping time.*

Proof. Fix Q equivalent to P such that W is a local martingale under Q. Note that W_t is a closable supermartingale, so there exists $W_\infty \in L^1(dQ)$ such that $W_t \to W_\infty$ almost surely. Also, since S is a nonnegative local martingale under the risk neutral measure, $\lim_{t\to\infty} S_t = S_\infty$ exists a.s. (cf., e.g., [128, Theorem 10, p. 8]). The fundamental wealth process is one's best guess of future wealth, given today's knowledge: $W_t^* = E_Q(W_\infty|\mathcal{F}_t)$. Note that analogously, W_∞^* exists, and $W_\infty = W_\infty^*$. Let

$$\beta_t' = W_t - E_Q[W_\infty|\mathcal{F}_t] = W_t - W_t^*. \tag{6}$$

Then β_t' is a (non-negative) local martingale since it is a difference of a local martingale and a uniformly integrable martingale. It is simple to check that

$$E_Q[W_\infty|\mathcal{F}_t] = E_Q[W_\infty^*|\mathcal{F}_t] + E[S_\infty|\mathcal{F}_t] = W_t^* + E[S_\infty|\mathcal{F}_t]. \tag{7}$$

By the definition of wealth processes and (6), (7):

$$\begin{aligned}\beta_t &= S_t - S_t^* \\ &= W_t - W_t^* \\ &= \left(E_Q[W_\infty|\mathcal{F}_t] + \beta_t'\right) - \left(E_Q[W_\infty|\mathcal{F}_t] - E_Q[S_\infty|\mathcal{F}_t]\right) \\ &= \beta_t' + E_Q[S_\infty|\mathcal{F}_t].\end{aligned} \tag{8}$$

If $\tau < T$ for $T \in \mathbb{R}_+$, then $S_\infty = 0$. A bubble $\beta_t = \beta_t' = 0$ for $t \geq \tau$ and in particular $\beta_T = 0$. If β_t is a martingale,

$$\beta_t = E[\beta_T|\mathcal{F}_t] = 0 \qquad \forall t \leq T \tag{9}$$

It follows that $\beta.$ is a strict local martingale. This proves (1). For (2) assume that β_t is uniformly integrable martingale. Then by Doob's optional sampling theorem, for any stopping time $\tau_0 \leq \tau$,

$$\beta_{\tau_0} = E_Q[\beta_\tau|\mathcal{F}_{\tau_0}] = 0 \tag{10}$$

and since β is optional, it follows from (for example) the section theorems of P.A. Meyer (see for example [39]) that $\beta = 0$ on $[0, \tau]$. Therefore the bubble does not exist. For (3), $E_Q[S_\infty|\mathcal{F}_t]$ is a uniformly integrable martingale and the claim holds. □

As indicated, there are three types of bubbles that can be present in an asset's price. Type 1 bubbles occur when the asset has infinite life with a payoff at $\{\tau = \infty\}$.

Type 2 bubbles occur when the asset's life is finite, but unbounded. Type 3 bubbles are for assets whose lives are bounded.

Of the three types of bubbles, the most interesting are those on a compact time interval, $[0, T]$. In this case we are dealing exclusively with Type 3 bubbles, and as seen in Theorem 2 we have that β will be a strict local martingale. Since $S^\star$ is a true martingale, and $\beta = S - S^\star$, we have that β being a strict local martingale is equivalent to the price process S being a strict local martingale. Indeed, we see that:

Corollary 1. *We have a bubble on* $[0, T]$ *if and only if the price process* S *is a strict local martingale.*

For the important special case of a bounded horizon (that is, we are working on a compact time interval, $[0, T]$), we can summarize as follows:

Theorem 3. *Any non-zero asset price bubble* β *on* $[0, T]$ *is a strict* Q*-local martingale with the following properties:*

1. $\beta \geq 0$,
2. $\beta_\tau = 0$,
3. *if* $\beta_t = 0$ *then* $\beta_u = 0$ *for all* $u \geq t$, *and*
4. *if no cash flows, then*

$$S_t = E_Q\left(\frac{S_T}{B_T}\middle|\mathcal{F}_t\right) B_t + \beta_t - E_Q\left(\frac{\beta_T}{B_T}\middle|\mathcal{F}_t\right) B_t$$

for any $t \leq T \leq \tau \leq T^\star$.

This theorem states that the asset price bubble β is a strict Q- local martingale. Condition (1) states that bubbles are always non-negative, i.e. the market price can never be less than the fundamental value. Condition (2) states that the bubble must burst on or before τ. Condition (3) states that if the bubble ever bursts before the asset's maturity, then it can never start again. Alternatively stated, condition (3) states that in the context of our model, bubbles must either exist at the start of the model, or they never will exist. And, if they exist and burst, then they cannot start again. Requiring bubbles to exist since the beginning of the modeling period is clearly a weak spot of the theory; fortunately this can be resolved within the context of incomplete markets, which allow for the concept of bubble birth. For this reason complete market models are ill suited to the study of bubbles, at least using our models of them. We will return to this subject in Sect. 6.

4 Examples of Bubbles

Of course it is of interest to know if such phenomena as bubbles occur, both in reality and in our models. We deal with our models first. Because we are working on a compact time interval, the fundamental value $S^\star$ will be a martingale as soon as $\Delta \in L^1(dQ)$, assuming no dividends and zero interest rate. In the presence of dividends and interest rates, other assumptions on integrability with respect to a given risk neutral measure enter the picture. Therefore the existence of a bubble becomes equivalent to the stock price process being a local martingale, which is not a martingale. (The space of all local martingales includes martingales as a subspace.) However it is easy to generate local martingales. Let us make the reasonable assumption that S follows a stochastic differential equation with a unique strong solution of the rather general form

$$dS_t = \sigma(S_t)dB_t + b(S_t)dt \tag{11}$$

where B is standard Brownian motion. Since Brownian motion has martingale representation, it generates complete markets (see, e.g., [83]). Therefore in this Brownian paradigm there is only one risk neutral measure Q. Under mild hypotheses on σ and b, including that σ never vanishes, (11) under Q becomes

$$dS_t = \sigma(S_t)dB_t, \tag{12}$$

and we have that S is a strict local martingale if and only if

$$\int_{\epsilon}^{\infty} \frac{x}{\sigma(x)^2}dx < \infty. \tag{13}$$

for some $\epsilon > 0$. This (and much more) is proved in detail in the papers [101, 116]. The idea goes back to Delbaen and Shirakawa [38]; see also [69]. However this is also easy to prove directly, using Feller's test for explosions. We have the following results, which are based on remarks made to us by Dmitry Kramkov [102]. Theorem 4 is classic:

Theorem 4. *Let S be a nonnegative Q local martingale with $S_0 = 1$. Then S is a true martingale if and only if there exists a probability measure R, with $R \ll Q$, and $\frac{dR}{dQ}|_{\mathcal{F}_t} = S_t$; otherwise S is a strict local martingale.*

The intuition behind why Theorem 4 is true, is that S has to have an expectation constant in time (and equal to one) in order to be a true martingale. Since it is nonnegative, this turns out also to be sufficient. If the expectation decreases with time, then R would be a sub probability measure, but not a true probability measure: some "mass would escape to ∞."

Following Jacod and Shiryaev [76, pp. 166ff], for a stopping time ν we let P_ν denote the restriction of P to the sigma algebra $\mathcal{F}_\nu$, and we define $R \ll_{\text{loc}} Q$ if there exists a sequence of stopping times τ_n of stopping times such that $\tau_n \nearrow \infty$ a.s. and $R_{\tau_n} \ll P_{\tau_n}$ for each n. With the hypotheses of Theorem 4 one has automatically that $R \ll_{\text{loc}} Q$. Indeed, we have a true martingale when $R \ll Q$ without the "local" caveat. Using Feller's test for explosions for one dimensional diffusions (see [97, 109] as in the recent treatment in [117]), we find the criterion of Mijatovic and Urusov [116, Corollary 4.3].

Theorem 5. *Let B be a Q Brownian motion and let S be of the form*

$$dS_t = S_t a(S_t) dB_t, \text{ under } Q. \tag{14}$$

Then S is a martingale if and only if

$$\int_1^\infty \frac{1}{xa(x)^2} dx = \int_1^\infty \frac{x}{\sigma(x)^2} dx = \infty$$

where $\sigma(x) = xa(x)$.

Proof. By Girsanov's theorem, (14), under R becomes

$$dS_t = S_t a(S_t) d\beta_t + S_t a(S_t)^2 dt, \quad S_0 = 1$$

for an R Brownian motion β. Mijatovic and Urusov show (though they are not the first to do something like this) that S is a true martingale if and only if $\int_0^t a(S_s)^2 ds < \infty$ a.s. (dQ). This implies S cannot explode. To use Feller's test to see if S explodes, we use the notation of Karatzas and Shreve [97]. Simple calculations show that in this case their scale function p is given by $p(x) = -\frac{1}{x} + C$, and their speed measure m is

$$m(dx) = \frac{2dx}{p'(x)\sigma^2(x)} = \frac{2x^2}{\sigma^2(x)} dx.$$

Finally their function $v(x) = \int_c^x (p(x) - p(y))\, m(dy)$ equals

$$= \int_c^x \left(-\frac{1}{x} + \frac{1}{y}\right) \cdot \frac{2y^2}{\sigma^2(y)} dy$$

$$= 2\int_c^x \frac{x-y}{xy} \cdot \frac{y^2}{\sigma^2(y)} dy$$

$$= 2\int_c^x \frac{y}{\sigma^2(y)} dy - \frac{2}{x} \int_c^x \frac{y^2}{\sigma^2(y)} dy$$

and since in the second integral we have $\frac{y}{x} < 1$ we get that $v(+\infty) = +\infty$ if and only if $\int_c^\infty \frac{y}{\sigma^2(y)} dy = +\infty$. Taking $\sigma(x) = xa(x)$ means in this context

$$\int_c^\infty \frac{y}{y^2 a(y)^2} dy = \int_c^\infty \frac{1}{ya(y)^2} dy = +\infty.$$

Therefore we see that by Feller's test S does not explode if and only if $\int_1^\infty \frac{1}{xa(x)^2} dx = +\infty$, and we are done. □

We end this discussion by noting that we do not really need to use Feller's test, but could have instead used the local time-space formula of stochastic calculus (see for example [128]). Namely we have that

$$\int_0^T a(S_s)^2 ds = \int_0^\infty a(x)^2 L_T^x dx$$

where L_T^x is the local time in x at time T of S. Since for almost all ω we have $x \mapsto L_T^x(\omega)$ is a continuous function of x that vanishes off a compact set, and if the function a never vanishes, we can conclude $0 < \epsilon(\omega) \leq L_T^x(\omega) < K(\omega) < \infty$ and once again we can deduce the result. This approach is developed in detail in [116].

Of course one can ask for examples of bubbles coming from the markets. For economy wide bubbles there are many, as we mentioned in the introduction. In the case of individual assets, we detail examples in Sect. 10 later in this paper. A recent paper of X. Li, M. Lipkin, and R. Sowers [106] has shown a way in which bubbles can arise as a consequence of short squeezes related to bankruptcy stocks. There are of course many more examples, as a simple Google Scholar search will exhibit. Strict local martingales have received attention in the mathematical literature irrespective of their connection to models of financial bubbles. See for example [46, 137, 146].

Simulations for the Inverse Bessel Process

The inverse Bessel process is perhaps the most famous (or infamous) strict local martingale. It goes back at least to the renowned 1963 paper of Johnson and Helms [93] who gave it to provide an example of a nonnegative supermartingale which is uniformly integrable but is not of "Class D", the class proposed by P.A. Meyer when he solved Doob's decomposition conjecture, by showing it did not hold in full generality, but that it did nevertheless hold for supermartingales of Class D (the theorem is now known as the Doob–Meyer Decomposition Theorem of Supermartingales). The construction of Johnson and Helms is now classical: Let W be a standard three dimensional Brownian motion starting from the point $(1, 0, 0)$. Let $u(p) = 1/r$, where $r = \| p \|$, the Euclidean distance of $p \in \mathbb{R}^3$ to the origin. Define a process X by $X_t = u(W_t)$ for $t \geq 0$. That is,

$$X_t = \frac{1}{\| W_t \|}. \tag{15}$$

Then X is a uniformly integrable nonnegative process, with finite values a.s. because W never hits the origin with probability 1, and Itô's formula shows that X is a local martingale, because u is the Newtonian potential and therefore a harmonic function for Brownian motion in $\mathbb{R}^3$. However simple calculations show that $t \mapsto E(X_t)$ is not constant (these calculations are given in detail in the little book of Chung and Williams [29]) and indeed $E(X_0) = 1$ while $\lim_{t\to\infty} E(X_t) = 0$. An alternate representation for the inverse Bessel process is as a solution to a stochastic

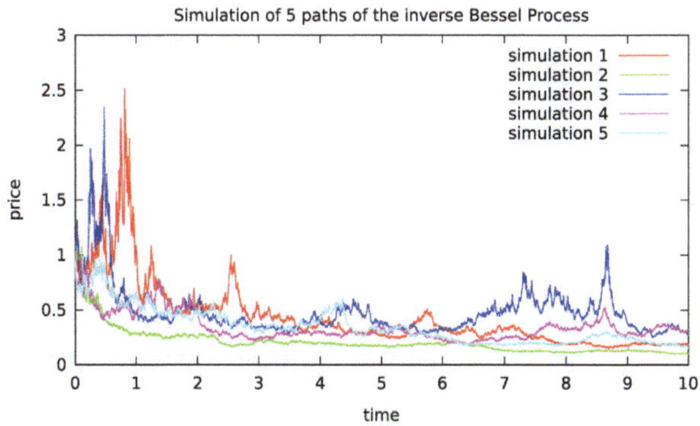

Fig. 1 Five simulated sample paths of the inverse Bessel process

differential equation of the form

$$dX_t = -X_t^2 dB_t; \quad X_0 = 1 \tag{16}$$

where B is a standard one dimensional Brownian motion, and therefore since (16) is of the form of Corollary 5, we know from the Mijatovic–Urysov theorem that X is a strict local martingale. Nevertheless, it is easier to simulate paths of X using the representation given in (15), so it is nice to have both methods of representing X. One can see the two representations [(15) and (16)] of X are equivalent by applying Itô's formula to the X given in (15).

To show that the inverse Bessel process has paths that can behave as if they are paths of a stock price with bubbles, we have the following simulations[11] (Fig. 1).

Note that a roughly half of the simulations of the sample paths of the inverse Bessel could reasonably represent a history of the price of a stock that underwent bubble pricing. For clarity, we isolate one of these paths in Fig. 2:

Simulations for Stochastic Differential Equations

It is nice to go beyond the canonical case of the inverse Bessel process, and to consider other simple models of local martingales, to see if their simulations agree with one's expectations for a model of a bubble price process. The theory tells us that they should, but one can always ask: Do simulations back up the theory? In this respect we are grateful to Jing Guo, who (at our request) simulated solutions of SDEs of the form

---

[11]We thank Etienne Tanre of INRIA for making these simulations of paths of the inverse Bessel process.

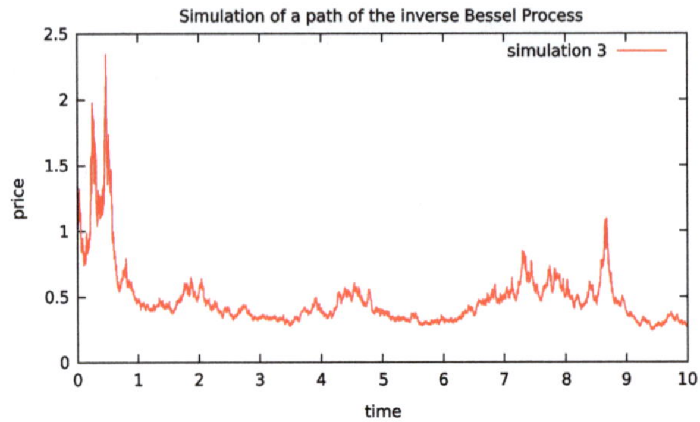

Fig. 2 An inverse Bessel sample path

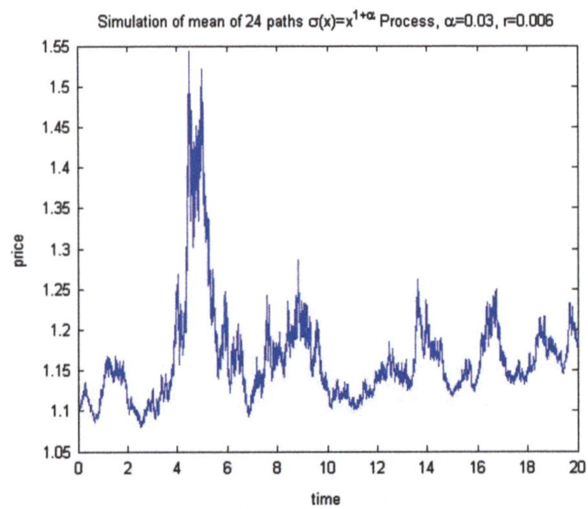

Fig. 3 Average of 24 paths with $\alpha = 0.03$

$$dX_t = X_t^{1+\alpha} dB_t$$

for various values of α, with of course $\alpha > 0$ always. One of his observations is that as α grows, the bubble peaks get more peaked: that is, they both get higher, and they also get narrower. Figure 3 below illustrates what happens, with a graph of the average of 24 paths, for $\alpha = 0.3$:

Note that we have not included the drift in the models used for these pictures, and yet certainly in practice there is a drift, as far as the data is concerned. (The dynamics under the risk neutral measure removes the drift, but the data should reflect the dynamics under the objective measure, not the risk neutral measure.) When a drift is present, it should diminish the future peaks that the simulations show occur after the initial primary peak, but we are not including here even more simulations in order to illustrate that.

The Case of Stochastic Volatility

While the examples provided by equations of type (11) form a wide and useful class of equations, several examples that include stochastic volatility already exist in the literature. They provide examples of strict local martingales (and hence bubbles on a compact time interval $[0, T]$) for models with stochastic volatility.

Theorem 6 (Sin). *Assume there are no cash flows on the underlying asset, B is as in (2), that (W^1, W^2) is a standard two dimensional Brownian motion, and let (S_t, v_t) satisfy*

$$\frac{dS_t}{S_t} = r_t dt + v_t^{\alpha}\left(\sigma_1 dW_t^1 + \sigma_2 dW_t^2\right)$$
$$\frac{dv_t}{v_t} = \rho(b - v_t)dt + a_1 dW_t^1 + a_2 dW_t^2$$

under the risk neutral measure Q where $S_0 = x$, $v_0 = 1$, $\alpha > 0$, $\rho \geq 0$, $b > 0$, a_1, σ_1, a_2, σ_2 are constants. Then, $\frac{S_t}{B_t}$ is a strict local martingale under Q if and only if $a_1\sigma_1 + a_2\sigma_2 > 0$.

For another example in this vein the reader can consult the work of B. Jourdain [94]. Also, L. Andersen and V. Piterbarg [3], of Bank of America and Barclay's Capital respectively, consider a class of stochastic volatility models of the form

$$dX_t = \lambda X_t \sqrt{V_t} dW_t^1 \tag{17}$$
$$dV_t = \kappa(\theta - V_t)dt + \epsilon V_t^p dW_t^2$$

where (W^1, W^2) is a two dimensional Brownian motion with correlation coefficient ρ.

Note that this is a generalization of the model of Sin above, and adds the feature that the correlation coefficient of the noise processes plays an important role. Anderson and Piterbarg in [3] are not trying to determine if a process is in a bubble or not, but rather their main thrust is to determine if extensions of what is known as the Heston model, a simple model using stochastic volatility, are reasonable in a financial context or not; they find that it depends on a range of parameters. And almost in passing, they discover a characterization of when the model forms a true martingale, or is a strict local martingale. This *inter alia* provides a simple test to determine if a process in their context is a strict local martingale, or a true martingale. They establish the following result, among many others.

Theorem 7 (Andersen–Piterbarg). *For the model* (17) *above, if* $p \leq \frac{1}{2}$ *or* $p > \frac{3}{2}$ *then* X *is a true martingale, and if* $\frac{3}{2} > p > \frac{1}{2}$, X *is a true martingale for* $\rho \leq 0$ *and it is a strict local martingale for* $\rho > 0$. *For the case* $p = \frac{3}{2}$, X *is a true martingale for* $\rho \leq \frac{1}{2}\epsilon\lambda^{-1}$, *and* X *is a strict local martingale for* $\rho > \frac{1}{2}\epsilon\lambda^{-1}$

Perhaps the most definitive result already existing in the literature is that of P.L. Lions and M. Musiela [107]. Indeed, in their interesting paper they prove the results in Theorem 8 and in the more general result Theorem 9.

Theorem 8 (Lions–Musiela). *Let* $Z_t = \rho W_t + \sqrt{1-\rho^2} B_t$ *where* (W_t, B_t) *is a standard two dimensional Brownian motion, and let* (F, σ) *solve*

$$dF_t = \sigma_t F_t dW_t, \quad F_0 = F > 0 \tag{18}$$

$$d\sigma_t = \mu(\sigma_t) dZ_t + b(\sigma_t) dt, \quad \sigma_0 - \sigma \geq 0 \tag{19}$$

with

$$\mu(0) = 0, \quad b(0) \geq 0 \tag{20}$$

$$\mu(\xi) > 0, \quad \text{for } \xi > 0, \text{ and } \mu \text{ is Lipschitz on } [0, \infty) \tag{21}$$

$$b(\xi) \leq C(1+\xi) \text{ on } [0,\infty), \text{ for some } C \geq 0 \tag{22}$$

Suppose in addition that the following condition holds

$$\limsup_{\xi\to\infty} (\rho\mu(\xi)\xi + b(\xi))\xi^{-1} < \infty \tag{23}$$

then $E(F_t|\ln F_t|) < \infty$, $E(\sup_{0\leq s\leq t} |F_s|) < \infty$ *for all* $t \geq 0$ *and* F_t *is an integrable nonnegative martingale. On the other hand, if the following holds:*

$$\liminf_{\xi\to\infty} (\rho\mu(\xi)\xi + b(\xi))\frac{1}{\phi(\xi)} > 0 \tag{24}$$

for some smooth, positive, increasing function ϕ *such that* $\int_\epsilon^\infty \frac{1}{\phi(\xi)} d\xi < \infty$, *for all* $\epsilon > 0$, *then* F_t *is a strict local martingale, and we have* $E(F_t) < F_0$ *for all* $t > 0$.

We observe that in the special case that $b = 0$ and $\mu(\xi) = \alpha\xi$ with $\alpha > 0$, then (23) is equivalent to $\rho \leq 0$, while (24) is equivalent to $\rho > 0$ (take $\phi(\xi) = \xi^2$). Therefore we see that the correlation coefficient ρ plays an important role in determining whether or not the process $(F_t)_{t\geq 0}$ is a strict local martingale.

Lions and Musiela go on to consider a more general case than that of Theorem 8. Instead of (18) and (19), they consider the equations

$$dF_t = \sigma_t^\delta F_t dW_t, \quad F_0 = F > 0 \tag{25}$$

$$d\sigma_t = \gamma\sigma_t^\gamma dZ_t + b(\sigma_t) dt, \quad \sigma_0 = \sigma > 0 \tag{26}$$

and again they want conditions under which F_t is a martingale, and conditions under which F_t is a strict local martingale. Their reasons for such an analysis are again not really related to bubble detection, but instead address the important issue as to whether or not certain stochastic volatility models are "well posed or not." As with Anderson and Piterbarg, in Theorem 7, they provide, *inter alia*, a parametric framework for detecting whether or not a process is a martingale or a local martingale, based an a range of parameter values. We have the following theorem:

Theorem 9 (Lions–Musiela). *With* $(F_t)_{t\geq 0}$ *given by* (25) *and* (26), *and* W *and* Z *given as in Theorem 8,*

1. *If* $\rho > 0$ *and if* $\gamma + \delta > 1$, *we assume that* b *satisfies*

$$\limsup_{\xi\to\infty} \frac{b(\xi) + \rho\alpha\xi^{\gamma+\delta}}{\xi} < \infty. \tag{27}$$

   Then $(F_t)_{t\geq 0}$ *is an integrable nonnegative martingale and*

$$E(F_t|\ln F_t|) < \infty, \quad E(\sup_{0\leq s\leq t} |F_s|) < \infty \quad \text{for all } t \geq 0.$$

2. *If* $\rho > 0, \gamma + \delta > 1$ *and* b *satisfies*

$$\liminf_{\xi\to\infty} \frac{b(\xi)_\rho\alpha\xi^{\gamma+\delta}}{\phi(\xi)} > 0 \tag{28}$$

   for some smooth, positive, increasing function ϕ *such that* $\int_\epsilon^\infty \frac{1}{\phi(\xi)} d\xi < \infty$, *then* $(F_t)_{t\geq 0}$ *is a strict local martingale (and not a true martingale), and we have* $E(F_t) < E(F_0)$ *for all* $t > 0$.

Removal of Drift in the Multidimensional Case, and Strict Local Martingales

The multidimensional case is intrinsically interesting, since it is easy to imagine contagion within bubbles. The most obvious case might be that instead of an individual stock undergoing bubble pricing, the phenomenon might apply to an entire financial sector, such as technology stocks, automotive stocks, telecommunications, etc. Therefore it is interesting to understand some examples of multidimensional bubbles.

Since we know from the one dimensional case that strict local martingales are more likely if the coefficient σ increases quickly to ∞, we assume that σ is only locally Lipchitz. This guarantees existence and uniqueness of solutions up to an explosion time ξ, which can be ∞ but need not be in general. Let $J = (0, \infty)$ and J_i be the ith copy of J, and let $I = \Pi_{i=1}^d J_i$, a subset of $\mathbb{R}^d$. We let

$$\begin{aligned} \mu &: I \to \mathbb{R}^d \\ \sigma &: I \to \mathbb{R}^d \star \mathbb{R}^d \end{aligned} \tag{29}$$

where μ and σ are locally Lipschitz functions. We let W denote a d dimensional Brownian motion, and then our stochastic differential equation takes the usual form

$$dS_t = \mu(S_t)dt + \sigma(S_t)dW_t, \text{ for } t < \xi, \text{ where } \xi \text{ is a possibly infinite explosion time.} \tag{30}$$

We make the hypotheses that the solution process S *lives in the positive orthant.*

The simplest case is to assume the square matrix σ is invertible. Then we can find a vector δ such that $\sigma \times \delta = -\mu$. We also assume that δ is locally bounded. Our candidate Radon Nikodym process will as usual be an exponential local martingale:

$$Z_t = e^{\int_0^{\xi \wedge t} \delta(S_s) dW_s - \frac{1}{2} \int_0^{\xi \wedge t} \|\delta(S_s)\|^2 ds}, \tag{31}$$

where we set $Z_t = 0$ on $\{t \geq \xi\}$.

We assume that $\int_0^{\xi \wedge t} \| \delta(S_s) \|^2 ds < \infty$ on the event $\{t < \xi\}$, so that Z is well defined. Z is of course a nonnegative local martingale (since it solves a multidimensional exponential equation, with driving term being a continuous stochastic integral), hence (by Fatou's Lemma) a supermartingale, and since the time horizon T is fixed, we have

$$Z = (Z_t)_{0 \leq t \leq T} \text{ is a martingale, if and only if } E(Z_T) = 1.$$

Note that since we are in a multidimensional Brownian paradigm, by (for example) the Kunita–Watanabe version of the martingale representation theorem, we know that all local martingales have continuous paths, and cannot therefore jump to 0, even at the time T. (See for example [128, Theorem 43, p. 188].)

We next use a technique present in the book by Karatzas and Shreve [97, Exercise 5.38, p. 352] for one dimension, and developed in much more generality and for multiple dimensions in Cheridito et al. [27]. We repeat it here since for our case, the argument is perhaps easier to follow than the more general one treated in [27]. We let

$$\tau_n = \inf\{t > 0 | S_t \notin [\frac{1}{n}, n]^d\},$$

the first exit time from the solid $[\frac{1}{n}, n]^d$. Note that $\tau_n \nearrow\nearrow \xi$ as $n \to \infty$, where $\tau_n < \xi$ for each n. We next modify μ and σ, calling the new coefficients μ_n and σ_n, where μ_n and σ_n agree with μ and σ on $[\frac{1}{n}, n]^d$, and also are globally Lipschitz, and σ_n is also invertible. We then have that there exists a unique, everywhere defined, *and nonnegative* solution S^n of the auxiliary equation

$$dS_t^n = \mu_n(S_s^n)ds + \sigma_n(S_s^n)dB_s, \tag{32}$$

where B is again a Brownian motion. Next we define δ_n such that $\sigma_n \times \delta_n = -\mu_n$, and define

$$L_t^n = \int_0^{\tau_n \wedge t} -\delta_n(S_s^n)dB_s$$

which is well defined globally since L^n is a local martingale with

$$[L^n, L^n]_t = \int_0^{t\wedge\tau_n} \| \delta \|^2 (S_s^n)ds \leq \| \delta \|^2_{L^\infty,[\frac{1}{n},n]} t < \infty$$

and hence $[L^n, L^n]_t \in L^1$ and L^n is actually a (true) square integrable martingale. However by Novikov's criterion (see for example [128]) we also have that the stochastic (also known as the Doléans–Dade) exponential $\mathcal{E}(L^n)$ is a martingale. We let

$$D_t^n = D_0\mathcal{E}(L_t^n) \text{ for } t < \xi, \text{ and } D_\xi^n = \lim_{n\to\infty} D_{\tau_n}^n$$

and again, D^n is a (nonnegative) martingale, so there is no problem in asserting the limit above exists. D^n so defined is a supermartingale, by Fatou's Lemma. We next relate it to the process Z defined in (31). For $n \geq m$ we have $D_t^n = D_t^m$ for $t \leq \tau_m$, and hence for $t < \xi$ we define $D_t = \lim D_t^n \geq 0$, as $n \to \infty$. Note that $D_t > 0$ on $\{t < \xi\} \cap \{D_0 > 0\}$. Finally, for $t < \xi$ we have $D_t = D_0 \frac{Z_t}{Z_0}$. All this is preamble to defining a sequence of new measures:

$$\frac{dQ^n}{dP}|_{\mathcal{F}_{\tau_n}} = \frac{D_{\tau_n}}{D_0}.$$

Using Girsanov's theorem we have that $W_t^n = W_t + \int_0^{t\wedge\tau_n} \delta(S_s^n)ds$ is a Q^n Brownian motion up to τ_n, giving rise to the SDE system (up to time τ_n):

$$dW_t^n = dW_t + \delta(S_t^n)dt$$
$$dS_t^n = \sigma(S_t^n)dW_t^n$$

and using the uniqueness in law of the solutions we have that the Q^n measures are compatible and give an über measure Q with $Q^n = Q|_{\mathcal{F}_{\tau_n}}$ for each n, with

$$\frac{dQ}{dP}|_{\mathcal{F}_{\tau_n}} = \frac{D_{\tau_n}}{D_0} \text{ and } Q^n_{|\mathcal{F}_{\tau_n}} = Q_{|\mathcal{F}_{\tau_n}}. \tag{33}$$

Theorem 10. *With Z, Q, as defined above, and ξ the explosion time of S, we have*

$$E_P(Z_T 1_{\{\xi > T\}}) = Q(\xi > T).$$

Proof. Let $A \in \mathcal{F}_T$. Recalling that $D = Z$ a.s. on the event $\{T < \xi\}$, we have:

$$\begin{aligned} E_P(Z_T 1_{\{T<\xi\}} 1_A) &= E_P(\frac{D_T}{D_0} 1_{\{T<\xi\}} 1_A) \\ &= E_P(\frac{D_T}{D_0} \lim_{n\to\infty} 1_{\{T<\tau_n\}} 1_A) \\ &= \lim_{n\to\infty} E_P(\frac{D_T}{D_0} 1_{\{T<\tau_n\}} 1_A) \end{aligned}$$

by the monotone convergence theorem; and using that

$D_T = D_{T\wedge\tau_n}$ on $1_{\{T<\tau_n\}}$, the above equals

$$\begin{aligned} &= \lim_{n\to\infty} E_Q(1_{\{T<\tau_n\}} 1_A) \\ &= Q(A \cap T < \xi) \end{aligned}$$

again by the monotone convergence theorem. The theorem follows by taking $A = \Omega$. □

Corollary 2. *Let S be as given in* (30) *and Z be as given in* (31)*. With the notation and assumptions of Theorem 10, if S does not explode under Q, then Z is a true martingale. If S does not explode under P, then Z is a martingale if and only if S does not explode under Q.*

Proof. Let us first assume that S does not explode under Q. But Z is a martingale if and only if $E(Z_T) = 1$, and this happens if and only if $Q(\xi > T) = 1$. Next we suppose that S does not explode under P. Then Z is a supermartingale, so $E_P(Z_T) \leq 1$. Therefore if S does not explode under Q, we have $E_P(Z_T 1_{\{\xi>T\}}) = 1$. However since $E_P(Z_T) \leq 1$, and $Z_T = 0$ on $\{T \geq \xi\}$ a.s., we deduce the result. □

Why do we care whether or not Z is a martingale or only a local martingale? We know that the solution S of (30) is nonnegative and let us suppose it does not explode under P. We know that under a risk neutral measure the drift disappears and S is always a vector of at least local martingales, and it is a vector of martingales if and only if S does not explode in each of every component, and as we have seen by Corollary 2, this is tied to whether or not Z is a martingale. This is nice to know, but it is not much help in analyzing whether or not a given system is a martingale or a strict local martingale, the key property for telling whether or not we have a financial bubble.

We next give a criterion to determine whether or not the system is a strict local martingale through the use of Hellinger Processes. We use freely results about

Hellinger Processes from the book of Jacod and Shiryaev [76]. First we note that if Q and P are two probabilities, we can define $R = \frac{P+Q}{2}$ and then $P \ll R$ and $Q \ll R$. We let $X = \frac{dP}{dR}$ and $Y = \frac{dQ}{dR}$, with $X = (X_t)_{t \geq 0}$ and $Y = (Y_t)_{t \geq 0}$ being their respective martingale versions, through projections onto the filtration. We set $U_t = \frac{X_t}{Y_t}$, and define

$$\alpha_t = \begin{cases} \frac{X_t}{Y_t} & \text{if } \ 0 < U_{t-} < \infty \\ 0 & \text{if } \ U_{t-} = 0 \\ \infty & \text{if } \ U_{t-} = \infty \end{cases} \tag{34}$$

While we do not reproduce the proof here, Younes Kchia has shown (2011, private communication):

Theorem 11. *Let the process Z be given as in* (31)*, the process α be as given in* (34)*, and the probability Q be as given in* (33)*. We then have that Z is a true martingale if and only if $Q(h(\frac{1}{2})_T < \infty) = 1$ and $Q(\sup_{0 \leq t \leq T} \alpha_t < \infty) = 1$. Here $h(\frac{1}{2})_T$ is the Hellinger process of order $\frac{1}{2}$ between P and Q.*

We note that in the case considered above, if all processes are continuous and using $R = \frac{P+Q}{2}$, we have

$$h(\frac{1}{2}) = \frac{1}{8}\left(\frac{1}{X} + \frac{1}{Y}\right)^2 \cdot [X, X].$$

(See for example [76, p. 236].) We also note that these are much less practical conditions to check than those we have in the one dimensional case. We will see later that the one dimensional case presents its own formidable problems if we want to check if a condition such as (13) holds, in order to determine whether or not S is a strict local martingale.

For more ways to generate strict local martingales, as well as a study of their asymptotic behaviors, we refer the interested reader to [122]. Related papers involving strict local martingales include [11,15,30,50,96,108], as well as the recent book [124].

5 Incomplete Markets: Choosing a Risk Neutral Measure

When we consider incomplete markets we immediately have a problem: How do we choose a risk neutral measure so that we can well define the fundamental value of a risky asset? The Second Fundamental Theorem of Finance states that a market is incomplete if and only if there exists an infinite number of equivalent risk neutral measures (see, e.g., [37], or [83]), so the question is not a trivial one. Many different methods have been proposed to solve this question, including (with sample references) indifference pricing (see for example the volume edited by R. Carmona [20]), choosing a risk neutral measure by choosing one that minimizes the entropy (or alternatively the "distance") between the objective measure and the class of risk neutral measures (see for example the excellent paper of Grandits and Rheinlander [60]), by minimizing the variance of certain terms in the semimartingale decomposition, known as choosing the minimal variance measure (see for example Föllmer–Schweizer [53], or the subsequent results of Monat and Stricker [119]). Each of these methods works but they all give the uneasy feeling of arbitrariness, whose main value is a canonical procedure to choose a risk neutral measure. Instead, and as an alternative, we will sketch here a procedure due to Jacod and this author [73], which gives conditions under which it is apparent that the market has itself chosen a unique risk neutral measure. A similar approach (with a similar result) was taken in Schweizer and Wissel [141, 142], albeit in a more restrictive case (i.e., restricted to the Brownian paradigm). When sufficient conditions hold for the uniqueness of a risk neutral measure compatible with all market prices, it seems intuitively reasonable to use that risk neutral measure for pricing purposes, since it is the one the market itself is using.

The basic idea of the article [73] is to take an inherently incomplete market, and to complete it artificially by including option prices. This is accomplished by modeling the market price S of our risky asset together with a family of traded options. In this way, the options can in theory "complete" the market, rendering the choice of a compatible risk neutral measure unique. This idea is not new with [73], and its beginnings can be traced to the late 1990s, with the works of Dengler and Jarrow [40], Dupire [43], Derman and Kani [102], and also Schönbucher [140]. Note that if one ignores the options, the model depending only on the risky asset price remains incomplete, with an infinite choice of risk neutral measures, and we call this set $\mathbb{Q}_S$. Therefore if the option prices change, for whatever reason, they could become compatible with a different choice of risk neutral measure in $\mathbb{Q}_S$, and it is this flexibility that allows us to include bubble birth in our model, in the incomplete case.

We assume the following model for the stock price X. First, in the *continuous case* we suppose that

$$X_t = X_0 + \int_0^t a_s ds + \sum_{i \in I} \int_0^t \sigma_s^i dW_s^i. \tag{35}$$

In the general case, when there are jumps, we suppose that

$$X_t = X_0 + \int_0^t a_s ds + \sum_{i \in I} \int_0^t \sigma_s^i d W_s^i + (\psi 1_{\{|\psi| \leq 1\}}) * (\mu - \nu)_t + (\psi 1_{\{|\psi| > 1\}}) * \mu_t. \tag{36}$$

Here we are using established notation for stochastic integrals with respect to Brownian motions W^i and random measure μ, or compensated random measure $\mu - \nu$, see for example the book of Jacod and Shiryaev [76]. We assume also that ν factors: $\nu(dt, dx) = dtF(dx)$. The index set I is assumed finite. In (36) $X_0 > 0$

is non-random and the coefficients a, σ^i and ψ are such that the integrals and sums above make sense: that is, a and σ^i are predictable and ψ is $\tilde{\mathcal{P}}$-measurable, and

$$\int_0^t \left(|a_s| + \sum_{i \in I} |\sigma_s^i|^2 + \int (\psi(s,x)^2 \wedge 1)\, F(dx) \right) ds < \infty \qquad \text{a.s.} \tag{37}$$

for all t. (We use $\tilde{\mathcal{P}}$ to denote the product σ algebra $\mathcal{P} \otimes \mathcal{R}$ on $\Omega \times \mathbb{R}_+ \times \mathbb{R}$.)

Of course these coefficients should also be such that $X_t > 0$: this amounts to saying that they factor as $a_t = X_{t-}\bar{a}_t$ and $\sigma_t^i = X_{t-}\bar{\sigma}_t^i$ and $\psi(t,x) = X_{t-}\bar{\psi}(t,x)$ with $\bar{\psi} > -1$ identically, with $\bar{a}, \bar{\sigma}^i$ and $\bar{\psi}$ satisfying (37), but it is more convenient to use the form (36). Note that this represents the most general semimartingale driven by μ and the W^i's that has a chance to satisfy the hypotheses NFLVR (No Free Lunch with Vanishing Risk) of Delbaen and Schachermayer.

For options, we consider a fixed pay-off function g on $(0,\infty)$ which is *non-negative and convex*, and we denote by $P(T)_t$ the price at time $t \in [0,T]$ of the option with pay-off $g(X_T)$ at expiration date T. We also assume that g is not affine, otherwise $P(T)_t = g(X_t)$ and we are in a trivial situation.

We denote by $\mathcal{T}$ the set of expiration dates T corresponding to tradable options (always with the same given pay-off function g), and by $T_\star$ the time horizon up to when trading may take place. Even when $T_\star < \infty$, there might be options with expiration date $T > T_\star$, so we need to specify the model up to infinity.

In practice $\mathcal{T}$ is a finite set, although perhaps quite large. For the mathematical analysis it is much more convenient to take $\mathcal{T}$ to be an interval, or perhaps a countable set which is dense in an interval. We consider the case where $T_\star < \infty$ and $\mathcal{T} = [T_0, \infty)$, with $T_0 > T_\star$.

Apart from the fact that $P(T)_T = g(X_T)$, the prices $P(T)_t$ are so far unspecified, and the idea is to model them on the basis of the same W^i and μ, rather than with X. However, since these are option prices, they should have some internal compatibility properties.

Indeed, if the option prices were derived in the customary way, we would have a measure $\mathbb{Q}$ which is equivalent to $\mathbb{P}$, and under which X is a martingale and $g(X_T)$ is $\mathbb{Q}$-integrable and $P(T)_t = \mathbb{E}_{\mathbb{Q}}(g(X_T)|\mathcal{F}_t)$ for $t \leq T$. Then of course $P(T)$ is a $\mathbb{Q}$-martingale indexed by $[0,T]$. But we can also look at how $P(T)_t$ varies as a function of the expiration date T, on the interval $[t,\infty)$. That is, we are taking the non customary step of fixing t, and considering $P(T)_t$ *as a process where T varies.* Since X is a quasi-left continuous martingale and g is convex, then $T \mapsto g(X_T)$ is a quasi-left continuous submartingale relative to $\mathbb{Q}$, and this implies that $T \mapsto P(T)_t$ is *non-decreasing and continuous* for $T \geq t$. Observe that this property holds $\mathbb{Q}$-almost surely, hence $\mathbb{P}$-almost surely as well because $\mathbb{P}$ and $\mathbb{Q}$ are equivalent.

Remark 12. We wish to emphasize that, for example in the case of European call options, the usual theory calls for $P(T)_t = \mathbb{E}_{\mathbb{Q}^\star}((X_T - K)_+|\mathcal{F}_t)$ for some risk neutral measure $Q^\star$. *We do not make this assumption here.* Indeed, the previous paragraph is simply motivation for us to *assume a priori* that $T \mapsto P(T)_t$ is non-decreasing and continuous. This seems completely reasonable from the viewpoint of practice, where (in the absence of dividends or interest rate changes and anomalies) it is always observed that $T \mapsto P(T)_t$ is nondecreasing. In the language of practitioners, if it were not it would imply a "negative pricing of the calendar," which makes no economic sense Lipkin, American stock exchange, 2007, private communication. Nevertheless we warn the reader that there are pathological examples where this assumption does not hold: for example if X is the reciprocal of a three dimensional Bessel process starting at $X_0 = 1$, then X is a local martingale for its natural filtration, but $T \mapsto P(T)_0$ is not increasing, since $P(0)_0 = 0$, $P(T)_0 > 0$ for $T \in (0, \infty)$, but $\lim_{T\to\infty} P(T)_0 = 0$, hence $T \mapsto P(T)_0$ cannot be increasing for $T \geq 0$. Thus our assumption $T \mapsto P(T)_t$ is increasing in T rules out the possibility of the market being governed by such price processes. This is an important exception, since the inverse Bessel process is the classic example of a strict local martingale, going back to the paper of Johnson and Helms [93]. The inverse Bessel process is of course a canonical example of a strict local martingale, fitting into the theory of when there are bubbles, so it would seem that this particular theory is excluding precisely the case where there are bubbles in call options, a topic treated in Sect. 7. Note however that in the proofs presented in [73], the assumption that $T \mapsto P(T)_t$ is increasing in T is not essential, and could be replaced simply with $T \mapsto P(T)_t$ is absolutely continuous as a function of T. This change allows us to apply this theory to the more general case where bubbles in option prices are included.

We write

$$P(T)_t = P(T_0)_t + \int_{T_0}^{T} f(t,s)ds. \tag{38}$$

In this case, the inverse Bessel process and other local martingales are included. The function f has the representation

$$\begin{aligned} f(t,s) = f(0,s) &+ \int_0^t \alpha(u,s)du + \sum_{i\in I}\int_0^t \gamma^i(u,s)dW_u^i \\ &+ (\phi(.,s)1_{\{|\phi(.,s)|\leq 1\}}) * (\mu-\nu)_t + (\phi(.,s)1_{\{|\phi(.,s)|>1\}}) * \mu_t. \end{aligned} \tag{39}$$

We further assume that the process $P(T_0)$ is given for $t \leq T_\star$ by

$$P(T_0)_t = P(T_0)_0 + \int_0^t \overline{\alpha}_s ds + \sum_{i\in I}\int_0^t \overline{\gamma}_s^i dW_s^i \tag{40}$$

in the *continuous case*, and in the general case by

$$P(T_0)_t = P(T_0)_0 + \int_0^t \overline{\alpha}_s ds + \sum_{i\in I}\int_0^t \overline{\gamma}_s^i dW_s^i + (\overline{\phi}1_{\{|\overline{\phi}|\leq 1\}}) * (\mu-\nu)_t + (\overline{\phi}1_{\{|\overline{\phi}|>1\}}) * \mu_t, \tag{41}$$

where the above coefficients are predictable and satisfy

$$\int_0^{T_\star} \left(|\overline{\alpha}_t| + \sum_{i\in I} |\overline{\gamma}_t^i|^2 + \int (\overline{\phi}(t,x)^2 \wedge 1) F(dx) \right) dt < \infty \tag{42}$$

a.s., and further the (non-random) initial condition $P(T_0)_0$ and these coefficients are such that we have identically

$$t \in [0, T_\star] \quad \Rightarrow \quad P(T_0)_t \geq g(X_t). \tag{43}$$

Finally we assume that we have $\int_{T_0}^T \chi(s)_{T_\star} ds < \infty$ a.s. for all $T > T_0$, where

$$\chi(s)_t = \int_0^t \left(|\alpha(u,s)| + \sum_{i\in I} |\gamma^i(u,s)|^2 + \int (\phi(u,x,s)^2 \wedge 1)\, F(dx) \right) du.$$

An example of the type of results obtained in [73] is when trading takes place up to time $T_\star$, and the expiration dates of the options are all $T \geq T_0$, where $T_0 > T_\star$. We denote $\mathcal{M}_{\text{loc}}(T_\star, T_0)$ the collection of risk neutral measures for X that are compatible with the option structure so that no arbitrage opportunities exist. The following result is shown in [73]:

Theorem 13. *Consider a $(T_\star, T_0)$ partial fair model such that the set $\mathcal{M}_{loc}(T_\star, T_0)$ is not empty. Then this set is a singleton if and only if, for a good version of the coefficients of the model, we have the following property: the system of linear equations*

$$\sum_{i\in I} \sigma_s^i(\omega)\beta_i + \int \psi(\omega,s,x)y(x)\,dx = 0, \tag{44}$$

$$\sum_{i\in I} \overline{\gamma}_s^i(\omega)\beta_i + \int \overline{\phi}(\omega,s,x)y(x)\,dx = 0, \tag{45}$$

$$T \geq T_0 \;\Rightarrow\; \sum_{i\in I} \alpha^i(s,T)(\omega)\beta_i + \int \phi(\omega,s,x,T)y(x)\,dx = 0, \tag{46}$$

where $((\beta_i), y) \in \Upsilon'(\omega, s)$, has for its only solution $\beta_i = 0$ and $y = 0$ up to a Lebesgue-null set.

A consequence is that we see when conditions such as those in Theorem 13 above are met, the market prices for the options have uniquely determined a risk neutral measure. Also, should the market change its collective mind about the pricing of options, it could still choose a unique risk neutral measure, but a new one. Such phenomena have been noticed by economists, and it is referred to colloquially as the sun spot theory, since occasionally the sun gets sun spots, and they appear to happen randomly and without explanation (see for example [7,22]).

6 Incomplete Markets: Bubble Birth

We use the idea of the previous section to extend our model of the economy to allow for the possibility of bubble "birth" after the model starts. A modification involves the market exhibiting different local martingale measures across time. We note that this is different from the usual paradigm of choosing an initial equivalent local martingale measure, and remaining with it fixed as our choice for all time, but we will see it is not that different from the standard notion of regime change. Indeed, shifting local martingale measures corresponds to regime shifts in the underlying economy (in any of the economy's endowments, beliefs, risk aversion, institutional structures, or technologies). For pedagogical reasons we choose a simple and intuitive structure consistent with this extension.

To begin this extension, we need to define the regime shifting process. Let $(\sigma_i)_{i\geq 0}$ denote an increasing sequence of random times with $\sigma_0 = 0$. The random times $(\sigma_i)_{i\geq 0}$ represent the times of regime shifts in the economy. It is important that these times σ_i be totally inaccessible stopping times. (See for example [128] for definitions and properties of totally inaccessible stopping times.) For if they were to be predictable, traders could see the regime shifts coming and develop arbitrage strategies around the shifts.[12] If we are working within a minimal Brownian paradigm, then there are no totally inaccessible stopping times, so we would need to consider a larger space that supports such times.

We let $(Y^i)_{i\geq 0}$ be a sequence of random variables characterizing the state of the economy at those times (the particular regime's characteristics) such that $(Y^i)_{i\geq 0}$ and $(\sigma)_{i\geq 0}$ are independent of each other. Moreover, we further assume that both $(Y^i)_{i\geq 0}$ and $(\sigma)_{i\geq 0}$ are also independent of the underlying filtration $\mathbb{F}$ to which the price process S is adapted.

Define two stochastic processes $(N_t)_{t\geq 0}$ and $(Y_t)_{t\geq 0}$ by

$$N_t = \sum_{i\geq 0} \mathbf{1}_{\{t\geq\sigma_i\}} \quad \text{and} \quad Y_t = \sum_{i\geq 0} Y^i \mathbf{1}_{\{\sigma_i\leq t<\sigma_i+1\}}. \tag{47}$$

N_t counts the number of regime shifts up to and including time t, while Y_t identifies the characteristics of the regime at time t. Let $\mathbb{H}$ be a natural filtration generated by N and Y and define the enlarged filtration $\mathbb{G} = \mathbb{F} \vee \mathbb{H}$ (for example see [128] or [120] for a discussion of some of the general theory of filtration enlargement). By the definition of $\mathbb{G}$, $(\sigma_i)_{i\geq 0}$ is an increasing sequence of $\mathbb{G}$ stopping times.

Since N and Y are independent of $\mathbb{F}$, every $(Q,\mathbb{F})$-local martingale is also a $(Q,\mathbb{G})$-local martingale. By this independence, changing the distribution of N and/or Y does not affect the martingale property of the wealth process W. To discuss a collection of ELMMs, however, it is prudent to work on a finite horizon ($[0,T]$), and not on the infinite half line $[0,\infty)$. Therefore, we do not speak of the

[12] We thank a referee for suggesting we include this remark.

mathematically appealing set of ELMMs defined on $\mathcal{G}_\infty$, but rather we fix a (non random) horizon time $T < \infty$ and speak of the ELMMs defines on $\mathcal{G}_T$, and that is *a priori* larger than the set of ELMMs defined on $\mathcal{F}_T$. We are not concerned with this enlarged set of ELMMs. We will, instead, focus our attention on the $\mathcal{F}_T$ ELMMs and sometimes write $\mathcal{M}_{loc}^{\mathbb{F}}(W)$ to recognize explicitly this restriction. With respect to this restricted set, given the Radon Nikodym derivative $Z_T = \frac{dQ}{dP}|_{\mathcal{F}_T}$, we define its density process by $Z_t = E[Z_T|\mathcal{F}_t]$. Of course, Z is an $\mathbb{F}$-adapted process. Note that this construction implies that the distribution of Y and N is invariant with respect to a change of ELMMs in $\mathcal{M}_{loc}^{\mathbb{F}}(W)$.

We will henceforth always be working in this section on the finite horizon case $[0, T]$ *with the non random time* T *chosen* a priori *and fixed. We will no longer make special mention of this implicit assumption.*

The independence of the filtration $\mathbb{H}$ from $\mathbb{F}$ gives this increased randomness in our economy the interpretation of being *extrinsic uncertainty*. It is well known that extrinsic uncertainty can affect economic equilibrium as in the sunspot equilibrium of Cass and Shell [7, 22]. This form of our information enlargement, however, is not essential to our arguments. It could be relaxed, making both N and Y pairwise dependent, and dependent on the original filtration $\mathbb{F}$ as well. This generalization would allow bubble birth to depend on *intrinsic uncertainty* (see Froot and Obstfeld [54] for a related discussion of intrinsic uncertainty). However, this generalization requires a significant extension in the mathematical complexity of the notation and proofs, so we leave it aside.

We are now ready to discuss the fundamental price of a risky asset in the incomplete market context. Of course to do this, we need to select a risk neutral measure from an infinite selection of possibilities. We do this with the aid of Theorem 13. Because the unique measure specified in Theorem 13 can change as the regime shifts, so too might the fundamental value of the asset. Since the selection of the risk neutral measure affects the fundamental value, and this can change as the regime shifts, we can have the birth of price bubbles. More formally, we let the local martingale measure in our extended economy depend on the state of the economy at time t as represented by the original filtration $(\mathcal{F}_t)_{t\geq 0}$, the state variable(s) Y_t, and the number of regime shifts N_t that have occurred. Suppose $N_t = i$. Denote $Q^i \in \mathcal{M}_{loc}(W)$ as the ELMM "selected by the market" at time t given Y^i.

As in the complete market case, the fundamental price of an asset (or portfolio) represents the asset's expected discounted cash flows.

Definition 2 (Fundamental Price). Let $\phi \in \Phi$ be an asset with maturity ν and payoff (Δ, Ξ^ν). The *fundamental price* $\Lambda_t^\star(\phi)$ of asset ϕ is defined by

$$\Lambda_t^\star(\phi) = \sum_{i=0}^{\infty} E_{Q^i}\left[\int_t^\nu d\Delta_u + \Xi^\nu \mathbf{1}_{\{\nu<\infty\}}\middle|\mathcal{F}_t\right]\mathbf{1}_{\{t<\nu\}\cap\{t\in[\sigma_i,\sigma_{i+1})\}} \tag{48}$$

$\forall t \in [0, \infty)$ where $\Lambda_\infty^\star(\phi) = 0$.

In particular the fundamental price of the risky asset $S_t^\star$ is given by

$$S_t^\star = \sum_{i=0}^{\infty} E_{Q^i}\left[\left.\int_t^\tau dD_u + X_\tau \mathbf{1}_{\{\tau<\infty\}}\right|\mathcal{F}_t\right]\mathbf{1}_{\{t<\tau\}\cap\{t\in[\sigma_i,\sigma_{i+1})\}}. \tag{49}$$

To understand this definition, let us focus on the risky asset's fundamental price. At any time $t < \tau$, given that we are in the ith regime $\{\sigma_i \le t < \sigma_{i+1}\}$, the right side of expression (49) simplifies to:

$$S_t^\star = E_{Q^i}\left[\left.\int_t^\tau dD_u + X_\tau \mathbf{1}_{\{\tau<\infty\}}\right|\mathcal{F}_t\right].$$

Given the market's choice of the ELMM is $Q^i \in \mathcal{M}_{loc}^{\mathbb{F}}(W)$ at time t, we see that the fundamental price equals its expected future cash flows. Note that the payoff of the asset at infinity, $X_\tau \mathbf{1}_{\{\tau=\infty\}}$, does not contribute to the fundamental price. This reflects the fact that agents cannot consume the payoff $X_\tau \mathbf{1}_{\{\tau=\infty\}}$. Furthermore note that at time τ, the fundamental price $S_\tau^\star = 0$. We emphasize that a fundamental price is not necessarily the same as the market price S_t. Under NFLVR the market price S_t equals the arbitrage-free price,[13] but this need not equal the fundamental price $S_t^\star$.

For notational simplicity, we can alternatively rewrite the fundamental price in terms of an equivalent probability measure, indexed by time t, that is not a local martingale measure because of this time dependence.

Theorem 14. *There exists an equivalent probability measure $Q^{t\star}$ such that*

$$\Lambda_t^\star(\phi) = E_{Q^{t\star}}\left[\left.\int_t^\nu d\Delta_u + \Xi^\nu \mathbf{1}_{\{\nu<\infty\}}\right|\mathcal{F}_t\right]\mathbf{1}_{\{t<\nu\}} \tag{50}$$

Proof. Let $Z^i \in \mathcal{F}_T$ be a Radon Nykodym derivative of Q^i with respect to P and $Z_t^i = E[Z^i|\mathcal{F}_t]$. Define

$$Z_T^{t*} = \sum_{i=0}^{\infty} Z^i \mathbf{1}_{\{t\in[\sigma_i,\sigma_{i+1})\}} \tag{51}$$

[13]What we mean by this is that if NFLVR holds, then one can neither find not exploit an arbitrage opportunity in the short run by strategies of buying and selling the asset, or by using financial derivatives

Then $Z_T^{t*} > 0$ almost surely and

$$\begin{aligned} EZ_T^{t*} &= E\left[\sum_{i=0}^{\infty} Z^i \mathbf{1}_{\{t\in[\sigma_i,\sigma_{i+1})\}}\right] = \sum_{i=0}^{\infty} E[Z^i \mathbf{1}_{\{t\in[\sigma_i,\sigma_{i+1})\}}] \\ &= \sum_{i=0}^{\infty} E[Z^i]E[\mathbf{1}_{\{t\in[\sigma_i,\sigma_{i+1})\}}] \\ &= \sum_{i=0}^{\infty} P(\sigma_i \leq t < \sigma_{i+1}) \\ &= 1 \end{aligned} \tag{52}$$

Therefore we can define an equivalent measure Q^{t*} on $\mathcal{F}_T$ by $dQ^{t*} = Z_T^{t*}dP$. The Radon Nykodim density Z_t^{t*}on $\mathcal{G}_t$ is

$$\begin{aligned} Z_t^{t*} &= \left.\frac{dQ^{t*}}{dP}\right|_{\mathcal{G}_t} = E[Z^{t*}|\mathcal{F}_t] = \sum_{i=0}^{\infty} E[Z^i \mathbf{1}_{\{t\in[\sigma_i.\sigma_{i+1})\}}|\mathcal{G}_t] \\ &= \sum_{i=0}^{\infty} E[Z^i|\mathcal{G}_t]\mathbf{1}_{\{t\in[\sigma_i.\sigma_{i+1})\}}. \end{aligned} \tag{53}$$

Then

$$\begin{aligned} \Lambda_t^*(\phi) &= \sum_{i=0}^{\infty} E_{Q^i}\left[\int_t^{\nu} d\Delta_u + \Xi^{\nu}\mathbf{1}_{\{\nu<\infty\}}\middle| \mathcal{F}_t\right]\mathbf{1}_{\{t<\nu\}\cap\{t\in[\sigma_i,\sigma_{i+1})\}} \\ &= \sum_{i=0}^{\infty} E_{Q^i}\left[\int_t^{\nu} d\Delta_u + \Xi^{\nu}\mathbf{1}_{\{\nu<\infty\}}\middle| \mathcal{G}_t\right]\mathbf{1}_{\{t<\nu\}\cap\{t\in[\sigma_i,\sigma_{i+1})\}} \\ &= E\left[\left(\sum_{i=0}^{\infty}\frac{Z^i}{Z_t^i}\mathbf{1}_{\{t\in[\sigma,\sigma_{i+1})\}}\right)\left(\int_t^{\nu} d\Delta_u + \Xi^{\nu}\mathbf{1}_{\{\nu<\infty\}}\right)\middle| \mathcal{G}_t\right]\mathbf{1}_{\{t<\nu\}} \end{aligned} \tag{54}$$

and observing that

$$\frac{Z^i}{Z_t^i}\mathbf{1}_{\{t\in[\sigma_i,\sigma_{i+1}]\}} = \frac{Z^i\mathbf{1}_{\{t\in[\sigma_i,\sigma_{i+1})\}}}{\sum_{i=0}^{\infty} Z_t^i\mathbf{1}_{\{t\in[\sigma_i,\sigma_{i+1})\}}},$$

we can continue:

$$
\begin{aligned}
&= E\left[\left(\frac{\sum_{i=0}^{\infty} Z^i \mathbf{1}_{\{t\in[\sigma_i,\sigma_{i+1})\}}}{\sum_{i=0}^{\infty} Z_t^i \mathbf{1}_{\{t\in[\sigma_i,\sigma_{i+1})\}}}\right)\left(\int_t^{\nu} d\Delta_u + \Xi^{\nu}\mathbf{1}_{\{\nu<\infty\}}\right)\middle|\mathcal{G}_t\right]\mathbf{1}_{\{t<\nu\}} \\
&= E\left[\left(\frac{Z_T^{t*}}{Z_t}\right)\left(\int_t^{\nu} d\Delta_u + \Xi^{\nu}\mathbf{1}_{\{\nu<\infty\}}\right)\middle|\mathcal{G}_t\right]\mathbf{1}_{\{t<\nu\}} \\
&= E_{Q^{t*}}\left[\int_t^{\nu} d\Delta_u + \Xi^{\nu}\mathbf{1}_{\{\nu<\infty\}}\middle|\mathcal{G}_t\right]\mathbf{1}_{\{t<\nu\}} \\
&= E_{Q^{t*}}\left[\int_t^{\nu} d\Delta_u + \Xi^{\nu}\mathbf{1}_{\{\nu<\infty\}}\middle|\mathcal{F}_t\right]\mathbf{1}_{\{t<\nu\}}
\end{aligned}
\tag{55}
$$

□

We call $Q^{t\star}$ the *valuation measure* at t, and the collection of valuation measures $(Q^{t\star})_{t\geq 0}$ the *valuation system*.

In our new model with regime change, there is no single risk neutral measure generating fundamental values across time. The valuation measures $Q^{s\star}$ and $Q^{t\star}$ at times $s < t$ are usually two different measures, and neither is an ELMM. The $\star$ superscript is used to emphasize that $Q^{t\star}$ is the measure *chosen by the market*, and the superscript t is used to indicate that it is selected at time t. In the ith regime $\{\sigma_i \leq t < \sigma_{i+1}\}$, the valuation measure coincides with $Q^i \in \mathcal{M}_{loc}^{\mathbb{F}}(W)$. Since $Q^{t\star}$ is a family of ELMMs and not one that is fixed, $Q^{t\star} \notin \mathcal{M}_{loc}^{\mathbb{F}}(W)$ in general, unless the system is static.[14]

Given the definition of an asset's fundamental price, we can now define the fundamental wealth process.

For subsequent usage, we see that the fundamental wealth process of the risky asset is given by

$$
W_t^{\star} = S_t^{\star} + \int_0^{\tau\wedge t} dD_u + X_{\tau}\mathbf{1}_{\{\tau\leq t\}}. \tag{56}
$$

Then,

$$
W_t^{\star} = \sum_{i=0}^{\infty} E_{Q^i}\left[\int_0^{\tau} dD_u + X_{\tau}\mathbf{1}_{\{\tau<T\}}\middle|\mathcal{F}_t\right]\mathbf{1}_{\{t\in[\sigma_i,\sigma_{i+1})\}} \tag{57}
$$

$\forall t \in [0,\infty)$ and $W_{\infty}^{\star} = \int_0^{\tau} dD_u + X_{\tau}\mathbf{1}_{\{\tau<T\}}$.

[14]Although the definition of the fundamental price as given depends on the construction of the extended economy, one could have alternatively used expression (50) as the initial definition. This alternative approach relaxes the extrinsic uncertainty restriction explicit in our extended economy.

Alternatively, we can rewrite $W_t^\star$ by

$$W_t^\star = \sum_{i=0}^{\infty} E_{Q^i}\left[W_T^\star | \mathcal{F}_t\right] \mathbf{1}_{\{t \in [\sigma_i, \sigma_{i+1})\}} \qquad \forall t \in [0, \infty). \tag{58}$$

In general, the choice of a particular ELMM affects fundamental values. But, for a certain class of ELMMs, when $\tau < \infty$ the fundamental values are invariant. This invariant class is characterized in the following lemma. We let $\mathcal{M}_{UI}(W)$ denote the collection of equivalent measures that render W a uniformly integrable martingale. In contrast, $\mathcal{M}_{NUI}(W)$ denotes those equivalent measures that render W at least a sigma martingale, but not a uniformly integrable martingale.

Lemma 1. *Suppose $\tau < T$ almost surely. In the ith regime $\{\sigma_i \le t < \sigma_{i+1}\}$, if the market chooses $Q^i \in \mathcal{M}_{UI}^{\mathbb{F}}(W)$, then the fundamental price of the risky asset $S_t^\star$ and fundamental wealth $W_t^\star$ do not depend on the choice of the measure Q^i almost surely.*

Proof. Fix $Q^*, R^* \in \mathcal{M}_{UI}^{\mathbb{F}}(W)$. $\tau < T$ implies that $W_T = W_T^*$. Let $W_t^{Q^*}$ and $W_t^{R^*}$ be the fundamental prices on $\{\sigma_i \le t < \sigma_{i+1}\}$ when $Q^i = Q^*$ and R^* respectively. Since W is uniformly integrable martingale under Q^* and R^*,

$$\begin{aligned} W_t^{Q^*} &= E_{Q^*}[W_T^* | \mathcal{F}_t] = E_{Q^*}[W_T | \mathcal{F}_t] \\ &= W_t = E_{R^*}[W_T | \mathcal{F}_t] \\ &= E_{R^*}[W_T^* | \mathcal{F}_t] \\ &= W_t^{R^*} \qquad \text{a.s. on } \{\sigma_i \le t < \sigma_{i+1}\} \end{aligned} \tag{59}$$

The difference of $W_t^{Q^*}$ and $S_t^{Q^*}$ does not depend on the choice of measure. Therefore $W_t^{Q^*} = W_t^{R^*}$ implies $S_t^{Q^*} = S_t^{R^*}$ on $\{\sigma_i \le t < \sigma_{i+1}\}$. □

This lemma applies to the risky asset only. If the measure shifts from $Q^i \in \mathcal{M}_{UI}^{\mathbb{F}}(W)$ to $R^i \in \mathcal{M}_{UI}^{\mathbb{F}}(W)$, then the fundamental price of other assets can in fact change.

The next lemma describes the relationship between the fundamental prices of the risky asset when two measures are involved, one being a measure $R^\star \in \mathcal{M}_{\mathrm{NUI}}^{\mathbb{F}}(W)$.

Lemma 2. *Suppose $\tau < T$. In the ith regime $\{\sigma_i \le t < \sigma_{i+1}\}$, consider the case where $Q^i \in \mathcal{M}_{UI}(W)$ and $R^i \in \mathcal{M}_{NUI}(W)$. Then,*

$$W_t^{R\star} \le W_t^{Q\star}, \qquad \text{a.s. on } \{\sigma_i \le t < \sigma_{i+1}\}. \tag{60}$$

That is, the fundamental price based on a uniformly integrable martingale measure is greater than that based on a non-uniformly integrable martingale measure.

Proof. Pick $Q^* \in \mathcal{M}_{\mathrm{UI}}(W)$ and $R^* \in \mathcal{M}_{NUI}(W)$. Since $\tau < T$ almost surely, $W_T = W_T^*$. Under R^*, W is not a uniformly integrable non-negative martingale and $W_t \geq E_{R^*}[W_T|\mathcal{M}_t]$. Therefore

$$\begin{aligned} W_t^{Q^*} - W_t^{R^*} &= E_{Q^*}[W_T^*|\mathcal{M}_t] - E_{R^*}[W_T^*|\mathcal{M}_t] \\ &= E_{Q^*}[W_R|\mathcal{M}_t] - E_{R^*}[W_T|\mathcal{M}_t] \\ &= W_t - E_{R^*}[W_R|\mathcal{M}_t] \\ &\geq 0. \end{aligned} \tag{61}$$

□

We can now finally define what me mean by a price bubble in an incomplete market. As is standard in the economics literature,

Definition 3 (Bubble). An asset price bubble β for S is defined by

$$\beta = S - S^\star. \tag{62}$$

Recall that S_t is the market price and $S_t^\star$ is the fundamental value of the asset. Hence, a price bubble is defined as the difference in these two quantities. Within a fixed regime, the theory simplifies to a complete market case where there is only one risk neutral measure, since the measure chosen by the market is fixed. Thus we have:

Theorem 15. *Within a fixed regime, S admits a unique (up to an evanescent set) decomposition*

$$S = S^\star + \beta = S^\star + (\beta^1 + \beta^2 + \beta^3), \tag{63}$$

where $\beta = (\beta_t)_{t\geq 0}$ is a càdlàg local martingale and

1. *β^1 is a càdlàg non-negative uniformly integrable martingale with $\beta_t^1 \to X_\infty$ almost surely,*
2. *β^2 is a càdlàg non-negative non-uniformly integrable martingale with $\beta_t^2 \to 0$ almost surely,*
3. *β^3 is a càdlàg non-negative supermartingale (and strict local martingale) such that $E\beta_t^3 \to 0$ and $\beta_t^3 \to 0$ almost surely. That is, β^3 is a potential.*

Furthermore, $(S^\star + \beta^1 + \beta^2)$ is the greatest submartingale bounded above by W.

As in the previous Theorem 2, β^1, β^2, β^3 correspond to the type 1, 2 and 3 bubbles, respectively. First, for type 1 bubbles with infinite maturity, we see that the β^1 bubble component converges to the asset's value at time ∞, X_∞. This time ∞ value X_∞ can be thought of as analogous to fiat money, embedded as part of the asset's price process. Indeed, it is a residual value to an asset that pays zero dividends for all finite times. Second, this decomposition also shows that for finite maturity assets, $\tau < \infty$, the critical threshold is that of uniform integrability. This is due to the fact that when $\tau < \infty$, the β^2, β^3 bubble components converge to 0 almost surely, while they need not converge in L^1. Finally, the β^3 bubble components are strict local martingales, and not martingales.

As a direct consequence of this theorem, we obtain the following corollary.

Corollary 3. *Within a fixed regime, any asset price bubble β has the following properties:*

1. $\beta \geq 0$,
2. $\beta_\tau \mathbf{1}_{\{\tau<\infty\}} = 0$,
3. *if* $\beta_t = 0$ *then* $\beta_u = 0$ *for all* $u \geq t$, *and*
4. $S_t = E_{Q^\star}[S_T | \mathcal{F}_t] + \beta_t^3 - E_{Q^\star}[\beta_T^3 | \mathcal{F}_t]$ *for any* $t \leq T \leq \tau$.

As in the complete market case, we still have that bubbles must be nonnegative, even without regard to the regime being fixed or not:

Theorem 16. *Bubbles are nonnegative. That is, if β denotes a bubble, then $\beta_t \geq 0$ for all $t \geq 0$.*

Proof. Fix $t \geq 0$. On $\{\sigma_i \leq t < \sigma_{i+1}\}$, the market chooses Q^i as a valuation measure and the fundamental price S_t^* is given by

$$\begin{aligned} S_t^* \mathbf{1}_{\{\sigma_i \leq t < \sigma_i+1\}} &= E_{Q^i}\left[\int_t^\tau dD_u + X_\tau \mathbf{1}_{\{\tau<\infty\}} \mid \mathcal{F}_t\right] \mathbf{1}_{\{t<\tau\}} \mathbf{1}_{\{\sigma_i \leq t < \sigma_i+1\}} \\ &= S_t^{\star i} \mathbf{1}_{\{\sigma_i \leq t < \sigma_i+1\}}, \end{aligned} \tag{64}$$

where $S_t^{\star i}$ denotes a fundamental price with valuation measure $Q^i \in \mathcal{M}_{loc}(W)$ and

$$S_t^* = \sum_i S_t^{\star i} \mathbf{1}_{\{\sigma_i \leq t < \sigma_i+1\}} \tag{65}$$

and

$$\beta_t^* = \sum_i \beta_{i,t} \mathbf{1}_{\{\sigma_i \leq t < \sigma_i+1\}} \tag{66}$$

By Corollary 3, $\beta_i = S - S^{\star i} \geq 0$ for each i and hence $\beta^* \geq 0$. □

The next example illustrates how we can model bubble birth.

Example 1. Suppose that the measure chosen by the market shifts at time σ_0 from $Q \in \mathcal{M}_{\text{UI}}(W)$ to $R \in \mathcal{M}_{\text{NUI}}(W)$. To avoid ambiguity, we denote a fundamental price based on valuation measures Q and R by $W^{Q\star}$ and $W^{R\star}$, respectively. By Lemma 2, we can choose Q, R and σ such that the difference of fundamental prices based on these two measures,

$$W_{\sigma_0}^{Q\star} - W_{\sigma_0}^{R\star} \geq 0, \tag{67}$$

is strictly positive with positive probability. Then, the fundamental price and the bubble are given by

$$W_t^\star = W_t^{Q\star}\mathbf{1}_{\{t<\sigma_0\}} + W_u^{R\star}\mathbf{1}_{\{\sigma_0\leq t\}} \tag{68}$$

$$\beta_t = \beta_t^R\mathbf{1}_{\{\sigma_0\leq t\}}. \tag{69}$$

And, a bubble is born at time σ_0.

As shown in Lemma 1, a switch from one measure Q to another measure Q' such that $Q, Q' \in \mathcal{M}_{\text{UI}}(W)$ does not change the value of $W^\star$. Therefore, if a bubble does not exist under Q, it also does not exist under Q'. Bubble birth occurs only when a valuation measure changes from a uniformly integrable martingale $Q \in \mathcal{M}_{\text{UI}}(W)$ to a non-uniformly integrable martingale $R \in \mathcal{M}_{\text{NUI}}(W)$.

Remark 17. The reader may well wonder if it is even possible that such a phenomenon happens: that there exists a framework with a process X that is a uniformly integrable martingale under one probability, and is a non uniformly integrable martingale under an equivalent martingale measure. The answer is yes, and it is provided in the work of Delbean and Schachermayer [36]. See alternatively [14].

We next wish to mention an alternative idea to treat the concept of bubble birth, although it complicates the model. It is often believed that bubbles arise due to "easy money," when speculators have access to large pools of funds to invest. This is reflected in the market by its having a high degree of liquidity. Therefore it seems reasonable to try to combine the ideas of high liquidity and bubbles to see if the former can help us understand the birth of the latter. A first mathematical attempt in this direction is attempted in the research paper of R. Jarrow et al. [90]. See also the Ph.D. thesis of A. Roch [133].

In Jarrow et al. [90, 135] the authors combine ideas for bubble birth with mathematical models of liquidity issues presented for example in the work of Çetin et al. [23, 24] and Blais and Protter [16]. See also [136]. The idea, loosely put, is to use a liquidity risk model developed in [133, 134] for highly liquid stocks with a supply curve identified in [16], in order to gain insight into how liquidity can affect bubble births and bubble bursts. Instead of an instant return to the price takers' general asset price, in this model each trade engenders a short exponential decay of its return time; in times of high liquidity these decays can overlap one upon the other, thereby mounting and artificially raising the price above its fundamental value. Whether or not this happens depends on whether or not key parameter values reach certain ranges.

Remark 18. In very recent work of Biagini et al. [14], a concept of "slow bubble birth" is developed. This differs from the regime change idea, which ultimately is an abrupt change at a random time, but rather contains a slow and continuous transition from one probability measure in $\mathcal{M}_{\text{UI}}(W)$ to another in $\mathcal{M}_{\text{NUI}}(W)$.

7 Calls, Puts, and Bubbles

Bubbles have surprising implications for financial derivatives, and these implications indicate that the standard no arbitrage assumption of NFLVR is ever so slightly too weak. This was first noticed, to our understanding, by Heston et al. [63], and underlined by A.M.G. Cox and David Hobson [30]. This also creates problems with the numerical solutions of option prices under the risk neutral measure (see for example [44,45]).

We consider three standard derivative securities all on the same risky asset: a forward contract, a European put option, and a European call option. Each of these derivative securities is defined by its payoff at its maturity date. A *forward contract* on the risky asset with strike price K and maturity date T has a payoff $[S_t - K]$. We denote its time t market price as $V_t^f(K)$. A *European call option* on the risky asset with strike price K and maturity T has a payoff $[S_t - K]^+$, with time t market price denoted as $C_t(K)$. Finally, a *European put option* on the risky asset with strike price K and maturity T has a payoff $[K - S_t]^+$, with time t market price denoted as $P_t(K)$.[15] Finally, let $V_t^f(K)^\star$, $C_t(K)^\star$, and $P_t(K)^\star$ be the fundamental prices of the forward contract, call option and put option, respectively.

A straightforward implication of the definitions is the following theorem.

Theorem 19 (Put-Call Parity for Fundamental Prices).

$$C_t^\star(K) - P_t^\star(K) = V_t^{f\star}(K). \tag{70}$$

Proof. The proof follows from the linearity of conditional expectation. At maturity T,

$$(S_T - K)^+ - (K - S_T)^+ = S_T - K \tag{71}$$

Since a fundamental price of a contingent claim with payoff function H is $E_{Q^{t*}}[H(S)_T|\mathcal{F}_t]$,

$$\begin{aligned} C_t^*(K) - P_t^*(K) &= E_{Q^{t*}}[(S_T - K)^+|\mathcal{F}_t] - E_{Q^{t*}}[(K - S_T)^+|\mathcal{F}_t] \\ &= E_{Q^{t*}}[S_T - K|\mathcal{F}_t] \\ &= V_t^{f\star}(K). \end{aligned} \tag{72}$$

□

Note that put-call parity for the fundamental prices holds regardless of whether or not there are bubbles in the asset's market price.

[15] To be precise, we note that the strike price is quoted in units of the numéraire for all of these derivative securities.

As noted by Heston et al. [63], put-call parity in market prices has been seen to be violated in the presence of bubbles. Examples are provided by the work of Ofek et al. [121] and that of Lamont and Thaler [105] who, in the words of Heston et al., "provide evidence that options on Palm and other stocks violated put-call parity at the same time the stocks clearly had bubbles."

We give an example to show what can happen mathematically under NFLVR.

Example 2. Let B_t^i, $i = \{1,2,3,4,5\}$ be independent Brownian motions. Let M_t^i satisfy

$$M_t^1 = \exp\left(B_t^1 - \frac{t}{2}\right), \qquad M_t^i = 1 + \int_0^t \frac{M_s^i}{\sqrt{T-s}} dB_s^i \qquad 2 \le i \le 5. \tag{73}$$

Consider a market with a finite time horizon $[0,T]$. The market is complete for all five processes M^i with respect to the filtration generated by $\{(M_t^i)_{t\ge 0}\}_{i=1}^5$ in the sense that martingale representation holds, and hence all contingent claims in L^2 are replicable in theory. M_t^1 is a uniformly integrable martingale on $[0,T]$. The processes $\{M_t^i\}_{i=2}^5$ are non-negative strict local martingales that converge to 0 almost surely as $t \to T$. Let $S_t^* = \sup_{s\le t} M_s^1$. Suppose the market prices in this model are given by

- $S_t = S_t^* + M_t^2$
- $C_t(K) = C_t^*(K) + M_t^3$
- $P_t(K) = P_t^*(K) + M_t^4$
- $V_t^f(K) = V_t^{f,\star}(K) + M_t^5$

All of the traded securities in this example have bubbles. To take advantage of any of these bubbles $\{M_t^i\}_{i=2}^4$ based on the time T convergence, an agent must short sell at least one asset. However, to do this one would need to short an asset with a type 3 bubble, and this is not an admissible strategy. Therefore such strategies are not a free lunch with vanishing risk.

For a general contingent claim H, if we let $V_t(H)$ denote its market price at time t, and $V_t^\star$ denote its fundamental price, then the bubble in a contingent claim is defined by

$$\delta_t = V_t(H) - V_t^\star(H) \tag{74}$$

We now have that, as seen by Example 2, NFLVR is not a strong enough assumption to eliminate the possibility of (a fortiori Type 3) bubbles in contingent claims. And, given the existence of bubbles in calls and puts, we get various possibilities for put-call parity in market prices.

- $C_t(K) - P_t(K) = V_t^f(K)$ if and only if $\delta_t^{V^f} = \delta_t^c - \delta_t^p$.
- $C_t(K) - P_t(K) = S_t - K$ if and only if $\delta_t^S = \delta_t^c - \delta_t^p$.

This example validates the following important observation. In the well studied Black Scholes economy (a complete market under the standard NFLVR structure), contrary to common belief, the Black–Scholes formula need not hold! Indeed, if there is a bubble in the market price of the option (M_t^3), then the market price ($C_t(K)$) can differ from the option's fundamental price ($C_t^*(K)$)—the Black–Scholes formula. This insight has numerous ramifications, for example, it implies that the implied volatility (from the Black–Scholes formula) does not have to equal the historical volatility. In fact, if there is a bubble, then the implied volatility should exceed the historical volatility, and yet there exist no arbitrage opportunities. (Note that this is with the market still being complete.) This possibility, at present, is not commonly understood. However, not all is lost. *One additional assumption* returns the Black–Scholes economy to normalcy. This is the assumption of *No Dominance*.

We have seen that put call parity need not hold in practice (as observed in [105, 121] as mentioned before), and that it need not hold mathematically under NFLVR. Nevertheless it is rare that it does not hold in practice, and it is distressing that the situation can invalidate (in some sense) the usual beliefs about the Black–Scholes paradigm. The observations of Ofek et al. and Lamont and Thaler notwithstanding, they are the exception, not the rule. The usual mathematical proof of put-call parity is that of Theorem 19 above, since the usual model does not account for bubbles and market prices, but simply implicitly assumes that market prices and what we call fundamental prices, are the same. The NFLVR assumption allows for market price put-call parity to be violated, but if one wants a model where that cannot happen, then one needs to add an assumption, and the assumption that is usually added is that of *No Dominance*. It dates back to R.C. Merton who proposed it in 1973 (see [114]), although he proposed it only with a verbal description. Jarrow et al. [88] first proposed a mathematical formulation of Merton's idea, and it has since been refined by Sergio Pulido [131], whose definition we give here.

Definition 4. A *Price Operator* is a (not necessarily linear) operator Λ such that

$$\Lambda : L^\infty(dP) \to \mathbb{R} \tag{75}$$

Definition 5. A price operator Λ satisfies the *No Dominance* condition *ND* if for all $f, g \in L^\infty(dP)$ such that $P(f \geq g) = 1$ and $P(f > g) > 0$ we have that $\Lambda(f) > \Lambda(g)$. We further say that the price operator Λ satisfies *No Dominance at* 0, denoted ND_0, if Λ is positive; that is, if for all $f \in L_+^\infty(dP)$ with $P(f > 0) > 0$ we have $\Lambda(f) > 0$.

Jarrow et al. [88, 89] show that No Dominance implies NFLVR. This formulation of the result is taken from Pulido [131], where S denotes the market price of our risky asset. The sets $\mathcal{K}$ and $\mathcal{C}$ defined below are the now standard notations from the formulation of NFLVR given by Delbaen and Schachermayer [34, 35] and also given in their book [37].

Theorem 20. *Suppose a price operator Λ is lower semi continuous on $L^\infty(dP)$, satisfies ND_0, and $\Lambda(f) \leq 0$ for all $f \in \mathcal{C}$, where*

$$\begin{aligned}
\mathcal{A} &= \textit{the set of admissible strategies relative to } S \\
\mathcal{K} &= \{(H \cdot S)_T : H \in \mathcal{A}\} \\
\mathcal{C} &= (\mathcal{K} - L^0_+(dP)) \cap L^\infty(dP) \qquad (76) \\
&= \{g \in L^\infty(dP) : g = f - h \textit{ for some } f \in \mathcal{K} \textit{ and } h \in L^0_+(dP).
\end{aligned}$$

Then NFLVR holds.

Proof. First we observe that NFLVR does not hold if and only if there exists a sequence H^n of processes in $\mathcal{A}$, and a sequence of bounded random variables f_n and a bounded random variable f such that $H^n \cdot S_T \geq f_n$ for all n, and f_n converges to $f \in L^\infty(dP)$, with $P(f \geq 0) = 1$ and $P(f > 0) > 0$. Therefore suppose that NFLVR does not hold. By the preceding observation, we can find a sequence of elements of $\mathcal{C}$, call them $(f_n)_{n\geq 1}$, and an $f \in L^\infty_+(dP)$ such that $f_n \to f$ in $L^\infty(dP)$ and $P(f > 0) > 0$. By hypothesis however,

$$0 < \Lambda(f) \leq \liminf_{n\to\infty} \Lambda(f_n) \leq 0,$$

which gives us a contradiction. So NFLVR must hold. □

With this assumption of No Dominance, we can prove the following useful lemma.

Lemma 3. *Assume No Dominance and NFLVR hold. Let J be a payoff function of a contingent claim such that $V_t(J) = V_t^*(J)$. Then for every contingent claim with payoff H such that $H(S)_T \leq J(S)_T$, $V_t(H) = V_t^*(H)$.*

Proof. Since contingent claims have bounded maturity, we only need to consider type 3 bubbles. Let $\mathcal{L}$ be a collection of stopping times on $[0, T]$. Then for all $L \in \mathcal{L}$, $V_L(H) \leq V_L(J)$ by No Dominance. Since $\{V_t(J)\}_{t\in[0,T]}$ is a martingale it is uniformly integrable martingale and of class (D) on $[0, T]$. Then $\{V_t(H)\}$ is also of class (D) and it is a uniformly integrable martingale on $[0, T]$. (See Jacod and Shiryaev [76, Definition 1.46, Proposition 1.47 in page 11]). Therefore type 3 bubbles do not exist for this contingent claim. □

This lemma states that if we have a contingent claim with no bubbles, and this contingent claim dominates another contingent claim's payoff, then the dominated contingent claim will not have a bubble as well. Immediately, we get the following corollary.

Corollary 4. *If $H(S)_T$ is bounded, then $V_t(H) = V_t(H^*)$. In particular a put option does not have a bubble.*

Proof. Assume that $H(S)_T < \alpha$ for some $\alpha \in \mathbb{R}_+$. Then applying Lemma 3 for $H(x) = \alpha$, we have desired result. □

Theorem 21 (European Put Price). *For all $K \geq 0$,*

$$P_t(K) = P_t^{\star}(K). \tag{77}$$

The proof of this theorem is contained in Corollary 4. Hence, European put options always equal their fundamental values, regardless of whether or not the underlying asset's price has a bubble.

We next consider the put call parity of market prices. We have already seen this is violated occasionally in practice, and that it is not implied by the no arbitrage assumption NFLVR. It is trivial algebraically that $C_T(K) - P_T(K) = V_T^f(K) = S_T - K$; what we want is for this relation to hold at intermediate times $t, 0 \leq t \leq T$.

Theorem 22. *Under NFLVR and No Dominance, we have put call parity of market prices. That is,*

$$C_t(K) - P_t(K) = V_t^f(K) = S_t - K \tag{78}$$

Proof. We re-write equation (78) at time 0 as

$$C = P + V^f = P + S - K, \tag{79}$$

and we see that the left side and right side of (79) have the same cash flows. Therefore if the left side is larger at time 0, the right side dominates the call. If the left side is larger at time 0, then the call dominates the right side of (79). Because we are assuming No Dominance, these phenomena cannot happen, so the two sides must be the same. The same argument works at intermediate times t. (Note that this cannot follow from NFLVR alone, because one would need to use a short selling argument, and it would not be an admissible strategy, due to theoretically potential unlimited losses.) □

Theorem 23 (European Call Price). *For all $K \geq 0$,*

$$C_t(K) - C_t^{\star}(K) = S_t - E_{Q^{t\star}}[S_T|\mathcal{F}_t]. \tag{80}$$

Proof.

$$\begin{aligned} V_t^f(K) &= S_t - K \\ &= (S_t - E_{Q^{t*}}[S_T|\mathcal{F}_t]) + (E_{Q^{t*}}[S_T|\mathcal{F}_t] - K) \\ &= V_t^{f*}(K) + (S_t - E_{Q^{t*}}[S_T|\mathcal{F}_t]). \end{aligned} \tag{81}$$

Using put-call parity in fundamental prices:

$$C_t^*(K) - P_t^*(K) = V_t^{f*}(K) \tag{82}$$

Using put-call parity in market prices,

$$C_t(K) - P_t(K) = V_t^f(K) \tag{83}$$

By subtracting (82) from (83),

$$\begin{aligned}[C_t(K) - C_t^*(K)] - [P_t(K) - P_t^*(K)] &= V_t^f(K) - V_t^{f*}(K) \\ &= S_t - E_{Q^{t*}}[S_T|\mathcal{F}_t] \qquad (84) \\ &= \delta_t,\end{aligned}$$

since the put option has a bounded payoff, $P_t(K) = P_t^*(K)$ and $C_t(K) - C_t^*(K) = \delta_t$. □

Since call options have finite maturity, call option bubbles must be of type 3, if they exist. The magnitude of such a bubble is independent of the strike price and it is related to the magnitude of the asset's price bubble. In a static market, Corollary 3 shows that

$$S_t - E_{Q^{t*}}[S_T|\mathcal{F}_t] = \beta_t^3 - E_{Q^{t*}}\left[\beta_T^3 \middle| \mathcal{F}_t\right]$$

where β_t^3 is the type 3 bubble component in the underlying stock.[16] Here, the call option's bubble equals the difference between the type 3 bubble in the underlying stock less the expected type 3 bubble remaining at the option's maturity.

American Options

The issue of American options is quite interesting, because one finds a surprise: we will see that American call options do not have bubbles, even if there is a bubble in the underlying asset. This is due to the special nature of American calls where early exercise is possible. We will assume throughout our treatment of American options that we are in one regime that does not change, so we will be dealing with one fixed risk neutral measure. Also, because the time value of money plays an important role in the analysis of the early exercise decision of American options, we need to modify our notation to make explicit the numéraire. We denote the time t value of a money market account as

$$A_t = \exp\left(\int_0^t r_u du\right) \tag{85}$$

[16]In an analogous theorem in Jarrow et al. [89], they used the implicit assumption that $T = \tau$ which would imply that $E_{Q^{t*}}\left[\beta_T^3 \middle| \mathcal{F}_t\right] = 0$.

where r is the non-negative adapted process representing the default free spot rate of interest. To simplify comparison with the previous, we still let S_t denote the risky asset's price in units of the numéraire. We choose and fix a risk neutral measure Q. In the terminology of Sect. 6 the measure $Q = Q^{t*}$ lies within a fixed period for all t in this period, in between possible regime shifts.

Definition 6 (The Fundamental Price of an American Option). The fundamental price $V_t^{A^\star}(H)$ of an American option with payoff function H and maturity T is given by

$$V_t^{A^\star}(H) = \sup_{\eta \in [t,T]} E_Q[H(S_\eta)|\mathcal{F}_t] \tag{86}$$

where η is a stopping time and the market selected $Q \in \mathcal{M}_{loc}(S)$.

This definition is a straightforward extension of the standard formula for the valuation of American options in the classical literature. It is also equivalent to the *fair price* as defined by Cox and Hobson [30] when the market is complete. We apply this definition to a call option with strike price K and maturity T. Letting $C_t^{A\star}(K)$ denote the American call's fundamental value, the definition yields

$$C_t^{A^\star}(K) = \sup_{\eta \in [t,T]} E_Q[(S_\eta - \frac{K}{A_\eta})^+|\mathcal{F}_t]. \tag{87}$$

Let $C^A(K)_t$ be the market price of this same option, and $C^E(K)_t$ the market price of an otherwise identical European call.

Before we continue, we establish some technical results of which we will have need. They are taken from [89].

Lemma 4. *Let M_u be a non-negative càdlàg local martingale. Assume that there exists some function f and a uniformly integrable martingale X such that*

$$\triangle M_u \leq f(\sup_{t \leq r < u} M_r)(1 + X_u), \tag{88}$$

where $\triangle M_u = M_u - M_{u-}$. Then for $U_m = \inf\{u > t : M_u \geq x_m\}$,

$$\lim_{m \to \infty} E_Q\left[M_{U_m} 1_{\{U_m \in (t,T)\}}|\mathcal{F}_t\right] = M_t - E_Q[M_T|\mathcal{F}_t] \tag{89}$$

Proof. To simplify the notation, we omit the Q subscript on the expectations operator. Let T_n be a fundamental sequence of M_t. Then $M_t^{T_n} = E[M_T^{T_n}|\mathcal{F}_t]$ and hence

$$M_t^{T_n} = M_t^{T_n} 1_{\{U_m = t\}} + E[M_{U_m}^{T_n} 1_{\{U_m \in (t,T)\}}|\mathcal{F}_t] + E[M_T^{T_n} 1_{\{U_m = T\}}|\mathcal{F}_t] \tag{90}$$

By hypothesis $M_{U_m}^{T_n} \leq x_m + f(x_m)(1 + \triangle X_{U_m})$ and $M_T^{T_n} \leq x_m + f(x_m)(1 + X_T)$. By the bounded convergence theorem,

$$M_t = \lim_{n\to\infty} M_t^{T_n} = M_t 1_{\{U_m=t\}} + E[M_{U_m} 1_{\{U_m\in(t,T)\}}|\mathcal{F}_t] + E[M_T 1_{\{U_m=T\}}|\mathcal{F}_t] \quad (91)$$

Since X is a uniformly integrable martingale, it is in class D and $(X^\tau)_{\{\tau:\text{ stopping times}\}}$ is uniformly integrable. Fix m. Then $M_T^{T_n}$, $M_{U_m}^{T_n}$ are bounded by a sequence of uniformly integrable martingales. Therefore taking the limit with respect to n and interchanging the limit with the expectation yields:

$$M_t = \lim_{m\to\infty} E[M_{U_m} 1_{\{U_m\in(t,T)\}}|\mathcal{F}_t] + E[M_T|\mathcal{F}_t]. \quad (92)$$

□

Theorem 24. *Let M be a non negative local martingale with respect to $\mathbb{F}$ such that $\triangle M$ satisfies the condition* (88) *specified in Lemma 4. Let $G(x,t) : \mathbb{R}_+ \times [0,T] \to \mathbb{R}_+$ be a function such that*

- *$G(x,s) \leq G(x,t)$ for all $0 \leq s \leq t \leq T$*
- *For all $t \in [0,T]$, $G(x,t)$ is convex with respect to x.*
- *$\lim_{x\to\infty} \frac{G(x,t)}{x} = c$ for all $t \in [0,T]$,*

  then

$$\sup_{\tau\in[t,T]} E_Q[G(M_\tau,\tau)|\mathcal{F}_t] = E_Q[G(M_T,T)|\mathcal{F}_t] + (c\vee 0)(M_t - E_Q[M_T|\mathcal{F}_t]) \quad (93)$$

Proof of Theorem 24. To simplify the notation, we omit the Q subscript on the expectations operator. Suppose $c \leq 0$. Then by monotonicity with respect to t and Jensen's inequality applied to a convex function G and a non-negative local martingale M.,

$$\begin{aligned}\sup_{\tau\in[t,T]} E[G(M_\tau,\tau)|\mathcal{F}_t] \leq& \sup_{\tau\in[t,T]} E[G(M_\tau,T)|\mathcal{F}_t] \\ \leq& E[G(M_T,T)|\mathcal{F}_t] \\ \leq& \sup_{\tau\in[t,T]} E[G(M_\tau,\tau)|\mathcal{F}_t]\end{aligned} \quad (94)$$

and

$$\sup_{\tau\in[t,T]} E[G(M_\tau,\tau)|\mathcal{F}_t] = E[G(M_T,T)|\mathcal{F}_t]. \quad (95)$$

Suppose $c > 0$. Fix $\varepsilon > 0$. Then there exists $\xi > 0$ such that $\varepsilon > 0 \exists \xi > 0$ such that $\forall x > \xi$, $\frac{G(x,0)}{x} > c - \varepsilon$ and hence $\frac{G(x,u)}{x} > c - \varepsilon$ for all $u \in [0,T]$. Let $\{x_n\}_{n\geq 1}$ be a sequence in (ξ, ∞) such that $x_n \uparrow \infty$. Let

$$V_n = \inf\{u > t : M_u \geq x_n\} \wedge T. \tag{96}$$

Without loss of generality we can assume that $M_t < x_n$. Since $G(\cdot, t)$ is increasing in t,

$$\begin{aligned}\sup_{\tau\in[t,T]} E[G(M_\tau,\tau)|\mathcal{F}_t] &\geq E[G(M_{V_n}, S_n)|\mathcal{F}_t]\\ &= E[G(M_T,T)1_{\{V_n=T\}}|\mathcal{F}_t] + E[G(M_{V_n}, V_n)1_{\{V_n<T\}}|\mathcal{F}_t]\\ &\geq E[G(M_T,T)1_{\{V_n=T\}}|\mathcal{F}_t] + E[G(M_{V_n}, 0)1_{\{V_n<T\}}|\mathcal{F}_t]\end{aligned} \tag{97}$$

Since $M_{V_n} \geq x_n > \xi$, $G(M_{V_n}, 0) \geq (c-\varepsilon)M_{V_n}$. Next, let's take a limit of $n \to \infty$. By Lemma 4 applied with $\{V_n\}$ and the monotone convergence theorem,

$$\begin{aligned}&\lim_{n\to\infty}\sup_{\tau\in[t,T]} E[G(M_\tau,\tau)|\mathcal{F}_t]\\ &\geq \lim_{n\to\infty}\left\{E[G(M_T,T)1_{\{V_n=T\}}|\mathcal{F}_t] + (c-\varepsilon)E[M_{V_n}1_{\{V_n<T\}}|\mathcal{F}_t]\right\}\\ &\geq E[(G(M_T,T)|\mathcal{F}_t] + (c-\varepsilon)(M_t - E[M_T|\mathcal{F}_t]).\end{aligned} \tag{98}$$

Letting $\varepsilon \to 0$,

$$\sup_{\tau\in[t,T]} E[G(M_\tau,\tau)|\mathcal{F}_t] \geq E[G(M_T,T)|\mathcal{F}_t] + c\beta_t \tag{99}$$

To show the other direction, let $G^c(x,u) = cx - G(x,u)$. $G^c(x,\cdot)$ is a non-positive increasing concave function w.r.t x such that

$$\lim_{x\to\infty}\frac{G^c(\cdot,x)}{x} = 0 \tag{100}$$

By Jensen's inequality,

$$E[G^c(M_T,u)|\mathcal{F}_u] \leq G^c(E[M_T|\mathcal{F}_u],u) \leq G^c(M_u,u) \tag{101}$$

Therefore

$$\begin{aligned}G(M_u,u) &\leq c(M_u - E[G^c(M_T,u)|\mathcal{F}_u])\\ &= c\beta_u + E[G(M_T,u)|\mathcal{F}_u]\\ &\leq c\beta_u + E[G(M_T,T)|\mathcal{F}_u]\end{aligned} \tag{102}$$

Since this is true for all $u \in [t,T]$, $G(M_\tau,\tau) \leq c\beta_\tau + E[G(M_T,T)|\mathcal{F}_\tau]$ for all $\tau \in [t,T]$. By the tower property of martingales, and a supermartingale property,

$$E[G(M_\tau,\tau)|\mathcal{F}_t] \leq E[c\beta_\tau + E[G(M_T,T)|\mathcal{F}_\tau]|\mathcal{F}_t] \leq E[G(M_T,T)|\mathcal{F}_t] + c\beta_t. \tag{103}$$

Therefore

$$\sup_{\tau\in[t,T]} E[G(M_\tau,\tau)|\mathcal{F}_t] = E[G(M_T,T)|\mathcal{F}_t] + c\beta_t \tag{104}$$

□

This theorem extends Theorem B.2 of Cox and Hobson [30] in two ways: First, the assumption that a martingale M_t be continuous is dropped; and second, the payoff function $G(\cdot,x)$ permits a more general form and, in particular, an analysis of an American option in an economy with a non-zero interest rate.

Then, the following theorem is provable using standard techniques.

Theorem 25. *Assume NFLVR and No Dominance holds, and that the jump process of the asset's price, $\Delta S := (\Delta S_t)_{t\geq 0}$, where $\Delta S_t = S_t - S_{t-}$, satisfies the regularity conditions of Lemma 4. Then, for all K*

$$C_t^E(K) = C_t^A(K) = C_t^{A^\star}(K). \tag{105}$$

Proof. (i) By Theorem 24 with $G(x,u) = [x - K/A_u]^+$,

$$\begin{aligned} C^{A^\star}(K)_t &= \sup_{t\leq\tau\leq T} E[(S_\tau - K/A_\tau)^+|\mathcal{F}_t] \\ &= E[(S_T - K/A_T)^+|\mathcal{F}_t] + (S_t - E[S_T|\mathcal{F}_t]) \\ &= C_t^{E^\star}(K) + \beta_t^3 - E[\beta_T^3|\mathcal{F}_t] \\ &= C_t^E(K) \end{aligned} \tag{106}$$

The last equality is by Theorem 23. This equality implies, using Merton's original no dominance argument, that the American call option is not exercised early. The reason is that the European call's value is at least the value of a forward contract on the stock with delivery price K, and this exceeds the exercised value.

(ii) A unit of an American call option with arbitrary strike K is dominated by a unit of an underlying asset. Therefore by No Dominance (Definition 5),

$$C_t^A(K) \leq S_t. \tag{107}$$

Let $\gamma_t := C_t^A(K) - C_t^{A^\star}(K)$ be a bubble of an American call option with strike K. Since American options have finite maturity, γ_t is of type 3 and is a strict local martingale. Then by (i) and a decomposition of S_t,

$$\begin{aligned} C_t^{E*}(K) + \beta_t^3 - E[\beta_T^3|\mathcal{F}_t] + \gamma_t &= C_t^{A^\star}(K) + \gamma_t \\ &= C_t^A(K) \le S_t \\ &= S_t^\star + \beta_t^1 + \beta_t^2 + \beta_t^3, \end{aligned} \tag{108}$$

and therefore

$$\gamma_t \le [S_t^\star - C_t^{E*}(K) + \beta_t^1] + \beta_t^2 - E[\beta_T^3|\mathcal{F}_t]. \tag{109}$$

The right side of (109) is a uniformly integrable martingale on $[0, T]$. Hence γ is a non-negative local martingale dominated by a uniformly integrable martingale. Therefore $\gamma_t \equiv 0$. □

This theorem is the generalization of Merton's [114] famous no early exercise theorem, i.e. given the underlying stock pays no dividends, otherwise identical American and European call options have identical prices. This extension is the first equality in expression (105), applied to the options' *market prices*. Just as in the classic theory, this implies that an American call option on a stock with no dividends is not exercised early.

The second equality is particularly nice; if the reader has ever wondered what was the point of American call options, since they tend to behave similarly to European call options, the second equality gives a nice response: it implies that *American call option prices exhibit no bubbles, even if there is an asset price bubble!* This result follows because the stopping time associated with the American call's fundamental value (as distinct from the exercise strategy of the American call's market price) explicitly incorporates the price bubble into the supremum. Indeed, the fundamental value of the American call option is the minimal supermartingale dominating the value function. If there is a price bubble, then the stopping time associated with the American call option's fundamental value is stopped early with strictly positive probability. This is understood by examining the difference between the fundamental values of the European and American call. If stopping early had no value, then it must be true that $C_t^{A\star}(K) = C_t^{E\star}(K)$. However, By Theorem 23, an asset price bubble creates a difference between an American and European calls' *fundamental* prices, i.e.

$$C_t^{A\star}(K) - C_t^{E\star}(K) = \beta_t^3 - E_Q\left[\beta_T^3\middle|\, \mathcal{F}_t\right] > 0.$$

The intuition for the possibility of stopping early is obtained by recognizing that the market price equals the fundamental value plus a price bubble. The price bubble is a non-negative supermartingale that is expected to decline. Its effect on the market price of the stock is therefore equivalent to a continuous dividend payout. And, it is well known that continuous dividend payouts make early exercise of (the fundamental value of) an American call possible.

Indeed, in the presence of bubbles we need no longer have that the classic "no early exercise" theorem of Merton holds. S. Pal and P. Protter have shown the following in this regard:

Theorem 26 (Pal–Protter [122]). *Assume NFLVR holds. Suppose for a European option, the discounted pay-off at time T is given by a convex function $h(S_T)$ which is sub-linear at infinity, i.e., $\lim_{x\to\infty} h(x)/x = 0$. Then the price of the option is increasing with the time to maturity, T, whether or not a bubble is present in the market. In other words, $E(h(S_T))$ is an increasing function of T. For example, consider the put option with a pay-off $(K-x)^+$.*

However, for a European call option, the price of the option $E(S_T - K)^+$ with strike K might decrease as the maturity increases.

This feature may seem strange at first glance, but if we assume the existence of a financial bubble, the intuition is that it is advantageous to purchase a call with a short expiration time, since at the beginning of a bubble prices rise, sometimes dramatically. However in the long run it is disadvantageous to have a call, increasingly so as time increases, since the likelihood of a crash in the bubble taking place increases with time.

As observed in [122], pricing a European option by the usual formula when the underlying asset price is a strict local martingale is itself controversial. For example, Heston, Loewenstein, and Willard [63] observe that under the existence of bubbles in the underlying price process, put-call parity might not hold, American calls have no optimal exercise policy, and look-back calls have infinite value. Madan and Yor [110] have argued that when the underlying price process is a strict local martingale, the price of a European call option with strike price K should be modified as $\lim_{n\to\infty} E\left[(S_{T\wedge T_n} - K)^+\right]$, where $T_n = \inf\{t \geq 0 : S_t \geq n\}$, $n \in \mathbb{N}$, is a sequence of hitting times. This proposal does however, in effect, try to hide the presence of a bubble and act as if the price process is a true martingale under the risk neutral measure, rather than a strict local martingale.

American calls in the presence of bubbles have also recently been studied in a recent paper by Kardaras et al. [100]. They provide an analysis of the relation between bubbles and derivative pricing, incorporating and explaining previous work in the area. Also, using the approach pioneered by Fernholtz and Karatzas [50], Bayraktar et al. [9] show how to price an American call option in a market that does not necessarily admit an equivalent sigma martingale measure (i.e., in which the condition NFLVR for the absence of arbitrage does not hold everywhere). A subsequent work by Kardaras [99] studies exchange options, and here the mathematics becomes both complicated and interesting, with the possible presence of bubbles taken into account regarding the issue of put-call parity, a question originally raised by Cox and Hobson [30].

Finally, we remark here that we can also apply these ideas to a study of forwards and futures in the presence of bubbles. There are two unusual features that are worthy of note here for forwards and futures depending on an underlying risky commodity. First, a futures price can have its own bubble, one that is not present in the forward price. And second, when the underlying risky commodity asset has a bubble, the present value of the forward price is "equivalent" to the spot commodity, and therefore reflects all three types of bubbles, whereas the futures price is simply a bet on the market price S_T of the commodity at time T. When the futures price is viewed from time t the type 3 bubble component is excluded. For more, the interested reader can consult [84] where an explicit expression relating forward prices with futures prices, in the presence of bubbles and stochastic interest rates, is presented.

Another point worth mentioning is an implicit relationship between futures and bubbles. It is often believed that selling short should correct for bubbles, but we have explained that selling short is inadmissible as a strategy and thus cannot correct for bubbles. However just because selling short is too dangerous a strategy to be admissible certainly does not mean it is not pursued and does not exists; history is replete with examples of dangerous risks taken in the financial markets that lead sometimes to great riches, and sometimes to large financial catastrophes. Sometimes in the midst of a crash, such as the banking crisis of 2008, government imposed restrictions on short selling occur. On the face of it, this seems silly, since in most third world emerging markets, short selling is either not allowed or is not possible due to inadequate financial infrastructures (see [17, 25]), and we do not see more or longer lived bubbles in these markets. Nevertheless it is often said that short selling constraints on a given asset can be overcome by using trading strategies in futures contracts on that asset in order to replicate a short position. While this is not true in full generality, it is however largely true (see Jarrow et al. [91]). Therefore restrictions on short selling, in the presence of a lively futures market, are doomed to failure, even if in principle they could work. When bubbles crash, there appears to be no current effective palliative.

8 Foreign Exchange

A study of foreign currency bubbles is undertaken in Jarrow and Protter [85], and it is this approach we will follow here.[17] Of course this is a topic long studied in the academic literature, see for example the 1986 papers of Evans [48] and Meese [112]. Reasons for such bubbles to come into existence also have a long history in the economics literature; see Camerer [19] or Scheinkman and Xiong [139] for reviews. Using our martingale theory approach developed in this paper (with precedents in the work of Loewenstein and Willard [108], and Cox and Hobson [30]), and some of the resulting insights are the following:

1. A foreign currency exchange rate bubble is positive and its inverse exchange rate bubble is negative. This implies that, in contrast to asset price bubbles (financial

[17] We wish to thank Roy DeMeo of Morgan Stanley for stimulating discussion on bubbles and foreign exchange.

securities and commodities) that can only be positive, *foreign currency exchange rates can have negative bubbles*.

2. Foreign currency exchange rate bubbles are caused by price level bubbles in either or both of the relevant countries' currencies. Alternatively stated, foreign currency exchange rate bubbles reflect "distorted" inflation in either or both countries. By "distorted," we mean that the inflation is due to trading activity in the currency and not fundamental macroeconomic forces. This connection of bubbles to inflation has been recently studied with remarkable results by Carr et al. [21].
3. Domestic price level bubbles decrease the expected inflation rate in the relevant country. This counter intuitive result is due to the fact that bubbles, being supermartingales, are expected to decrease. Alternatively stated, bubbles are expected eventually to burst, thereby reducing the price level and the inflation rate.

Since we are dealing with foreign exchange, we need continually to specify the currency of which we are speaking. We will work with U.S. dollars ($) and Euros (€). To embed our foreign currency model in the previous model structure, we begin with our standard assumptions, assumed throughout this article: We have a filtered complete probability space $(\Omega, \mathcal{F}, \mathbb{F}, P)$ satisfying the usual hypotheses (see Footnote 5).

Let $\tau_{\$}$, τ_e be stopping times which represent the maturity (or life) of the U.S. and the European Union, respectively. Define $\tau = \min(\tau_{\$}, \tau_e)$, the economy's maturity date.

We assume trading in a dollar denominated money market account with value

$$A_t = \exp\left(\int_0^t r_u du\right) \tag{110}$$

where r_t is the dollar default free spot rate of interest, and we let $\hat{A}_t$ denote a euro denominated money market account with $\hat{r}_t$ the euro default free spot rate of interest. Next we let Y_t be the spot exchange rate of dollars per euro, and of course we assume that all of these processes are adapted with respect to the filtration $\mathbb{F}$.

The traded risky asset that we consider is the dollar value of the *euro money market account* (€ mma), i.e.

$$S_t = Y_t \hat{A}_t. \tag{111}$$

Note that using the notation from the previous section, we have that $D_t = 0$ for all t and $X_\tau = Y_\tau \hat{A}_\tau$. The dollar value of the euro money market has no cash flow and a terminal value equal to the dollar value of the € mma at the economy's maturity (which could be $+\infty$).

We assume that there are no arbitrage opportunities (NFLVR holds), hence, there exists an equivalent local martingale measure Q such that $\frac{S_t}{A_t}$ is a $Q-$ local martingale.[18] It is well known that the ELMM measure as identified herein depends crucially on using the dollar as the numéraire. A change in numéraire to the euro changes the perspective to a foreign investor, which in turn will change the martingale measure employed (see Amin and Jarrow [2], Sect. 5, pages 321–322.). In our context, fixing the numéraire determines the bubble's characterization. This dependency on the numéraire is necessary in the foreign currency context and it is related to the resolution of *Siegel's paradox*, a "paradox" that is well explained in the little book of Sondermann [146, pp. 74–84].

In order to characterize an exchange rate bubble we begin by defining the fundamental value of the dollar value of the € mma as

$$S_t^\star = E_Q\left(\frac{Y_\tau \hat{A}_\tau}{A_\tau}\middle|\mathcal{F}_t\right)A_t \tag{112}$$

The current market price is $S_t = Y_t\hat{A}_t$. Hence, the traded asset's price bubble (in dollars) is

$$\beta_t = S_t - S_t^\star \geq 0.$$

Because this is a traded asset, the price bubble must be nonnegative. But this is not the bubble in the exchange rate itself. To characterize the *exchange rate bubble*, we define the *fundamental (dollar/euro) exchange rate* as

$$Y_t^\star \equiv \frac{1}{\hat{A}_t}S_t^\star = E_Q\left(\frac{Y_\tau \hat{A}_\tau}{A_\tau}\middle|\mathcal{F}_t\right)\frac{A_t}{\hat{A}_t}. \tag{113}$$

The fundamental exchange rate is just the fundamental dollar value of the € mma divided by the euro value of the € mma. Hence, the *(dollar/euro) exchange rate bubble* is then

$$\beta_t^Y \equiv Y_t - Y_t^\star = Y_t - E_Q\left(\frac{Y_\tau \hat{A}_\tau}{A_\tau}\middle|\mathcal{F}_t\right)\frac{A_t}{\hat{A}_t} \geq 0. \tag{114}$$

We see that an exchange rate bubble exists if and only if the dollar value of the € mma has a price bubble. And, if it exists, the exchange rate bubble must be nonnegative. This is because we are using the dollar as the numéraire.

---

[18]To consider foreign currency derivatives, one would want to include trading in default free zero-coupon bonds in both dollars and euros. Then, the no arbitrage condition would be extended to include the discounted dollar values of the dollar zero-coupon bonds and the dollar value of the euro zero-coupon bonds (see Amin and Jarrow [2]).

Next, we consider the (euro/dollar) exchange rate $\frac{1}{Y_t}$. Defining the fundamental (euro/dollar) exchange rate to be $\frac{1}{Y_t^\star}$, we see that the bubble in the (euro/dollar) exchange rate is then given by

$$\beta_t^{\frac{1}{Y}} \equiv \left(\frac{1}{Y_t} - \frac{1}{Y_t^\star}\right) \leq 0,$$

which is negative. Hence, a negative bubble exists in this framework. Whether a bubble is positive or negative is a matter of perspective.

Remark 27. At first glance, it might seem as though combining $1/Y$ with the pricing measure associated to the dollar as numéraire seems artificial and does not quickly lend itself to an economic interpretation. However in this modern world it has immediate appeal. To give a banal example, imagine yourself as a world traveler. You might feel the dollar is over valued in relation to the euro. If you are right, this should reflect itself as a bubble in the dollar/euro exchange rate. Suppose you travel to the euro zone for a period of time for work and get a large payment in euros. When should you repatriate your euro earnings, by conversion into dollars? You now realize that the exchange rate $1/Y$, using your home currency the dollar as numéraire, is in a negative bubble, so you may choose to wait until that bubble ends. This applies analogously to businesses, of course. An example is that of the company Apple. According to many sources (see for example [13]) Apple has around \$1 trillion in profits sitting overseas. Apple would have a large U.S. tax bill were it to repatriate its profits, and claims to be waiting for the U.S. Congress to give a tax holiday to American multinational companies that wish to repatriate their foreign profits. Were the dollar to be in a bubble when such a holiday came (if it ever does), then presumably Apple would realize that its holdings in foreign currencies might be in a negative bubble, and the tax advantage of the tax holiday would be reduced or possibly eliminated by the negative bubble. This could affect Apple's actions.

Foreign Currency Price Bubbles and Inflation

We illustrate the ideas by considering an economy with a single consumption good, traded across economies. We let $\tau_\$$ and τ_e be stopping times which represent the maturity (or life) if the U.S. economy and the Euro zone economy, respectively. We let $\tau = \tau_\$ \wedge \tau_e$, the maturity of the joint economy. For interest rates, $r_\$(t)$ is the default free dollar spot rate of interest, and $B_\$(t)$ is the dollar value of a dollar money market account. We define a "real value" default rate free real spot rate of interest $r(t)$, and $B(t)$ is the "real value" of a money market account paying off in consumption goods. For convenience, we define

$$R(t) = \int_0^t r(s)ds.$$

By analogy for the dollar rates,

$$R_{\$}(t) = \int_0^t f_{\$}(s)ds, \tag{115}$$

and we let $\pi_{\$}(t)$ be the dollar price level of the consumption good. In essence, $\pi_{\$}(t)$ is the (dollar/cg) exchange rate (cg = consumption good). The inverse of the dollar price level, $\frac{1}{\pi_{\$}(t)}$, is the dollar deflator. The dollar deflator transforms dollars into consumption goods—real values. The rate of change in the dollar price level $\frac{d\pi_{\$}(t)}{\pi_{\$}(t)}$ is the dollar inflation rate. We define the same objects for the euro economy, r_e, R_e, and π_e analogously; these are of course denominated in euros.

The two traded assets of interest are the real value of the \$mma and the € mma, and these are

$$\frac{B_{\$}(t)}{\pi_{\$}(t)} \text{ and } \frac{B_e(t)}{\pi_e(t)}, \tag{116}$$

respectively. Given the trading of inflation protected bonds, the assumption of trading in these real-valued money market accounts is without loss of generality. Assuming we have NFLVR, we know there exists an equivalent probability measure Q such that

$$\frac{B_{\$}(t)}{\pi_{\$}(t)B(t)} \quad and \quad \frac{B_e(t)}{\pi_e(t)B(t)}$$

are Q sigma martingales, in this case local martingales, since the processes are nonnegative.

Note that when using the consumption good as the numéraire, the notion of no arbitrage takes on a new interpretation. No arbitrage in real values is the natural extension of *purchasing power parity*. Purchasing power parity states that the same consumption good has the same real price across all economies, after adjusting for the different currency exchange rates (see Taylor [150], Taylor and Taylor [151]). Also note that given the existence and frequency of trading in Treasury Inflation Protected Securities (TIPS), one can infer both the dollar and real term structure of interest rates from market data (see Jarrow and Yildirim [87]).

Dollar Price Bubbles

Let the traded asset be the real value of the dollar mma (\$mma) and its fundamental value $\left[\frac{B_{\$}(t)}{\pi_{\$}(t)}\right]^{\star}$, is equal to

$$\left[\frac{B_{\$}(t)}{\pi_{\$}(t)}\right]^{\star} = E_Q\left(\frac{B_{\$}(\tau)}{\pi_{\$}(\tau)B(\tau)}\middle|\mathcal{F}_t\right)B(t). \tag{117}$$

The traded asset's price bubble (in consumption goods) is

$$\beta_\$(t) = \frac{B_\$(t)}{\pi_\$(t)} - \left[\frac{B_\$(t)}{\pi_\$(t)}\right]^\star \geq 0. \tag{118}$$

We note that the bubble in the traded asset's price is nonnegative.

As before, this is not the bubble in the dollar price level. To derive this, we define the *fundamental dollar price level* as

$$\pi_\$^\star(t) \equiv \frac{B_\$(t)}{\left[\frac{B_\$(t)}{\pi_\$(t)}\right]^\star}. \tag{119}$$

The *dollar price level bubble* is then

$$\beta_\$^\pi(t) = \pi_\$(t) - \pi_\$^\star(t) \geq 0. \tag{120}$$

Note that the dollar price level bubble is with respect to the consumption good as the numéraire. It is nonnegative as well, since both $B_\$(t)$ and $\left[\frac{B_\$(t)}{\pi_\$(t)}\right]^\star$ are nonnegative.

The dollar inflation rate can be computed as

$$\frac{d\pi_\$(t)}{\pi_\$(t)} = \frac{\pi_\$^\star(t)}{\pi_\$(t)}\frac{d\pi_\$^\star(t)}{\pi_\$^\star(t)} + \frac{d\beta_\$^\pi(t)}{\pi_\$(t)}. \tag{121}$$

Taking expectations yields the expected dollar inflation rate

$$E_Q\left(\frac{d\pi_\$(t)}{\pi_\$(t)}\middle|\mathcal{F}_t\right) = \frac{\pi_\$^\star(t)}{\pi_\$(t)}E_Q\left(\frac{d\pi_\$^\star(t)}{\pi_\$^\star(t)}\middle|\mathcal{F}_t\right) + E_Q\left(\frac{d\beta_\$^\pi(t)}{\beta_\$(t)}\middle|\mathcal{F}_t\right).$$

Given a strictly positive dollar price level bubble, we have $\frac{\pi_\$^\star(t)}{\pi_\$(t)} < 1$. Given that the dollar price level bubble is a supermartingale, we have that

$$E_Q\left(\frac{d\beta_\$^\pi(t)}{\beta_\$(t)}\middle|\mathcal{F}_t\right) < 0.$$

Combined, we get the following result:

$$\textit{If}\ \ \beta_\$^\pi(t) > 0,\ \ \textit{then}\ \ E_Q\left(\frac{d\pi_\$(t)}{\pi_\$(t)}\middle|\mathcal{F}_t\right) < E_Q\left(\frac{d\pi_\$^\star(t)}{\pi_\$^\star(t)}\middle|\mathcal{F}_t\right). \tag{122}$$

That is, a dollar price level bubble decreases the dollar expected inflation rate from its fundamental level. Of course, there is nothing special here about the dollar, and the same analysis can be applied to the euro.

Currency Exchange Rate Bubbles

The dollar/euro exchange rate is given by

$$\frac{\pi_{\$}(t)}{\pi_{€}(t)}.$$

One can understand why this is true by considering the units of this ratio, where "cg" stands for consumption goods, i.e. $\frac{\frac{\text{dollar}}{\text{cg}}}{\frac{\text{euro}}{\text{cg}}} = \frac{\text{dollar}}{\text{euro}}$. The *fundamental dollar/euro exchange rate* is

$$\frac{\pi^{\star}_{\$}(t)}{\pi^{\star}_{€}(t)}.$$

The *dollar/euro exchange rate bubble* is

$$\beta_{\$/€}(t) = \left(\frac{\pi_{\$}(t)}{\pi_{€}(t)} - \frac{\pi^{\star}_{\$}(t)}{\pi^{\star}_{€}(t)}\right).$$

Recall that this is measured in consumption goods. In this context, we see that the dollar/euro exchange rate bubble can be either positive or negative, depending upon the magnitudes of the price level bubbles within each economy. However, if the dollar/euro exchange rate bubble is positive, then the euro/dollar exchange rate will be negative, and conversely. For much more on this subject, including the working out of illustrative examples, see [85].

9 Forwards and Futures

Futures have become an important element in modern day finance, especially if one judges by how much capital is tied up in them. The appeal of futures is that they reduce one's exposure to risk, since the accounts are settled in an ongoing and daily basis. Each future is of course intrinsically attached a risk asset, or basket of risky assets such as an index. Therefore it is interesting to examine whether or not they reflect a bubble in the underlying asset(s) should one occur, which is intuitively reasonable. However it might also be the case that futures themselves could develop their own bubbles, independent of the presence (or not) of a bubble in the underlying asset. This is perhaps less intuitive, but we will see that mathematically and theoretically it is indeed possible. Forwards and Futures are intimately related, and arose traditionally in relation to commodities, and for this reason we distinguish between cash settlement of a future and physical settlement, where the goods in question must be physically produced.[19] For this analysis, we rely on the published article [84].

We recall our usual framework, assumed throughout this article: Let $(\Omega, \mathcal{F}, \mathbb{F}, P)$ be a filtered complete probability space. We assume that the filtration $\mathbb{F} = (\mathcal{F}_t)_{t\geq 0}$ satisfies the "usual hypotheses."(See Footnote 5). Once again τ is a stopping time which represents the maturity (or life) of a risky asset, and $D = (D_t)_{0\leq t<\tau}$ is a (càdlàg) semimartingale adapted to $\mathbb{F}$, representing the cumulative cash flow process of the risky asset. ΔD_t can be positive or negative depending on the sign of the cash flows (e.g. storage costs are negative, dividends are positive). As before, $X_\tau \geq 0$ is an $\mathcal{F}_\tau$-measurable random variable representing the time τ terminal payoff or liquidation value of the asset. The *market price* of the risky asset is given by the non-negative semimartingale $S = (S_t)_{0\leq t\leq \tau}$. Note that for t such that $\Delta D_t \neq 0$, S_t denotes a price *ex-cash flows*, since S is càdlàg.

Let r_t be a non-negative semimartingale representing the default free spot rate of interest. We define a money market account A_t by

$$A_t = \exp\left(\int_0^t r_u du\right). \tag{123}$$

Note that $A_t \geq 1$ is continuous and non-decreasing.

One again W denotes a wealth process on $t \in [0, \infty)$ associated with the market price of the risky asset, i.e.

$$W_t = S_t \mathbf{1}_{\{t<\tau\}} + A_t \int_0^{t\wedge\tau} \frac{1}{A_u} dD_u + A_t \frac{X_\tau}{A_\tau} \mathbf{1}_{\{\tau\leq t\}}. \tag{124}$$

The market value of the wealth process is the position in the risky asset plus all accumulated cash flows, and the terminal payoff if $t \geq \tau$.[20] Note that the cash flows are invested in the money market account to keep the wealth process self-financing. We assume that (D, X_τ) are such that $W \geq 0$, i.e. holding the risky asset has non-negative value. This condition is needed to be consistent with the non-negativity of the risky asset's price process. Finally, we assume that there exists a probability measure Q *equivalent to* P *such that the wealth process* $\frac{W}{A}$ is a Q local martingale, so that NFLVR applies, by the First Fundamental Theorem of Asset Pricing.

Second, we do not assume such a Q is unique, hence the market is *incomplete*. Instead, in order to uniquely identify the price of a derivative security, we assume that the market selects a unique ELMM from the collection of all possible ELMMs.

---

[19] The author spent over 20 years at Purdue University in Indiana, and there he developed an appreciation for the importance of pork belly futures, for example.

[20] When considering non-financial commodities, this expression implicitly assumes that the risky asset is storable.

For example, this will be the case if enough static trading in call options exist as discussed in Sect. 5.

The Market Price Operator

To study forward and futures contracts, we need the concept of a *market price* operator. To do this we let $T < \infty$ represent some fixed future time that exceeds the maturity dates of all relevant forward and futures contracts. Also $\phi = A_T \int_t^T \frac{d\Delta_u}{A_u} + \Xi^T$ denotes a time T payoff, starting at time $t \leq T$, where: (a) $\Delta = (\Delta_t)_{0 \leq t \leq T}$ is an arbitrary semimartingale representing the asset's cumulative cash flow process, and (b) $\Xi^T \in \mathcal{F}_T$ is a random variable that represents the asset's terminal payoff at time T. Note that both of these quantities may be negative. The payoff ϕ is in $\mathcal{F}_T$. Then $\Phi_0(t)$ represents the collection of all these $\mathcal{F}_T$ measurable random variables, where one begins at time t when computing the payoff. Define $\Phi(t) \equiv \{\phi \in \Phi_0(t) : E_Q(|\phi|) < \infty\}$ where $E_Q(\cdot)$ denotes expectation under Q. By construction, $\Phi(t)$ is a linear space.

Define $\Phi_m(t) \subset \Phi(t)$ to be the linear combination of the random variables generated by all admissible and self-financing trading strategies involving the risky asset and money market account and all static trading strategies involving forward and futures contracts, and European call and put options on the risky asset. Note that both $W_T, A_T \in \Phi(0)$ where

$$W_T = 1_{\{T<\tau\}} S_T + A_T \int_0^{T \wedge \tau} \frac{1}{A_u} dD_u + A_T \frac{X_\tau}{A_\tau} \mathbf{1}_{\{\tau \leq T\}}.$$

As written, this expression extends the time domain of the risky asset wealth process beyond time τ.

We assume that we are given a unique *market price* operator[21] $\Lambda_t : \Phi_m(t) \to \mathbb{L}^0(\Omega, \mathcal{F}_t, P)$ that gives for each $\phi \in \Phi_m(t)$, its time t market price $\Lambda_t(\phi)$. Note that (in the presence of bubbles) the uniqueness of the market price operator is an additional assumption beyond the existence of an ELMM Q. We do not assume that Λ_t extends uniquely to the set $\Phi(t)$. For future reference, we note that by the definition of the market price operator, we have that both $\Lambda_t(A_T) = A_t$ and $\Lambda_t(W_T) = S_t$.

We need to impose two additional assumptions on the market price operator. Consistent with no arbitrage, the first is sometimes known as the "law of one price."

Assumption 28 (Linearity). *Given $\phi', \phi \in \Phi_m(t)$ and $a, b \in \mathbb{R}$, we have that $a\Lambda_t(\phi') + b\Lambda_t(\phi) = \Lambda_t(a\phi' + b\phi)$ for all t.*

That is, we assume that a portfolio of two assets trades for the same price as the cost of constructing the portfolio by trading in the individual assets themselves. We also

[21] $\mathbb{L}^0(\Omega, \mathcal{F}_t, P)$ is the collection of finite valued $\mathcal{F}_t$ measurable functions on Ω.

assume No Dominance, as defined in Sect. 7. In this framework, for $\phi', \phi \in \Phi_m(t)$, we say that ϕ' *dominates* ϕ if either of the following conditions holds

1. $Q(\phi' \geq \phi) = 1$ *and* $Q(\phi' > \phi) > 0$ *and* $\Lambda_t(\phi') \leq \Lambda_t(\phi)$ for some t almost surely.
2. $Q(\phi' = \phi) = 1$ *and* $\Lambda_t(\phi') < \Lambda_t(\phi)$ for some t almost surely.

If ϕ' were to dominate ϕ, then conceptually if one could short ϕ and go long ϕ', NFLVR would imply that no dominated assets exist in the economy. However, because of the admissibility condition, one cannot always short ϕ and hold it until time T. For example, one cannot short sell the risky asset and hold it until time T if the risky asset's price process is unbounded above. This is the reason that we need to assume no dominance directly.

Assumption 29 (No Dominance). *There are no dominated assets in the market.*

We can now define the fundamental price in terms of this market operator.

Definition 7 (Fundamental Price and Bubbles). Define the fundamental price $\Lambda_t^* : \Phi_m(t) \to \mathbb{L}^0(\Omega, \mathcal{F}_t, P)$ of $\phi = A_T \int_t^T \frac{d\Delta_u}{A_u} + \Xi^T \in \Phi_m(t)$ by

$$\Lambda_t^*(\phi) \equiv E_Q\left(\int_t^T \frac{d\Delta_u}{A_u} + \frac{\Xi^T}{A_T}\middle| \mathcal{F}_t\right) A_t, \text{ and} \tag{125}$$

define its bubble $\delta_t : \Phi_m(t) \to \mathbb{L}^0(\Omega, \mathcal{F}_t, P)$ *by*

$$\delta_t(\phi) \equiv \Lambda_t(\phi) - \Lambda_t^*(\phi). \tag{126}$$

Note that, by construction, δ_t is a linear function and $\delta_T(\phi) = 0$, i.e. any bubble disappears by time T. The linearity follows from the linearity of both Λ_t and Λ_t^*.

In an NFLVR economy, all discounted market prices under a risk neutral measure must be sigma martingales. [22] Hence, without loss of generality, we assume

Assumption 30 (Local Martingale Bubbles). $\frac{\delta_t(\phi)}{A_t}$ *is a Q sigma martingale.*

Theorem 31 (Bounded Assets). *If $\phi \in \Phi_m(t)$ is bounded, then $\delta_t(\phi) = 0$.*

Proof. If ϕ is bounded, then there exists $a > 0$ such that $\left|A_T \int_t^T \frac{d\Delta_u}{A_u} + \Xi^T\right| \leq a$. Then, investing a dollars in the money market account implies by no dominance that $\Lambda_t(\phi) \leq a\Lambda_t(A_T) = aA_t$. This implies that the Q sigma martingale $\frac{\Lambda_t(\phi)}{A_t}$ is bounded, and hence a martingale (see [128]). By expression (126), $\delta_t(\phi) = 0$. □

---

[22] Sigma martingales are defined and discussed for example in [76, 128]. When a sigma martingale is continuous, or bounded below, it is a local martingale. Otherwise, in general, local martingales are a proper subset of sigma martingales.

Forward Prices

A forward contract is a financial contract written on a risky asset S that obligates the owner (the long) to purchase the risky asset on the delivery date T for a predetermined price, called the *forward price*. If the contract is written at time t, denote the forward price by $f_{t,T}$. The payoff to the forward contract at delivery is $[S_T - f_{t,T}] \in \Phi_m(t)$. By market convention, the forward price is selected such that the forward contract has zero initial value. We consider forwards to commodities. For our analysis, we only consider underlying risky assets (commodities) whose liquidation dates exceed the maturity of the contract, e.g. gold, oil, a stock index. So, without loss of generality, we assume that $T < \tau$. We define

$$\mathrm{div}_{t,T} \equiv \Lambda_t(A_T \int_t^{T\wedge\tau} \frac{1}{A_u} dD_u). \tag{127}$$

where economically $\mathrm{div}_{t,T}$ represents the market price of the cash flow stream for the time interval $[t, T]$. We have in this context

$$S_t = \Lambda_t(S_T) + \mathrm{div}_{t,T} \tag{128}$$

and

$$S_t = E_Q\left(\frac{S_T}{A_T} + \int_t^T \frac{dD_u}{A_u}\middle| \mathcal{F}_t\right) A_t + \beta_t^3 - E_Q\left(\frac{\beta_T^3}{A_T}\middle| \mathcal{F}_t\right) A_t. \tag{129}$$

Consider $W_T = S_T \mathbf{1}_{\{T<\tau\}} + A_T \int_0^{T\wedge\tau} \frac{1}{A_u} dD_u + A_T \frac{X_\tau}{A_\tau} \mathbf{1}_{\{\tau\le T\}} \in \Phi_m(0)$. This represents the time T payoff from buying the risky asset at time t. Then,

$$S_t \equiv \Lambda_t\left(S_T \mathbf{1}_{\{T<\tau\}} + A_T \int_t^{T\wedge\tau} \frac{1}{A_u} dD_u + A_T \frac{X_\tau}{A_\tau} \mathbf{1}_{\{\tau\le T\}}\right).$$

Let us define some simpler notation. Let

$$\hat{S}_T \equiv S_T \mathbf{1}_{\{T<\tau\}} + A_T \frac{X_\tau}{A_\tau} \mathbf{1}_{\{\tau\le T\}}$$

and

$$\mathrm{div}_{t,T} \equiv \Lambda_t(A_T \int_t^{T\wedge\tau} \frac{1}{A_u} dD_u).$$

These represent the payoff to the risky asset at time T (less cash flows prior to T) and the market price of the cash flow stream between $[t, T]$, respectively. Then, using linearity of the market price operator, we obtain

$$S_t = \Lambda_t(\hat{S}_T) + \mathrm{div}_{t,T}. \tag{130}$$

Here, $\Lambda_t(\hat{S}_T) = S_t - \text{div}_{t,T}$ represents the time t market price of the payoff to the risky asset at time T.

Now, the payoff to the risky asset $\left(\hat{S}_T + A_T \int_t^{T\wedge\tau} \frac{1}{A_u} dD_u\right)$ has the bubble component given by

$$\begin{aligned}
&\delta_t\left(\hat{S}_T + A_T \int_t^{T\wedge\tau} \frac{1}{A_u} dD_u\right) \\
&= \Lambda_t\left(\hat{S}_T + A_T \int_t^{T\wedge\tau} \frac{1}{A_u} dD_u\right) - \Lambda_t^*\left(\hat{S}_T + A_T \int_t^{T\wedge\tau} \frac{1}{A_u} dD_u\right) \\
&= S_t - E_Q\left(\frac{\hat{S}_T}{A_T} + \int_t^{T\wedge\tau} \frac{1}{A_u} dD_u \middle| \mathcal{F}_t\right) A_t. \qquad (131)
\end{aligned}$$

We can relate the time t bubble component of $\left(\hat{S}_T + A_T \int_t^{T\wedge\tau} \frac{1}{A_u} dD_u\right)$ to our usual bubble of S_t, since under the assumption that $T < \tau$ we have that $\hat{S}_T = S_T$, leading to simplifications of the formulae. Using the fundamental price of the risky asset, S_t^*, we have

$$S_t^* = E_Q\left(\int_t^{\tau} \frac{1}{A_u} dD_u + \frac{X_\tau}{A_\tau} \mathbf{1}_{\{\tau<\infty\}} \middle| \mathcal{F}_t\right) A_t \qquad (132)$$

and the asset price bubble is β given by

$$\beta_t = S_t - S_t^*, \qquad (133)$$

when $S \geq 0$. Again, the above expressions simplify since $\hat{S}_T = S_T$.

Given these definitions, and using the notation established in Sect. 3, we have two simple theorems:

Theorem 32 (Forward Price).

$$f_{t,T} \cdot p(t,T) = S_t - div_{t,T} \qquad (134)$$

Proof. By definition of the contract $0 = \Lambda_t(S_T - f_{t,T})$. Linearity implies $0 = \Lambda_t(S_T) - f_{t,T}\Lambda_t(1_T)$. Using (128) and the notation for the zero coupon bond yields the final result. $0 = S_t - \text{div}_{t,T} - f_{t,T}\, p(t,T)$. □

Theorem 33 (Forward Price Bubbles).

$$f_{t,T} \cdot p(t,T) = S_t^* - div_{t,T} + \beta_t \textit{ where } \beta_t = S_t - S_t^*.$$

$$f_{t,T} \cdot p(t,T) = E_Q\left(\frac{S_T}{A_T}\middle| \mathcal{F}_t\right) A_t + \beta_t^3 - E_Q\left(\frac{\beta_T^3}{A_T}\middle| \mathcal{F}_t\right) A_t - \delta_t\left(A_T \int_t^T \frac{dD_u}{A_u}\right).$$

Proof. By (134), we obtain $f_{t,T} \cdot p(t,T) + \text{div}_{t,T} = S_t$, the first property follows. Finally, $f_{t,T} \cdot p(t,T) = \Lambda_t(S_T) = E_Q\left(\frac{S_T}{A_T}\middle|\mathcal{F}_t\right)A_t + \delta_t(S_T) = E_Q\left(\frac{S_T}{A_T}\middle|\mathcal{F}_t\right)A_t + \beta_t^3 - E_Q\left(\frac{\beta_T^3}{A_T}\middle|\mathcal{F}_t\right)A_t - \delta_t\left(A_T\int_t^T \frac{dD_u}{A_u}\right)$. The last equality uses the identity

$$\delta_t\left(\hat{S}_T + A_T\int_t^{T\wedge\tau}\frac{dD_u}{A_u}\right) = \beta_t^3 - E_Q\left(\frac{\beta_T^3}{A_T}\middle|\mathcal{F}_t\right)A_t. \tag{135}$$

This yields the second property. □

Futures Prices

A futures contract is similar to a forward contract. It is a financial contract written on the risky asset S,with a fixed maturity T. It represents the purchase of the risky asset at time T via a prearranged payment procedure. The prearranged payment procedure is called marking-to-market. Marking-to-market obligates the purchaser (long position) to accept a continuous cash flow stream equal to the continuous changes in the futures prices for this contract.

The time t *futures prices*, denoted $F_{t,T}$, are set (by market convention) such that newly issued futures contracts (at time t) on the same risky asset with the same maturity date T, have zero *market value*. Hence, futures contracts (by construction) have zero market value at all times, and a continuous cash flow stream equal to $dF_{t,T}$. At maturity, the last futures price must equal the asset's price $F_{T,T} = S_T$. Note that even with zero market value at all times, a futures contract can be worth a lot to an investor.

Let us construct a portfolio long one futures contract. The wealth process of this portfolio at time T is given by

$$A_T\int_0^T \frac{1}{A_u}dF_{u,T} \in \Phi_m(0). \tag{136}$$

Note that we do not a priori require futures prices $(F_{t,T})_{t\geq 0}$ to be non-negative.

Our definition of the Futures price below is a definition which depends on the processes themselves, and not (in the case of an incomplete market, where there are an infinite number of risk neutral measures) on the choice of a risk neutral measure. In this sense, we are following Definition 3.6 found in the book of Karatzas and Shreve [98, p. 45]. Of course, this is in contrast to the classical definition of the futures price, see Duffie [42, p. 143] or Shreve [144, p. 244], where futures price bubbles are excluded by fiat. Using our futures price process characterization, we can investigate the relationship between the futures price and the risky asset's price bubbles.

Definition (Futures Price). The futures price process $(F_{t,T})_{t\geq 0}$ is any càdlàg semimartingale process such that

$$\Lambda_t(A_T \int_t^T \frac{1}{A_u} dF_{u,T}) = 0 \textit{ for all } t \in [0,T] \quad \textit{and}$$

$$F_{T,T} = S_T.$$

Note that while this definition is the same as given in [84], it is different from the original definition in Jarrow et al. [89], where a futures price process is defined independently of the market price operator. The original definition does not explicitly use the fact that the futures price is that price which makes the futures contract have zero value. In contrast, the new definition does. The new definition nevertheless yields the same theorem as in Jarrow et al. [89], Theorem 7.3, that futures prices can have their own bubbles that are unrelated to any bubble in the underlying asset's price. In fact, a futures price bubbles can be positive or negative. This is in contrast to bubbles in the underlying asset's price process.

Theorem 34 (Futures Price Bubbles). *Let $(\gamma_u)_{u\geq t}$ be a local Q martingale with $\gamma_t = 0$. Then,*

$$F_{t,T} = E_Q\left(S_T \mid \mathcal{F}_t\right) + \gamma_T \tag{137}$$

is a futures price process.

Proof. We need to show that $\Lambda_t(A_T \int_t^T \frac{1}{A_u} dF_{u,T}) = 0$ for all $t \in [0,T]$ and $F_{T,T} = S_T$. The second condition is true by inspection. To facilitate the notation, let $F_t^* = E_Q\left(S_T \mid \mathcal{F}_t\right)$.

$$\begin{aligned}
0 &= \Lambda_t(A_T \int_t^T \frac{1}{A_u} dF_{u,T}) \\
&= \Lambda_t^*(A_T \int_t^T \frac{1}{A_u} dF_{u,T}) \frac{1}{A_t} + \delta_t(A_T \int_t^T \frac{1}{A_u} dF_{u,T}) \frac{1}{A_t} \\
&= E_Q\left(\int_t^T \frac{1}{A_u} dF_u^* \middle| \mathcal{F}_t\right) \frac{1}{A_t} + E_Q\left(\int_t^T \frac{1}{A_u} d\gamma_u \middle| \mathcal{F}_t\right) \frac{1}{A_t} + \delta_t(A_T \int_t^T \frac{1}{A_u} dF_{u,T}) \frac{1}{A_t}
\end{aligned}$$

But $E_Q\left(\int_t^T \frac{1}{A_u} dF_u^* \middle| \mathcal{F}_t\right) \frac{1}{A_t} = 0$. So,

$\delta_t(A_T \int_t^T \frac{1}{A_u} dF_{u,T}) = -E_Q\left(\int_t^T \frac{1}{A_u} d\gamma_u \middle| \mathcal{F}_t\right)$. This identity guarantees the value of the futures contract is always zero. □

We record the following useful corollary which is a slight generalization of Theorem 3.7, p. 45, of [98].

Corollary 5. *Let* $E_Q\left([F_{\cdot,T}, F_{\cdot,T}]_t^{\frac{1}{2}}\right) < \infty$ for all $0 \le t \le T$. *A futures contract has no bubbles if and only if*

$$F_{t,T} = E_Q\left(S_T \middle| \mathcal{F}_t\right). \tag{138}$$

Proof. If $F_{t,T} = E_Q\left(S_T \middle| \mathcal{F}_t\right)$, then $\gamma_t \equiv 0$ and the statement follows from the theorem. If there are no bubbles then $\delta_t(A_T \int_t^T \frac{1}{A_u} dF_{u,T})\frac{1}{A_t} = 0$. But,

$$\begin{aligned}0 &= \Lambda_t(A_T \int_t^T \frac{1}{A_u} dF_{u,T}) \\ &= \Lambda_t^*(A_T \int_t^T \frac{1}{A_u} dF_{u,T})\frac{1}{A_t} + \delta_t(A_T \int_t^T \frac{1}{A_u} dF_{u,T})\frac{1}{A_t} \\ &= E_Q\left(\int_t^T \frac{1}{A_u} dF_{u,T} \middle| \mathcal{F}_t\right).\end{aligned}$$

Hence, $\int_0^t \frac{1}{A_u} dF_{u,T} \equiv M_t$ is a martingale (compute the conditional expectation).

Then, $Y_t \equiv \int_0^t A_u dM_u = F_{t,T} - F_{0,T}$ is a martingale since $E_Q\left([Y,Y]_t^{\frac{1}{2}}\right) < \infty$ for all $0 \le t \le T$. (See [128].) This implies $F_{t,T} = E_Q\left(S_T \middle| \mathcal{F}_t\right)$ is a uniformly integrable $\mathcal{H}^1$ martingale on $[0,T]$. □

Corollary 6. *If a market is complete, futures processes price bubbles do not exist.*

Proof. Assuming No Dominance in a complete market, it is a consequence of the results of [88] that the process δ is zero. So Corollary (5) gives the result. □

Theorem 35 (Futures Price Bubbles).

$$\begin{aligned}F_{t,T} &= E_Q\left(A_T \middle| \mathcal{F}_t\right)\left(S_t^* - div_{t,T}\right) + cov_Q\left(\frac{S_T}{A_T}, A_T \middle| \mathcal{F}_t\right) \\ &\quad + \beta_t - \left[\beta_t^3 - E_Q\left(\frac{\beta_T^3}{A_T} \middle| \mathcal{F}_t\right) A_t - \delta_t\left(A_T \int_t^T \frac{dD_u}{A_u}\right)\right] + \gamma_t \end{aligned} \tag{139}$$

$$F_{t,T} = E_Q\left(A_T \middle| \mathcal{F}_t\right) E_Q\left(\frac{S_T}{A_T} \middle| \mathcal{F}_t\right) A_t + cov_Q\left(\frac{S_T}{A_T}, A_T \middle| \mathcal{F}_t\right) + \gamma_t \tag{140}$$

Proof. First, algebra yields

$E_Q\left(S_T \middle| \mathcal{F}_t\right) = E_Q\left(A_T \middle| \mathcal{F}_t\right) E_Q\left(\frac{S_T}{A_T} \middle| \mathcal{F}_t\right) + cov_Q\left(\frac{S_T}{A_T}, A_T \middle| \mathcal{F}_t\right)$.
This gives property (140).
Now, $\Lambda_t(S_T) = E_Q\left(\frac{S_T}{A_T} \middle| \mathcal{F}_t\right) A_t + \delta_t(S_T)$. Hence,
$E_Q\left(S_T \middle| \mathcal{F}_t\right) = E_Q\left(A_T \middle| \mathcal{F}_t\right)\left(\Lambda_t(S_T) - \delta_t(S_T)\right) + cov_Q\left(\frac{S_T}{A_T}, A_T \middle| \mathcal{F}_t\right)$.

But, $\Lambda_t(S_T) = S_t - \mathrm{div}_{t,T}$, $S_t = S_t^* + \beta_t$, and
$\delta_t(S_T) = \beta_t^3 - E_Q\left(\frac{\beta_T^3}{A_T}\middle|\mathcal{F}_t\right)A_t - \delta_t\left(A_T\int_t^T \frac{dD_u}{A_u}\right)$.
Substitution yields property (139). □

Property (139) shows that, modulo its own bubble γ_t, the futures price inherits the first two types of bubbles present in the risky asset price $\beta_t^1 + \beta_t^2$, but not the third β_t^3. It omits the type 3 bubble because the futures price is a bet on the market price of the risky asset S_T at time T. And, when viewed from time t, this market price already excludes $\left[\beta_t^3 - E_Q\left(\frac{\beta_T^3}{A_T}\middle|\mathcal{F}_t\right)A_t - \delta_t\left(A_T\int_t^T \frac{dD_u}{A_u}\right)\right]$. Property (140) is just the classical relationship between the futures and the spot price of the risky asset modified for the existence of the futures price bubble.

Forward vs Futures Prices

This section relates forward and futures prices. In the classical literature (see [31] and/or [82]) it is known that forward and futures prices are equal under deterministic interest rates, but unequal (in general) otherwise. To facilitate a comparison with the classical literature and to develop some intuition concerning forward and futures price bubbles, we first study an economy with deterministic interest rates before analyzing the general case.

Deterministic Interest Rates

For this subsection, we let the spot rate be a deterministic function of time. For this section only, we assume that $A_T(S_T - F_{0,T}) \in \Phi_m(0)$.

Theorem 36 (Deterministic Interest Rates).

$$F_{t,T} = f_{t,T} \text{ for all } t.$$

Proof. This logic is from Cox et al. [31].

Strategy 1: Let us consider the following trading strategy. At each time $t \in [0, T]$, go long $N(t)$ units of the futures contract. At each $t + dt$, invest the proceeds from the futures contract into the money market account (if negative, short). This implies we purchase $N(t)dF_{t,T}$ dollars of the money market account at time $t + dt$, or $\frac{N(t)dF_{u,T}}{A_{t+dt}}$ units. Note that A_t is continuous, so $A_{t+dt} = A_t$. Hold this position until time T. Because futures contracts always have zero value and reinvestment in the money market account has no cost, this strategy is self financing. Let the value of this portfolio be denoted $G(t)$. Note $G(0) = 0$. Then $G(t) = A_t\int_0^t \frac{N(u)}{A_u}dF_{u,T}$. Of course, we are interested in time T. Next choose $N(t) = A_t$. Then $G(T) = A_T\int_0^T \frac{A_u}{A_u}dF_{u,T} = A_T(F_{T,T} - F_{0,T})$. But $F_{T,T} = S_T$

whence $G(T) = A_T(S_T - F_{0,T})$. Note that by assumption, $A_T(S_T - F_{0,T}) \in \Phi_m(0)$.

Strategy 2: Consider the following trading strategy with a forward contract. At time 0 go long $\frac{1}{p(0,T)}$ forward contracts and hold until time T. This is self financing since it is a buy and hold position. Let the value of this portfolio be denoted $H(t)$. Note $H(0) = 0$. Then $H(T) = \frac{1}{p(0,T)}(S_T - f_{0,T})$. Now we apply the assumption of no dominance, and make a comparison at time T.

$$G(T) = A_T(S_T - F_{0,T}),\ H(T) = \frac{1}{p(0,T)}(S_T - f_{0,T}).$$

Under deterministic interest rates $\frac{1}{A_T} = p(0,T)$. Then both these strategies give the same payoff at time T. To avoid dominance, $0 = \Lambda_0(G(T)) = \Lambda_0(H(T))$. Linearity of Λ_0 implies that $F_{0,T} = f_{0,T}$. □

This implies that under deterministic interest rates, the classical relation holds.

Stochastic Interest Rates

We now consider the general case.

Theorem 37 (Stochastic Interest Rates).

$$f_{t,T} = F_{t,T} + cov_Q\left(S_T, \frac{1}{A_T}\middle| \mathcal{F}_t\right)\frac{A_t}{p(t,T)} - \gamma_t + \frac{\beta_t^3}{p(t,T)} - E_Q\left(\frac{\beta_T^3}{A_T}\middle| \mathcal{F}_t\right)\frac{A_t}{p(t,T)} - \delta_t\left(A_T\int_t^T \frac{dD_u}{A_u}\right). \quad (141)$$

Proof. Using expression (135), we get:

$$f_{t,T}\cdot p(t,T) = E_Q\left(S_T\middle| \mathcal{F}_t\right)E_Q\left(\frac{1}{A_T}\middle| \mathcal{F}_t\right)A_t + cov_Q\left(S_T, \frac{1}{A_T}\middle| \mathcal{F}_t\right)A_t + \beta_t^3 - E_Q\left(\frac{\beta_T^3}{A_T}\middle| \mathcal{F}_t\right)A_t - \delta_t\left(A_T\int_t^T \frac{dD_u}{A_u}\right).$$

Combine this with expression (140) to get:

$$f_{t,T}\cdot p(t,T) = (F_{t,T} - \gamma_t)\,p(t,T) + cov_Q\left(S_T, \frac{1}{A_T}\middle| \mathcal{F}_t\right)A_t + \beta_t^3 - E_Q\left(\frac{\beta_T^3}{A_T}\middle| \mathcal{F}_t\right)A_t - \delta_t\left(A_T\int_t^T \frac{dD_u}{A_u}\right).$$

Algebra generates the final result. □

This theorem relates forward prices to futures prices. The first covariance term is the classical difference between forward and futures prices. However, there are two additional differences. First, a futures price can have its own bubble γ_t not present in the forward price. Second, when the risky asset price has a bubble, there is an additional difference reflecting the type 3 bubble. The reason for this difference is that the (present value of the) forward price is "equivalent" to the spot commodity, and hence reflects all three types of bubbles. In contrast, the futures price is a bet on the market price S_T of the commodity at time T. When viewed from time t, this excludes the type 3 bubble component. Hence, expression (141).

We have not considered here how options interact with futures and bubbles. For this and more, we refer the reader to [84].

10 Testing for Bubbles in Real Time

No matter how many symptoms of the coming trouble there may have been, panics always come with a shock and a tremendous surprise and disappointment.
–President W.H. Taft, "The Panic of 1907," a speech given before the Merchants Association of Boston, Massachusetts, December 30, 1907; see [149, p. 212].

It might seem self evident that the presence of bubbles in the prices of risky financial assets is an important phenomenon to understand. Economists have studied it for a long time, but it is only within the last 10 years that the mathematical finance community has been trying to understand and analyze the phenomenon, and this paper is hopefully part of that effort. But going beyond understanding how it happens to the detection of *when it is happening* (if not necessarily why it happens, which is more properly the domain of economists [see for example [55] or more recently [67]]) seems especially timely, given the often disastrous consequences of the aftermath of large, economy or sector wide bubbles. But it also interesting on a more individual level, both for investors for the obvious reasons, but also for regulators for a more subtle reason. An example perhaps is that of banks and large financial institutions. After the banking crisis in the U.S. in 2008, and the banking crisis in much of Europe in 2011/2012, the detection of underlying bubbles is especially important. One reason, for example, is in the evaluation of capital reserves. Banks are required to hold capital reserves roughly in proportion to their capital at risk.[23] This is important for banking health, and helps to prevent runs on banks, but it does cut into profits, since capital reserves are not available for risky investment opportunities. Left to themselves, and in the presence of competition, banks would whittle away at their capital reserves until they were meaningless; thus it is important that government regulators ensure that proper capital reserves are maintained. To do this, regulators must evaluate capital reserves, and if some

[23] How one measures capital at risk (involving Value at Risk and the theory of risk measures) is another thorny issue that we do not even attempt to address in this article.

significant proportion of those reserves are in assets undergoing bubble pricing, then they are worth less than the face value at which they are undoubtedly evaluated, through the marked to market procedures.

This might help to explain why the US Federal Reserve is repeatedly questioned about what it plans to do on the subject of financial bubbles. Indeed, Federal Reserve Chairman Ben Bernanke said in 2009 at his confirmation hearings [12]:

> It is extraordinarily difficult in real time to know if an asset price is appropriate or not.

Dr. Bernanke is correct: Without a quantitative procedure, experts often have different opinions about the existence of price bubbles. A famous example is the oil price bubble of 2007/2008. Nobel prize winning economist Paul Krugman wrote in the New York Times that it was not a bubble, and 2 days later Ben Stein wrote in the same paper that it was.

William Dudley, the President of the New York Federal Reserve, in an interview with Planet Money in 2010 [59] stated

> ...what I am proposing is that we try to identify bubbles in real time, try to develop tools to address those bubbles, try to use those tools when appropriate to limit the size of those bubbles and, therefore, try to limit the damage when those bubbles burst.

A third example is from a report by Claire Baldwin of Reuters [8] of June 2, 2011:

> When LinkedIn shares jumped 109.4 % on their first day of trade, Chicago Fed president Charles Evans said he was withholding judgment over whether a new dot-com bubble was under way. “I have no way of knowing that those aren’t just exactly the right valuations,” Mr Evans told reporters after a speech in Chicago.

And a fourth example (that we found in [125]) comes from Donald Kohn, Federal Reserve Board Vice Chairman, who on March 24, 2010 declared:

> Federal Reserve policymakers should deepen their understanding about how to combat speculative bubbles to reduce the chances of another financial crisis.

S.M Davidoff, writing in 2011 in the *New York Times* [33] made a case for a gold bubble, and then in the same article made a case for there not being a gold bubble; this author found both of his arguments to be convincing(!). His article inspired the investigation [80].

Finally we note that the method proposed here for bubble detection is only one proposed of many. See for example [148] where the author (Matt Swayne, an *eHow* contributor) purports to be able to detect a gold bubble. What perhaps distinguishes the method presented in this paper from others such as that of Swayne is that it is mathematically and statistically based; although this is not to say it is without controversy. We discuss the leading two alternative methods, and some of this controversy, in Sect. 11.

For a risky asset such as a given stock, we need to be able to tell whether or not, under the risk neutral measure, the asset price is a martingale, or is only a local martingale which is not a true martingale (called a *strict local martingale*). This is incredibly hard to do, but there are a few situations where we have a chance to do so.

Indeed, we have already presented these situations in Sect. 4. We have three cases: that of the stock price following a stochastic differential equation of the form (where B is a standard Brownian motion):

$$dX_t = \sigma(X_t)dB_t + \mu_t dt; \quad X_0 = x \tag{142}$$

and the cases of the theorem of Andersen and Piterbarg (Theorem 7) and that of Lions and Musiela (Theorem 8) which handle situations that fit into what is known as the Heston paradigm of stochastic volatility. As far as we know, these last two situations have not been exploited for the purposes of bubble detection, and they might be quite difficult to analyze due to the precision required in order for the confidence intervals to be of reasonable size. However the framework of (142) has indeed been studied, and such an analysis appears in the articles [78–80]. We present a review of it here. We note that we do not require the stock price to follow (142) at all times, only during the period of investigation. One could have instead of (142) a regime change model (for example, see [62]), where during different periods of volatility the stock price might evolve according to different stochastic differential equations.

Due to the presence of the drift term $\mu_t dt$ in (142) we can assume we are dealing with an incomplete market model. However since all risk neutral measures in effect remove the drift, under any risk neutral measure Q the price process X will follow the same equation

$$dX_t = \sigma(X_t)dB_t; \quad X_0 = x \tag{143}$$

By the results presented in Sect. 4 under any of the risk neutral measures we have that X in (143) is a strict local martingale if and only if the non-random calculus integral

$$\int_\alpha^\infty \frac{x}{\sigma(x)^2}dx < \infty; \quad \text{any } \alpha > 0 \tag{144}$$

Therefore to determine whether or not X of (143) is a strict local martingale, we "only" need to know the function $x \mapsto \sigma(x)$, and in particular to know it for asymptotically large values of x. This is an impossible task. First of all, it is completely non-trivial to estimate accurately the function $\sigma(x)$ from data. The good news is that this is the subject of a fair amount of research, and Jean Jacod has effectively solved this issue in two important papers [71, 72]. We outline our own approach to this problem as well. The bad news is that one can only "know" the coefficient $\sigma(x)$ at those values x that the stock price X attains. Since any stock price is a fortiori bounded in range, in a finite time interval, we cannot know the asymptotic behavior of $\sigma(x)$ no matter how accurately we can estimate it for those x in the range of X. At this juncture, we could simply give up; but instead we try to do the best we can do, with the information we have. Therefore we smooth our estimate of σ where we can know it, and we analyze its behavior. It seems to be often the case that the behavior of σ is clear, and if it seems to be tending off to ∞ as $x \nearrow \infty$, then we make the leap that this behavior will continue even where we do not see σ. Thus the problem reduces to the issue of the asymptotic rate in which σ tends to ∞, when it does. We have tested this idea with data, and it seems to work in almost all of the cases in which we have tested it. By "seems to work" we mean that when the asset being tested went through a bubble, out test indicates that it did so. When the asset did not go through a bubble, our test indicates that there was no bubble. And when it is not obvious whether or not the asset went through a bubble, our test gives any of three results: a bubble, no bubble, or the test fails to decide. So let us now proceed to the method.

Estimation of the Diffusion Coefficient in a Bounded Domain

In addition to the work of Jacod discussed above [71, 72], many authors have proposed estimators for the volatility function $\sigma(x)$. D. Florens–Zmirou [52] proposed a non parametric estimator based on the local time of the diffusion process. We present this estimator later in this section, when we treat the example of Infospace. (See Theorems 44 and 45.) V. Genon Catalot and J. Jacod [56] proposed an estimation procedure for parameterized volatility functions. M. Hoffmann [64] constructs a wavelets based estimator.

In the article [78] we introduce a smooth kernel estimator, in the same spirit as that of Jacod in [72]. The estimator is constructed from the two quantities:

$$V_n^x = \frac{1}{nh_n}\sum_{i=0}^{n-1}\phi(\frac{S_{\frac{i}{n}} - x}{h_n})n(S_{\frac{i+1}{n}} - S_{\frac{i}{n}})^2 \tag{145}$$

$$L_n^x = \frac{1}{nh_n}\sum_{i=0}^{n-1}\phi(\frac{S_{\frac{i}{n}} - x}{h_n}) \tag{146}$$

The kernel function ϕ is a C^6 positive function with compact support and such that $\int_{\mathbb{R}_+}\phi = 1$. We are interested in the convergence of V_n^x and L_n^x to $\sigma^2(x)L^x$ and L^x respectively, where h_n satisfies $nh_n^2 \to \infty$. The following theorem is established in [78], where h_n is a sequence of positive real numbers converging to 0 and satisfying some constraints:

Theorem 38. *If $nh_n^2 \to \infty$ then $S_n^x = \frac{V_n^x}{L_n^x}$ converges in probability to $\sigma^2(x)$ and provides a consistent estimator of $\sigma^2(x)$.*

Remark 39. In [72] Jacod is able to take $h_n = \frac{1}{\sqrt{n}}$ and he also obtains a rate of convergence and an associated Central Limit Theorem. His method of proof is a bit more complicated that the one presented in [78].

Estimation of the Diffusion Coefficient's Asymptotic Behavior

Again we let h_n be a sequence of positive real numbers converging to 0 and satisfying some constraints. We construct an estimator of $\sigma(x)$ given by:

$$S_n(x) = \frac{\sum_{i=1}^{n} 1_{\{|S_{t_i}-x|<h_n\}} n(S_{t_i+1} - S_{t_i})^2}{\sum_{i=1}^{n} 1_{\{|S_{t_i}-x|<h_n\}}}. \tag{147}$$

The previous estimator for the volatility function $\sigma(x)$, presented in Theorem 38 is over a compact domain representing the observation interval. In this section, for the stochastic differential equation (143) we relax this boundedness assumption on the volatility function $\sigma(x)$. We now assume that $\sigma > 0$ on $I =]0, \infty[$, it is identically null elsewhere and it satisfies $\frac{1}{\sigma^2} \in L^1_{loc}(I)$.

This is the Engelbert Schmidt condition (see, e.g., [47] or [97]) under which the SDE has a unique weak solution S that does not explode to ∞. We let P be the law of the solution on the canonical space $\Omega = C([0, T], \mathbb{R})$ equipped with the canonical filtration $(\mathcal{F}_t)_{t\in[0,T]}$ and the canonical process $S = (S_t)_{t\in[0,T]}$. We also assume that σ is C^3 bounded and with bounded derivatives on every compact set. We add in passing that these hypotheses imply the existence of a strong solution, as well. Let $\tau_0(S)$ be the first time S hits zero. The next theorem is again taken from [78].

Theorem 40. *Suppose $\sigma(x)$ has three continuous derivatives. Assume that $nh_n^4 \to 0$ and $nh_n \to \infty$. Then conditional on $\{\tau_0(S) > T\}$, $S_n(x)$ given in (147) converges in probability to $\sigma^2(x)$. The same holds for our smooth kernel estimator under the constraint $nh_n^2 \to \infty$.*

Sketch of a Proof. Let $T_q = \inf\{t, S_t \geq q\}$ and $\tau_p = \inf\left\{t, S_t \leq \frac{1}{p}\right\}$. Then $\lim_{p\to\infty} \tau_p = \tau_0(S)$ and $\lim_{q\to\infty} T_q = \infty$ since S does not explode to ∞. We can take $\sigma_{p,q}$ to be a function bounded above and below away from zero with three bounded derivatives such that $\sigma_{p,q}(x) = \sigma(x)$ for all $\frac{1}{p} \leq x \leq q$. Let $(S_t^{p,q})_{t\in[0,T]}$ be the unique strong solution to the SDE $dS_t^{p,q} = \sigma_{p,q}(S_t^{p,q})dW_t$. Introduce now $S_n^{p,q}(x)$, the estimator computed on the basis of $(S_t^{p,q})_{t\in[0,T]}$ as in (147) or using our smooth kernel estimator. Then under suitable constraints on the sequence $(h_n)_{n\geq 1}$, $S_n^{p,q}(x)$ converges in probability to $\sigma_{p,q}^2(x)$. Moreover $S_n^{p,q}(x) = S_n(x)$ if $T < T_q \wedge \tau_p$. Then it follows that $S_n(x)$ converges in probability to $\sigma^2(x)$, in restriction to the set $\{T < \tau_0(S)\}$. □

Note that we have shown only the convergence of the estimators to the function σ. We can also obtain confidence intervals giving the accuracy of our predictions using the central limit results of Jacod, as mentioned in the above Remark 39, but we do not do so here. Such techniques are treated in detail in the recent book [74].

Bubble Detection

As we already discussed in the first part of this section (Sect. 10), while we can estimate σ reasonably accurately, we can only do so on the part of the domain of σ that is given by the range of X. But whether or not X is a strict local martingale under any and all risk neutral measures we need to determine whether or not the integral allows one to decide whether or not the following integral converges:

$$\int_{\alpha}^{\infty} \frac{x}{\sigma(x)^2} ds \quad \text{any } \alpha > 0. \tag{148}$$

We recall that if (148) is finite, then X is a strict local martingale; otherwise it is a true martingale. Therefore we need to know the behavior of $x \mapsto \sigma(x)$ as $|x| \to \infty$. Our procedure uses the theory of *Reproducing Kernel Hilbert Spaces (RKHS)* and it consists of two steps:

- We first interpolate an estimate of σ within the bounded interval where we have observations, and in this way we lose the irregularities of non parametric estimators.
- We next extrapolate our function σ by choosing a *RKHS* from a family of Hilbert spaces in such a way as to remain as close as possible (on the bounded interval of observations) to the interpolated function provided in the previous step.

This represents a new methodology which allows us to choose a *good* extrapolation method. We do this via the choice of a certain extrapolating RKHS, which—once chosen—determines the tail behavior of our volatility σ. If we let $(H_m)_{m\in\mathbb{N}}$ denote our family of RKHS, then any given choice of m, call it m_0, allows us to interpolate *perfectly* the original estimated points, and thus provides a valid *RKHS* H_m with which we extrapolate σ. But this represents a choice of m_0 and not an estimation. So if we stop at this point the method would be as arbitrary as parametric estimation. That is, choosing m_0 is analogous to choosing the parameterized family of functions which fits σ best. The difference is that we do not arbitrarily choose m_0. Instead we choose the index m *given the data available*. In this sense we are using the data twice. To do this we evaluate different RKHS's in order to find the most appropriate one *given the arrangement of the finite number of grid points* from our observations.

The *RKHS* method (see [65, 78]) is intimately related to the reconstruction of functions from scattered data in certain linear functional spaces. The reproducing kernel $Q(x, x')$ that is associated with an *RKHS* $H(\mathcal{D})$ in the spatial domain $\mathcal{D}$, over the coordinate x, is unique and positive and thus constitutes a natural basis for generic interpolation problems.

Reproducing Kernel Hilbert Spaces

Let $H(\mathcal{D})$ be a Hilbert space of continuous real valued functions $f(x)$ defined on a spatial domain $\mathcal{D}$. A reproducing kernel Q possesses useful properties for data interpolation and function approximation problems.

Theorem 41. *There exists a kernel function* $Q(x, x^{'})$*, the reproducing kernel, in* $H(\mathcal{D})$ *such that the following properties hold:*

(i) ***Reproducing property.*** *For all x and y, and for all* $f \in H(\mathcal{D})$*,*

$$f(x) = \langle f(x^{'}), Q(x, x^{'})\rangle^{'}$$
$$Q(x, y) = \langle Q(x, x^{'}), Q(y, x^{'})\rangle^{'}.$$

The prime indicates that the inner product $\langle\cdot,\cdot\rangle^{'}$ *is performed over* $x^{'}$*.*

(ii) ***Uniqueness.*** *The RKHS* $H(\mathcal{D})$ *has one and only one reproducing kernel* $Q(x, x^{'})$*.*

(iii) ***Symmetry and Positivity.*** *The reproducing kernel* $Q(x, x^{'})$ *is symmetric, i.e.* $Q(x^{'}, x) = Q(x, x^{'})$*, and positive definite, i.e.:*

$$\sum_{i=1}^{n}\sum_{k=1}^{n} c_i Q(x_i, x_k) c_k \geq 0$$

for any set of real numbers c_i *and for any countable set of points* $(x_i)_{i\in[1,n]}$.

For a proof of this theorem, we refer the reader to the classic works of N. Aronszajn [5,6].

In this framework, interpolation is seen as an inverse problem. The inverse problem is the following. Given a set of real valued data $(f_i)_{i\in[1,M]}$ at M distinct points $S_M = x_i, i \in [1, M]$ in a domain $\mathcal{D}$, and a *RKHS* $H(\mathcal{D})$, find a suitable function $f(x)$ that interpolates these data points. Using the reproducing property, this interpolation problem is reduced to solving the following linear inverse problem:

$$\forall i \in [1, M], f(x_i) = \langle f(x^{'}), Q(x_i, x^{'})\rangle^{'} \tag{149}$$

where we need to invert this relation and exhibit the function $f(x)$ in $H(\mathcal{D})$. We refer the reader to [65] for a detailed discussion.

We first present the normal solution that allows an exact interpolation, and second the regularized solution that yields quasi interpolative results, accompanied by an error bound analysis. Then in the next section, we will construct a family of *RKHS*'s that enable us to interpolate not $\sigma(x)$ but $\frac{1}{\sigma(x)^2}$. This transformation makes natural the choice of the family of *RKHS*'s. Note that for every choice of an *RKHS*, one can construct an interpolating function using the input data. For this reason, we define a family of Reproducing Kernel Hilbert Spaces that encapsulate different assumptions on the asymptotic forms and smoothness constraints. From this set, we choose that *RKHS* which best fits the input data in the sense explained below.

Normal Solutions: The most straightforward interpolation approach is to find the normal solution that has the minimal squared norm $||f||^2 = \langle f(x'), f(x')\rangle'$ subject to the interpolation condition (149).

That is, given a set of real valued data $\{f_i\}, 1 \leq i \leq K$ specified at K distinct points in a domain $\mathcal{D}$, we wish to find a function f that is the normal solution:

$$f(x) = \sum_{i=1}^{M} c_i Q(x_i, x)$$

where the coefficients c_i satisfy the linear relation:

$$\forall k \in [1, M], \sum_{i=1}^{M} c_i Q(x_i, x_k) = f_k. \tag{150}$$

If the matrix Q_M whose entries are $Q(x_i, x_k)$ is "well conditioned," then the linear algebraic system above can be efficiently solved numerically. Otherwise, we use regularized solutions.

Regularized Solutions: When the matrix Q_M is "ill conditioned," regularization procedures may be invoked for approximately solving the linear inverse problem. In particular, the Tikhonov regularization procedure produces an approximate solution f_α, which belongs to $H(\mathcal{D})$ and that can be obtained via the minimization of the regularization functional

$$||Qf - F||^2 + \alpha||f||^2$$

with respect to $f(x)$.[24] Note that here F is the data vector (f_i) and the residual norm $||Qf - F||^2$ is defined as:

$$||Qf - F||^2 = \sum_{i=1}^{M} (\langle f(x'), Q(x_i, x')\rangle' - f_i)^2.$$

The regularization parameter α is chosen to impose a proper balance between the residual constraint $||Qf - F||$ and the magnitude constraint $||f||$. The regularized solution has the form

$$f_\alpha(x) = \sum_{i=1}^{M} c_i^\alpha Q(x_i, x) \tag{151}$$

[24] See for example [65] for the details of how to go about this.

where the coefficients c_i^α satisfy the linear relation:

$$\forall k \in [1, M], \sum_{i=1}^{M} c_i^\alpha (Q(x_i, x_k) + \alpha\delta_{i,k}) = f_k \quad (152)$$

where $\delta_{i,k}$ is the Kronecker delta function. Note that for $\alpha > 0$, Q_M^α whose entries are $[Q(x_i, x_k) + \alpha\delta_{i,k}]$ is symmetric and positive definite and the problem can now be solved efficiently. Also, the *RKHS* interpolation method leads to an automatic error estimate of the regularized solution (see [65] for more details).

10.1 Construction of the Reproducing Kernels

We consider reciprocal power reproducing kernels that asymptotically behave as some reciprocal power of x, over the interval $[0, \infty[$. We are interested in this type of *RKHS* because this is a reasonable assumption for $f(x) = \frac{1}{\sigma^2(x)}$. The CEV model[25] $dS_t = S_t^\gamma dW_t$ where $\gamma > 0$ is a local volatility model proposed in the literature and satisfies this assumption, with $f_{cev}(x) = \frac{1}{x^{2\gamma}}$. We also assume that the function $f(x)$ possesses the asymptotic property

$$\lim_{x \to \infty} x^k f^{(k)}(x) = 0, \forall k \in [1, n-1].$$

for some $n \geq 1$ that controls the minimal required regularity. This property is often satisfied by the volatility functions used in practice. For instance, $x^k f_{cev}^{(k)}(x) = \frac{\prod_{i=0}^{k-1}(-2\gamma - i)}{x^{2\gamma}}$ converges to 0 as x tends to infinity, for all k. This is also satisfied by many volatility functions that explode faster than any power of x, for example $\sigma(x) = x^\gamma e^{\beta x}$, with $\gamma > 0$ and $\beta > 0$. The condition appears restrictive only when σ and its derivatives explode too slowly or when σ is bounded, however in these cases, it is likely that there is no bubble and no extrapolation using this *RKHS* theory will be required. We would like to emphasize that the asymptotic property satisfied by f is the key point for the whole method to work as this may be seen from Proposition 2 below.

Concerning the degree of smoothness, we usually take in practice n to be 1, 2 or 3. We can define now our Hilbert space

$$H_n = H_n([0, \infty[) = \left\{ f \in C^n([0, \infty[) \mid \lim_{x \to \infty} x^k f^{(k)}(x) = 0, \forall k \in [1, n-1] \right\}.$$

[25] "CEV" stands for constant elasticity of volatility.

We next need to define an inner product. A smooth reproducing kernel $q^{RP}(x, x')$ can be constructed via the choice:

$$< f, g >_{n,m}= \int_0^{\infty} \frac{y^n f^{(n)}(y)}{n!} \frac{y^n g^{(n)}(y)}{n!} \frac{dy}{w(y)}$$

where $w(y) = \frac{1}{y^m}$ is the asymptotic weighting function. From now on we consider the *RKHS* $H_{n,m} = (H_n, <, >_{n,m})$. The next proposition can be shown following the steps in [65].

Proposition 1. *The reproducing kernel is given by*

$$q_{n,m}^{RP}(x, y) = n^2 x_>^{-(m+1)} B(m + 1, n) F_{2,1}(-n + 1, m + 1, n + m + 1, \frac{x_<}{x_>})$$

where $x_>$ and $x_<$ are respectively the larger and smaller of x and y, $B(a, b)$ is the beta function and $F_{2,1}(a, b, c, z)$ is Gauss's hypergeometric function.

Remark 42. The integers $n - 1$ and $m + 1$ are respectively the order of smoothness and the asymptotic reciprocal power behavior of the reproducing kernel $q^{RP}(x, y)$. This kernel is a rational polynomial in the variables x and y and has only a finite number of terms, so it is computationally efficient.

As pointed out above, any choice of n and m creates an *RKHS* $H_{n,m}$ and allows one to construct an interpolating function $f_{n,m}(x)$ with a specific asymptotic behavior. The following result gives the exact asymptotic behavior.

Proposition 2. *For every x, $q^{RP}(x, y)$ is equivalent to $\frac{n^2}{y^{m+1}} B(m + 1, n)$ at infinity as a function of y and*

$$\lim_{x\to\infty} x^{m+1} f_\alpha(x) = n^2 B(m + 1, n) \sum_{i=1}^{M} c_i^\alpha$$

where f_α is defined as in (151) *and the constants c_i^α are obtained as in* (152). *Hence, if $\sum_{i=1}^{M} c_i^\alpha \neq 0$, then $f_\alpha(x)$ is equivalent to $\frac{n^2 B(m+1,n)}{x^{m+1}} \sum_{i=1}^{M} c_i^\alpha$.*

Choosing the Best *m*

The choice of m allows us to decide if the integral in (148) converges or diverges. If $m > 1$, there is a bubble. This section explains how to choose m. Let us first summarize the idea. We choose the *RKHS* by optimizing over the asymptotic weight m that allows us to construct a function that interpolates the input data points and remains as close as possible to the interpolated function on the finite interval $\mathcal{D}$. This optimization provides an $\overline{m}$ which allows us to construct $\sigma_{\overline{m}}(x)$. We employ a four step procedure:

Procedure 43. ***(i) Non-parametric estimation over $\mathcal{D}$:*** *Estimate $\sigma(x)$ using our non-parametric estimator on a fixed grid $x_1, \ldots, x_M$ of the bounded interval $\mathcal{D} = [\min S, \max S]$ where $\min S$ and $\max S$ are the minimum and the maximum reached by the stock price over the estimation time interval $[0, T]$. In our illustrative examples, we use the kernel $\phi(x) = \frac{1}{c} e^{\frac{1}{4x^2-1}}$ for $|x| < \frac{1}{2}$, where c is the appropriate normalization constant. The number of data available n and the restriction on the sequence $(h_n)_{n \geq 1}$ makes the number of grid points M relatively small in practice. In our numerical experiments, $7 \leq M \leq 25$.*

(ii) Interpolate $\sigma(x)$ over $\mathcal{D}$ using RKHS theory: *Use any interpolation method on the finite interval $\mathcal{D}$ to interpolate the data points $(\sigma(x_i))_{i \in [1,M]}$. Call the interpolated function $\sigma^b(x)$. For completeness, we provide a methodology to achieve this using the RKHS theory. However, any alternative interpolation procedure for a finite interval could be used.*

Define the Sobolev space: $H^n(\mathcal{D}) = \{u \in L^2(\mathcal{D}) \mid \forall k \in [1, n], u^{(k)} \in L^2(\mathcal{D})\}$ where $u^{(k)}$ is the weak derivative of u. The norm that is usually chosen is $||u||^2 = \sum_{k=0}^n \int_{\mathcal{D}} (u^{(k)})^2(x)dx$. Due to Sobolev inequalities, an equivalent and more appropriate norm is $||u|| = \int_{\mathcal{D}} u^2(x)dx + \frac{1}{\tau^{2n}} \int_{\mathcal{D}} (u^{(n)})^2(x)dx$. We denote by $K_{n,\tau}^{a,b}$ the kernel function of $H^n(]a, b[)$, where in this case $\mathcal{D} =]a, b[$. This reproducing kernel is provided for $n = 1$ and $n = 2$ in the following lemma.

Lemma 5.

$$K_{1,\tau}^{a,b}(x, y) = \frac{\tau}{\sinh(\tau(b-a))} \cosh(\tau(b - x_>)) \cosh(\tau(x_< - a))$$

$$K_{2,\tau}^{a,b}(x, y) = L_{x_>}(x_<)$$

and $L_x(t)$ is of the form $\sum_{i=1}^4 \sum_{k=1}^4 l_{ik} b_i(\tau t) b_k(\tau x)$.

We refer to [152, Eq. (22) and Corollary 3 on page 28] for explicit analytic expressions for l_{ik} and b_k, which while simple, are nevertheless tedious to write. In both equalities, $x_>$ and $x_<$ respectively stand for the larger and smaller of x and y. In practice, one should check the quality of this interpolation and carefully study the outputs by choosing different τ's before using the interpolated function $\sigma^b = \frac{1}{\sqrt{f^b}}$ in the algorithm detailed above, where $f^b(x) = \sum_{i=1}^M c_i^b K_{n,\tau}^{\mathcal{D}}(x_i, x)$, for all $x \in \mathcal{D}$ and for all $k \in [1, M]$, $\sum_{i=1}^M c_i^b K_{n,\tau}^{\mathcal{D}}(x_i, x_k) = f_k = \frac{1}{\sqrt{\sigma^{est}(x_k)}}$.

(iii) Deciding if an extrapolation is required: *If the interpolated estimate of $\sigma(x)$ appears to be a bounded function and not tending to $+\infty$ as $x \mapsto \infty$, or if the implicit extended form of the interpolated estimate of $\sigma(x)$ implies that the volatility does not diverge to ∞ as $x \to \infty$ and remains bounded on $\mathbb{R}^+$, no extrapolation is required. In such a case $\int_\epsilon^\infty \frac{x}{\sigma^2(x)} dx$ is infinite and the process is a true martingale. If one decides, however, that $\sigma(x)$ diverges to ∞ as $x \to \infty$, then the next step is required to obtain a "natural" candidate for its asymptotic behavior as a reciprocal power.*

(iv) Extrapolate $\sigma^b(x)$ ***to*** $\mathbb{R}^+$ ***using RKHS:*** *Fix* $n = 2$ *and define*

$$\overline{m} = \arg\min_{m \geq 0} \sqrt{\int_{[a,\infty[\cap \mathcal{D}} |\sigma_m - \sigma^b|^2 ds} \tag{153}$$

where $f_m = \frac{1}{\sigma_m^2}$ *is in the RKHS* $H_{2,m} = (H_{2,m}([0,\infty[), \langle,\rangle_{RP})$. *By definition, all* σ_m *will interpolate the input data points and* $\sigma_{\overline{m}}$ *has the asymptotic behavior that best matches our function on the estimation interval.* a *is the threshold determining closeness to the interpolated function. Choosing* a *too small is misleading since then it would account more (and unnecessarily) for the interpolation errors over the finite interval* $\mathcal{D}$ *than is desirable. We should choose a large* a *since we are only interested in the asymptotic behavior of the volatility function. In the illustrative examples below, the threshold* a *in* (153) *is chosen to be* $a = \max S - \frac{1}{3}(\max S - \min S)$.

Illustrative Examples from the Internet Dotcom Bubbles of 1998–2001

We illustrate our testing methodology for price bubbles using some stocks that are often alleged [111,159] as experiencing internet dot com bubbles. We consider those stocks for which we have high quality tick data. The data was obtained from WRDS [161]. We apply this methodology to four stocks: *Lastminute.com, eToys, Infospace, and Geocities*. The methodology performs well. The weakness of the method is the possibility of inconclusive tests as illustrated by *eToys*. For *Lastminute.com* and *Infospace* our methodology supports the existence of a price bubble. For *Infospace*, we reproduce the methodology step-by-step. Finally, the study of *Geocities* provides a stock commonly believed to have exhibited a bubble (see for instance [111,159]), but for which our method says it did not. We now provide our analyzes.

Lastminute.com: Our methodology confirms the existence of a bubble. The stock prices are given in Fig. 4.

The optimization performs as expected with the asymptotic behavior given by $\overline{m} = 8.26$, which means that $\sigma(x)$ is equivalent at infinity to a function proportional to x^α with $\alpha = 4.63$. We plot in Fig. 5 the different extrapolations obtained using different reproducing kernel Hilbert spaces $H_{2,m}$ and their respective reproducing kernels $q_{2,m}^{RP}$.

Figure 5 shows that m is between 7 and 9 as obtained by the optimization procedure. The orange curve labelled (sigma) is the interpolation on the finite interval $\mathcal{D}$ obtained from the non-parametric estimation procedure where the interpolation is achieved using the *RKHS* theory as described in step **(ii)** with the choice of the reproducing kernel Hilbert space $H^1(\mathcal{D})$ and the reproducing kernel $K_{1,6}^{\min S, \max S}$. Then m is optimized as in step **(iv)** so that the interpolating function $\sigma_{\overline{m}}(x)$ is as close as possible to the orange curve in the last third of the domain $\mathcal{D}$, i.e. the threshold a in (153) is chosen to be $a = \max S - \frac{1}{3}(\max S - \min S)$.

Fig. 4 Lastminute.com stock prices during the alleged dot com bubble

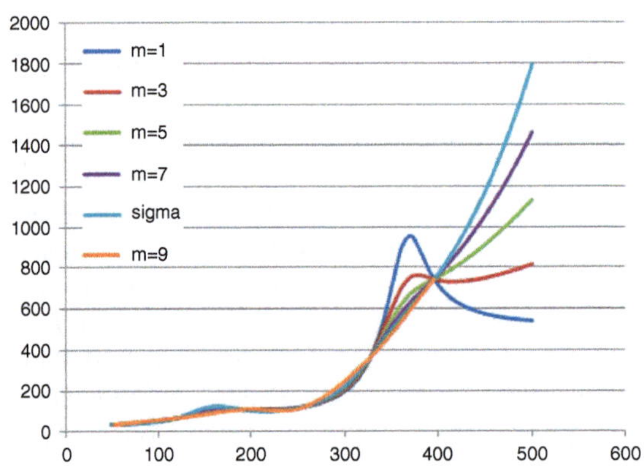

Fig. 5 Lastminute.com. RKHS estimates of $\sigma(x)$

eToys: While the graph of the stock price of eToys as given in Fig. 6 makes the existence of a bubble plausible, *the test nevertheless is inconclusive*. Different choices of m giving different asymptotic behaviors are all close to linear (see Fig. 7).

Because they are so close to being linear, we cannot tell with any level of assurance that the integral in question diverges, or converges. We simply cannot decide which is the case. If it were to diverge we would have a martingale (and hence no bubble), and were it to converge we would have a strict local martingale (and hence bubble pricing).

The estimated $\overline{m}$ is close to one. In Fig. 7, the powers α are given by $\frac{1}{2}(m+1)$ where m is the weight of the reciprocal power used to define the Hilbert space and its inner product. We plot the extrapolated functions obtained using different Hilbert spaces $H_{2,m}$ together with their reproducing kernels $q_{2,m}^{RP}$. Figure 8 shows that the extrapolated functions obtained using these different *RKHS* $H_{2,m}$ produce the same quality of fit on the domain $\mathcal{D}$.

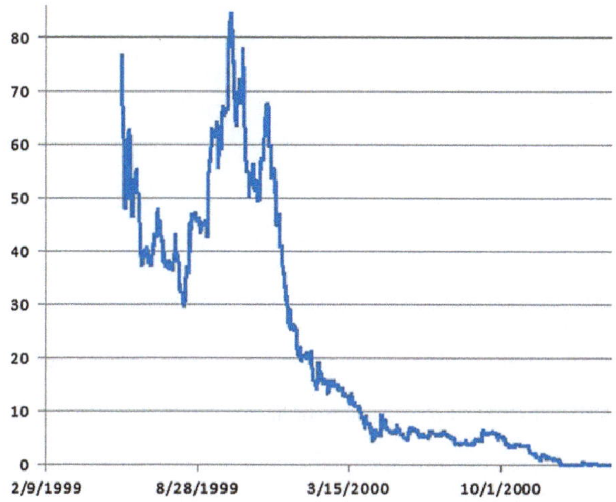

Fig. 6 Etoys.com Stock Prices during the alleged Dotcom Bubble

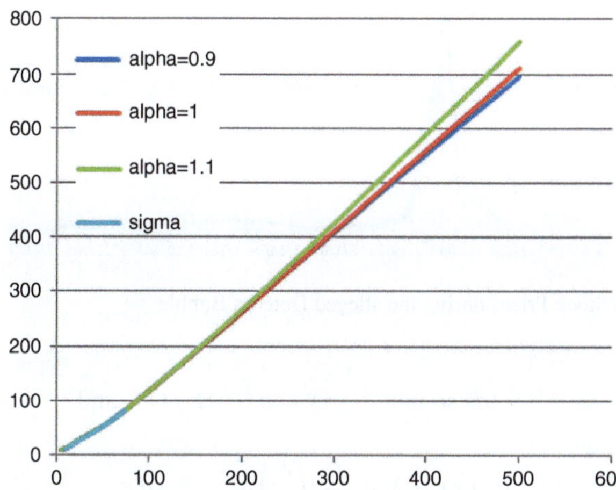

Fig. 7 eToys. RKHS estimates of $\sigma(x)$

Infospace: Our methodology shows that Infospace exhibited a price bubble. We detail the methodology step by step in this example. The graph of the stock prices in Fig. 9 suggests the existence of a bubble.

We present a summary of the estimator of Florens–Zmirou. Her estimator is based on the local time of a diffusion and is based on an analysis of local times. The local time is given by

$$\ell_T(x) = \lim_{\epsilon \to 0} \frac{1}{2\epsilon} \int_0^T 1_{\{|S_s - x| < \epsilon\}} d\langle S, S\rangle_s$$

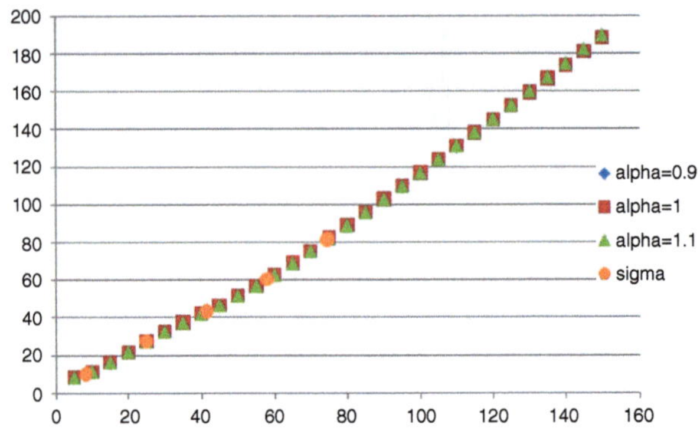

Fig. 8 eToys. RKHS estimates of $\sigma(x)$, quality of fit

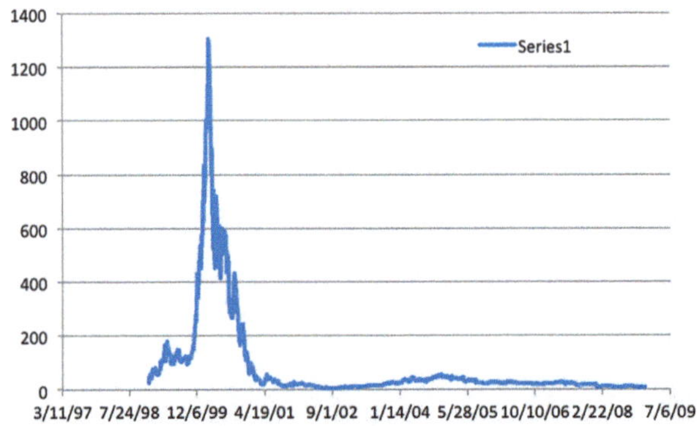

Fig. 9 Infospace Stock Prices during the alleged Dotcom Bubble

where $d\langle S, S\rangle_s = \sigma^2(S_s)ds$ so that $\ell_T(x) = \sigma^2(x)L_T(x)$, and

$$L_T(x) = \lim_{\epsilon\to 0} \frac{1}{2\epsilon} \int_0^T 1_{\{|S_s - x| < \epsilon\}} ds.$$

Hence, the ratio $\frac{\ell_T(x)}{L_T(x)} = \sigma^2(x)$ yields the volatility at x. These limits and integrals can be approximated by the following sums:

$$L_T^n(x) = \frac{T}{2nh_n} \sum_{i=1}^n 1_{\{|S_{t_i} - x| < h_n\}}$$

$$\ell_T^n(x) = \frac{T}{2nh_n} \sum_{i=1}^n 1_{\{|S_{t_i} - x| < h_n\}} n(S_{t_{i+1}} - S_{t_i})^2$$

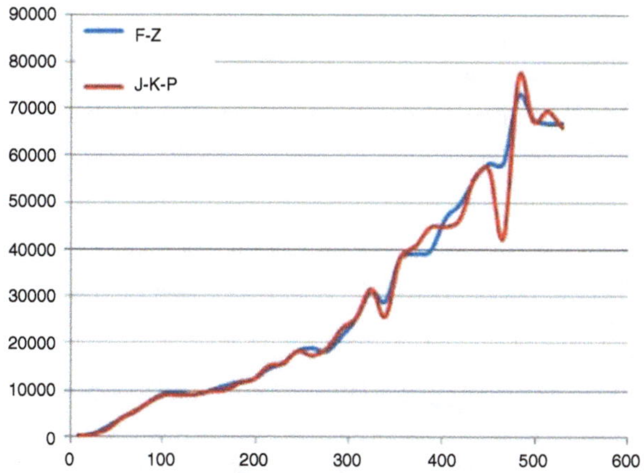

Fig. 10 Infospace. Non-parametric estimation using $h_n = \frac{1}{n^{\frac{1}{3}}}$

where h_n is a sequence of positive real numbers converging to 0 and satisfying some constraints. This allows us to construct an estimator of $\sigma(x)$ given by:

$$S_n(x) = \frac{\sum_{i=1}^{n} 1_{\{|S_{t_i}-x|<h_n\}} n(S_{t_{i+1}} - S_{t_i})^2}{\sum_{i=1}^{n} 1_{\{|S_{t_i}-x|<h_n\}}}. \tag{154}$$

Indeed, Florens–Zmriou [52] proves the following theorems.

Theorem 44. *If σ is bounded above and below from zero, has three continuous and bounded derivatives, and if $(h_n)_{n\geq 1}$ satisfies $nh_n \to \infty$ and $nh_n^4 \to 0$ then $S_n(x)$ is a consistent estimator of $\sigma^2(x)$.*

The proof of this theorem is based on the expansion of the transition density. The choice of a sequence h_n converging to 0 and satisfying $nh_n \to \infty$ and $nh_n^4 \to 0$ allows one to show that $L_T^n(x)$ and $\ell_T^n(x)$ converge in $L^2(dQ)$ to $L_T(x)$ and $\sigma^2(x)L_T(x)$, respectively. Hence $S_n(x)$ is a consistent estimator of $\sigma^2(x)$, for any x that has been visited by the diffusion.

Another result, developed in [78], is useful to obtain confidence intervals for the estimator $S_n(x)$ of $\sigma(x)$.

Theorem 45. *If moreover $nh_n^3 \to 0$ then $\sqrt{N_x^n}(\frac{S_n(x)}{\sigma^2(x)} - 1)$ converges in distribution to $\sqrt{2}Z$ where Z is a standard normal random variable and $N_x^n = \sum_{i=1}^{n} 1_{\{|S_{t_i}-x|<h_n\}}$.*

(i) We compute the Florens–Zmirou's estimator and our smooth kernel local time based estimator, using a sequence $h_n = \frac{1}{n^{\frac{1}{3}}}$. The result is not smooth enough as seen in Fig. 10.

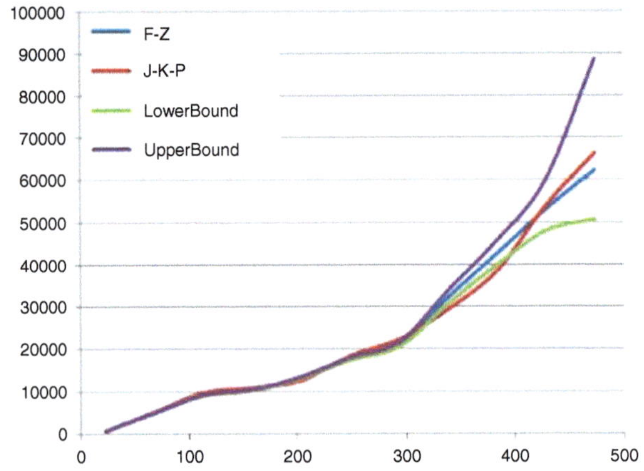

Fig. 11 Infospace. Non-parametric estimation using $h_n = \frac{1}{n^{\frac{1}{4}}}$

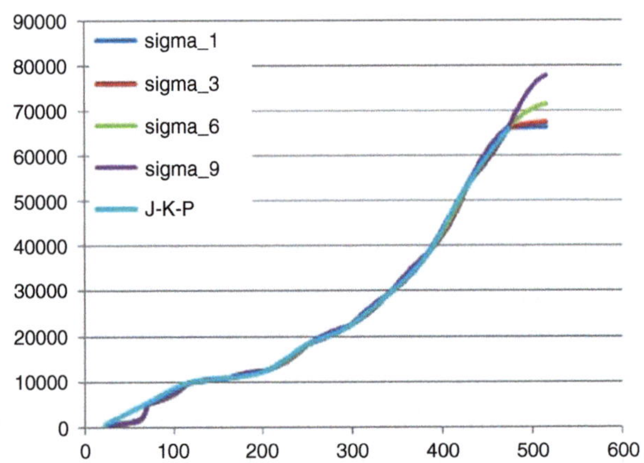

Fig. 12 Infospace. Interpolation $\sigma^b(x)$ on the compact domain

(ii) We use the sequence $h_n = \frac{1}{n^{\frac{1}{4}}}$ to compute our estimators (the number of points where the estimation is performed is smaller, M = 11). Theoretically, we no longer have the convergence of the Florens–Zmirou's estimator. However, as seen in Fig. 11, this estimator is robust with respect to the constraint on the sequence h_n. F-Z, LowerBound and UpperBound are Florens–Zmirou's estimator together with the 95 % confidence bounds her estimation procedure provides. J-K-P is our estimator.

(iii) We obtained in (ii) estimations on a fixed grid containing $M = 11$ points, and we now construct a function $\sigma^b(x)$ on the finite domain (see Fig. 12) which perfectly interpolates those points. Here the *RKHS* used is $H^1(\mathcal{D})$ where $\mathcal{D} = [\min S, \max S]$ together with the reproducing kernels $K^{\mathcal{D}}_{1,\tau}$, where τ takes the

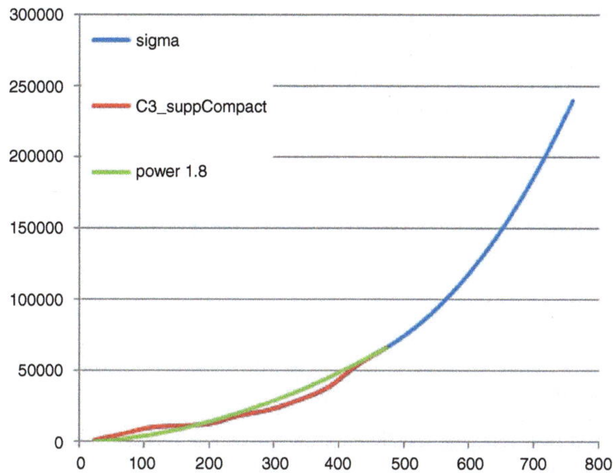

Fig. 13 Infospace. Final estimator and RKHS extrapolation

values 1, 3, 6 and 9. The functions obtained using these different reproducing kernels provide the same quality of fit within $\mathcal{D}$ and we can use any of the four outputs as the interpolated function, σ^b, over the finite interval $\mathcal{D}$.

(iv) Finally we optimize over m and find the *RKHS* $H_{2,m}$ that allows the best interpolation of the $M = 11$ estimated points and such that the extrapolated function $\overline{\sigma}(x)$ remains as close as possible to $\sigma^b(x)$ on the third right side of $\mathcal{D}$. Of course, the reproducing kernels used in order to construct the functions σ_m and minimize the target error as in (153) are $q_{2,m}^{RP}$. We obtain $\overline{m} = 6.17$ (i.e. $\alpha = \frac{\overline{m}+1}{2} = 3.58$) and we can conclude that there is a bubble.

Remark 46. One might expect $\alpha \approx 1.8$ as suggested by the green curve in Fig. 13. But this is different from what the *RKHS* extrapolation has selected. Why? In Fig. 13, we plot the *RKHS* extrapolation obtained when $\alpha = 1.8$. We have proved that

$$\lim_{x\to\infty} \frac{x^{m+1}}{\overline{\sigma}^2(x)} = 4B(m+1,2)\sum_{i=1}^{M} c_i.$$

The numerical computations give: $\overline{\sigma}(x) \approx \frac{x^{3.58}}{127009}$ when using optimization over m and $\overline{\sigma}(x) \approx \frac{x^{1.8}}{5.66}$ when fixing $\alpha = 1.8$. Independent of the power chosen, the c_i's and hence the constant of proportionality are automatically adjusted to interpolate the input points. But, as can be seen in Fig. 14, the power 3.58 is more consistent in terms of extending "*naturally*" the behavior of $\sigma^b(x)$ to $\mathbb{R}^+$.

Geocities: Our methodology shows that this stock did not have a price bubble. The stock prices are graphed in Fig. 15.

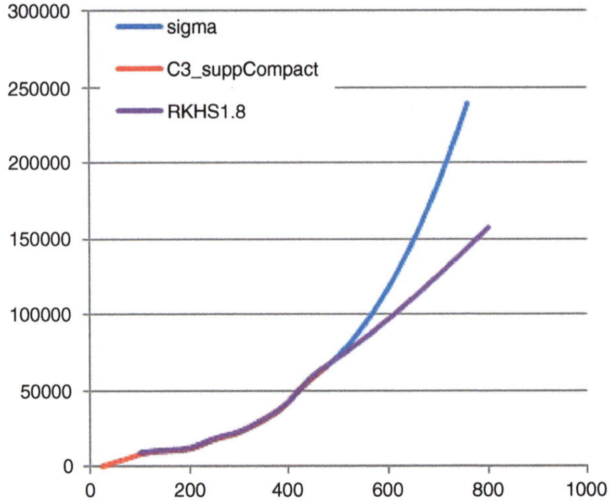

Fig. 14 Infospace

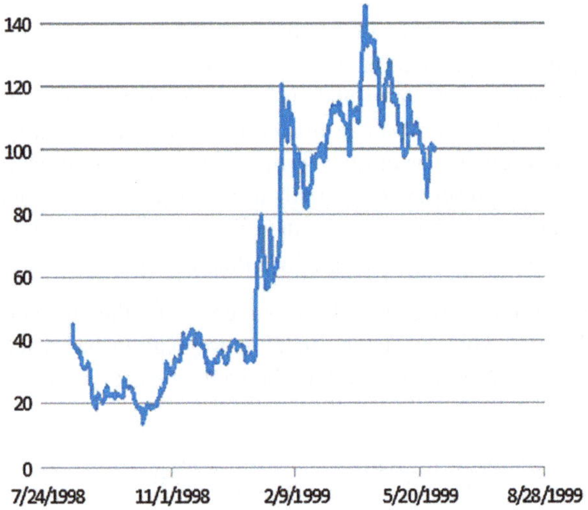

Fig. 15 Geocities Stock Prices during the alleged Dotcom Bubble

This is an example where we can stop at step (iii) of Procedure 43: we do not need to use *RKHS* theory to extrapolate our estimator in order to determine its asymptotic behavior. As seen from Fig. 16, the volatility is a nice bounded function, and any natural extension of this behavior implies the divergence of the integral $\int_\epsilon^\infty \frac{x}{\sigma^2(x)}dx$. Hence the price process is a true martingale.

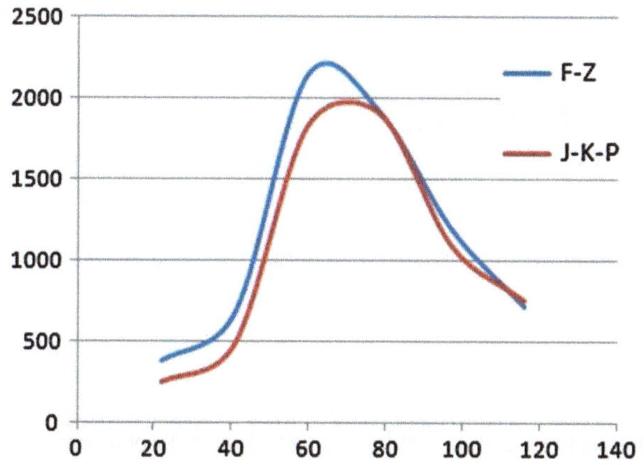

Fig. 16 Geocities. Estimates of σ

More Recent Examples

The Case of the IPO of LinkedIn

After giving talks based on the results of [78] with the examples of the dot com era, colleagues asked for examples from more recent and timely stocks. One of the times this occurred, at a conference in Ascona, Switzerland, May 23–27, 2011, we happened to read a New York Times article by Julie Creswell [32] discussing whether or not in the aftermath of the LinkedIn IPO the stock price had a bubble. Inspired by this controversy we obtained stock price tick data from Bloomberg.[26] And, we used our methodology to test whether LinkedIns stock price is exhibiting a bubble. We found, definitively, that there was indeed a price bubble in the opening days of the stock.

To perform our test, we obtained minute by minute stock price tick data for the 4 business days 5/19/2011 to 5/24/2011 from Bloomberg. There are exactly 1,535 price observations in this data set. The time series plot of LinkedIn's stock price is contained in Fig. 17. The prices used are the open prices of each minute but the results are not sensitive to using open, high or lowest minute prices instead.

The maximum stock price attained by LinkedIn during this period is $120.74 and the minimum price was $81.24. As evidenced in this diagram, LinkedIn experienced a dramatic price rise in its early trading. This suggests an unusually large stock price volatility over this short time period and perhaps a price bubble.

Let us recall from our treatment of the dot com bubbles that we just treated previously in this section, that our bubble testing methodology first requires us to estimate the volatility function σ using local time based non-parametric estimators.

[26]We thank Arun Verma of Bloomberg for quickly providing us with high quality tick data.

Fig. 17 LinkedIn Stock Prices from 5/19/2011 to 5/24/2011. (The observation interval is 1 min)

Zmirou's estimator					
x	sigma_Zmirou	lowerBound	upperBound	LocalTime	NbrePoints
84.665	19.0354	17.8579	20.4816	0.0393737	414
91.5149	24.0447	22.6762	25.6951	0.0472675	497
98.3648	22.4606	20.9575	24.3417	0.0330968	348
105.215	37.8995	34.9693	41.7162	0.0239666	252
112.065	86.192	68.0373	137.119	0.00199722	21
118.915	221.362	113.979	1e+006	9.51056e-005	1
125.764	0	0	1e+006	0	0

JKP Estimator		
x	sigma_JKP	LocalTime
84.665	13.4404	0.0619793
91.5149	19.1038	0.0259636
98.3648	27.7474	0.0223718
105.215	38.781	0.0229719
112.065	69.481	0.000708326
118.915	3.95e+014	0
125.764	3.95e+014	0

Fig. 18 Non-parametric volatility estimates

We use two such estimators. We compare the estimation results obtained using both Florens–Zmirou's[27] estimator (see Theorems 44 and 45) and the estimator developed in [78]. The implementation of these estimators requires a grid step h_n tending to zero, such that $nh_n \to \infty$ and $nh_n^4 \to 0$ for the former estimator, and $nh_n^2 \to \infty$ for the later one. We choose the step size $h_n = \frac{1}{n^{\frac{1}{3}}}$ so that all of these conditions are simultaneously satisfied. This implies a grid of seven points. The statistics are displayed in Fig. 18.

Since the neighborhoods of the grid points \$118.915 and \$125.764 are either not visited or visited only once, we do not have reliable estimates at these points. Therefore, we restrict ourselves to the grid containing only the first five points.

[27]Hereafter referred to as Zmirou's estimator.

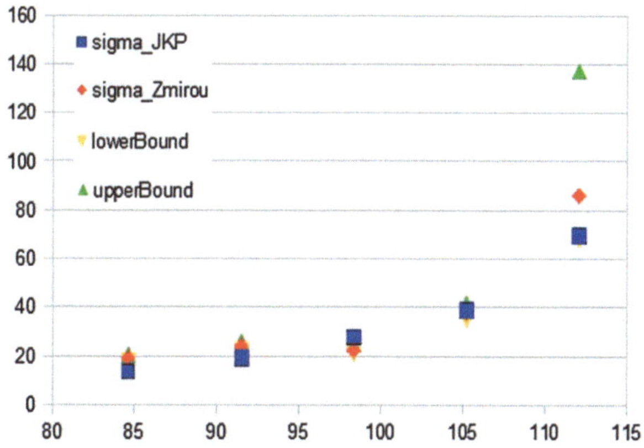

Fig. 19 Non parametric volatility estimation results

We note that the last point in the new grid \$112.065 still has only been visited very few times.

When using Zmirou's estimator, confidence intervals are provided. The confidence intervals are quite wide. Given these observations, we apply our methodology twice. In the first test, we use a five point grid. In the second test, we remove the fifth point where the estimation is uncertain and we use a four point grid instead. The graph in Fig. 19 plots the estimated volatilities for the grid points together with the confidence intervals.

The next step in our procedure is to interpolate the shape of the volatility function between these grid points. We use the estimations from our non parametric estimator with the five point grid case. For the volatility time scale, we let the 4 day time interval correspond to one unit of time. This scaling does not affect the conclusions of this paper. When interpolating one can use any reasonable method. We use both cubic splines and reproducing kernel Hilbert spaces as suggested in [78], Sect. 5.2.3 item (ii). The interpolated functions are in Fig. 20.

From these,we select the kernel function $K_{1,\tau}$ as defined in Lemma 10 in [78], and we choose the parameter $\tau = 6$.

The next step is to extrapolate the interpolated function σ^b using the RKHS theory to the left and right stock price tails. Here we refer the reader to our treatment given in Sect. 10.1, and do not repeat the necessities here. The reader desiring a detailed treatment for this specific example is referred to the published article [79]. We mention only that we take $f(x) = \frac{1}{\sigma^2(x)}$ and define the Hilbert space

$$H_n = H_n([0,\infty[) = \{f \in C^n([0,\infty[) \mid \lim_{x\to\infty} x^k f^{(k)}(x) = 0 \text{ for all } 0 \leq k \leq n-1\}$$

where n is the assumed degree of smoothness of f.

For $n \in \{1,2\}$ fixed, we construct our extrapolation $\sigma = \sigma_m$ as in [78], 5.2.3 item (iv), by choosing the asymptotic weighting function parameter m such that

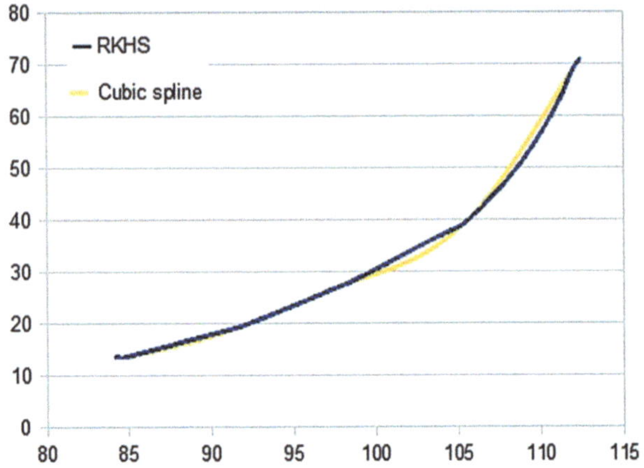

Fig. 20 Interpolated volatility using cubic splines and the RKHS theory

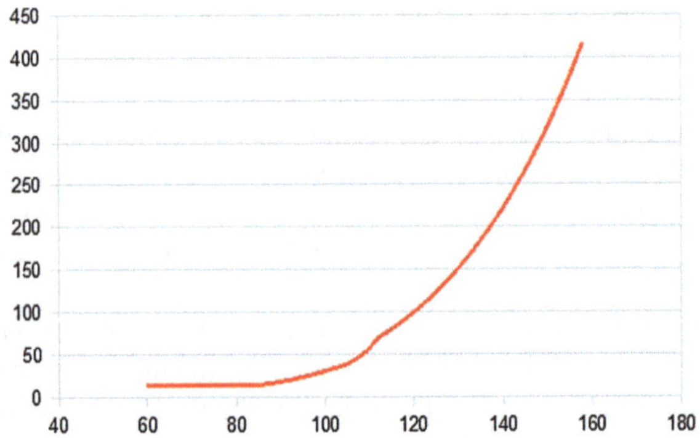

Fig. 21 RKHS based extrapolation of σ^b

$f_m = \frac{1}{\sigma_m^2}$ is in $H_{n,m}$, σ_m exactly matches the points obtained from the non parametric estimation, and σ_m is as close (in norm 2) to σ^b on the last third of the bounded interval where σ^b is defined. Because of the observed kink and the obvious change in the rate of increase of σ^b at the forth point, we choose $n = 1$ in our numerical procedure. The result is shown in Fig. 21.

We obtain $m = 9.42$.

From Proposition 3 in [78], the asymptotic behavior of σ is given by

$$\lim_{x \to \infty} x^{m+1} f(x) = n^2 B(m+1, n) \sum_{i=1}^{M} c_i$$

where $M = 5$ is the number of observations available, B is the Beta function, and the coefficients $(c_i)_{1 \le i \le M}$ are obtained by solving the system

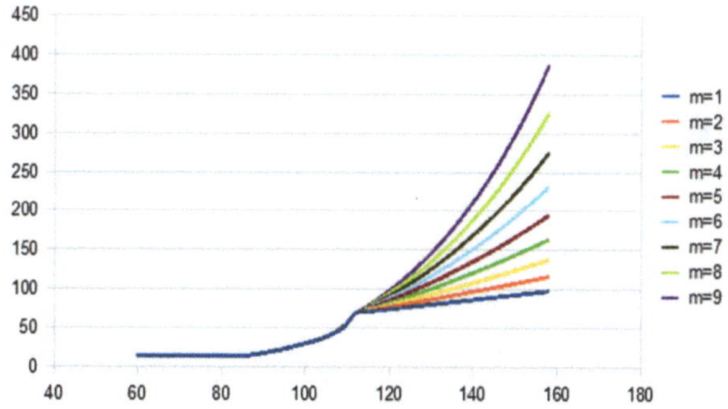

Fig. 22 Extrapolated volatility functions using different reproducing kernels

$$\sum_{i=1}^{M} c_i q_{n,m}^{RP}(x_i, x_k) = f(x_k) \text{ for all } 1 \leq k \leq M$$

where $(x_i)_{1 \leq i \leq M}$ is the grid of the non parametric estimation, $f(x_k) = \frac{1}{\sigma^2(x_k)}$ and $\sigma(x_k)$ is the value at the grid point x_k obtained from the non-parametric estimation procedure. This implies that σ is asymptotically equivalent to a function proportional to x^α with $\alpha = \frac{1+m}{2}$, that is $\alpha = 5.21$. This value appears very large, but the proportionality constant is also large. The c_i's are automatically adjusted to exactly match the input points $(x_i, f(x_i))_{1 \leq i \leq M}$.

We plot below the functions with different asymptotic weighting parameters m obtained using the RKHS extrapolation method, without optimization. All the functions exactly match the non-parametrically estimated points.

The asymptotic weighting function's parameter $m = 9.42$ obtained by optimization appears in Fig. 22 to be the estimate most consistent (within all the functions, in any Hilbert Space of the form $H_{1,m}$, that exactly match the input data) with a "natural" extension of the behavior of σ^b to $\mathbb{R}^+$. *The power $\alpha = 5.21$ implies then that LinkedIn stock price is currently exhibiting a bubble.*

Since there is a large standard error for the volatility estimate at the end point \$112.065, we remove this point from the grid and repeat our procedure. Also, the rate of increase of the function between the last two last points appears large, and we do not want the volatility's behavior to follow solely from this fact. Hence, we check to see if we can conclude there is a price bubble based only on the first 4 reliable observation points. We plot in Fig. 23 the function σ^b (in blue) and its extrapolation to $\mathbb{R}^+$, σ (in red).

Now $M = 4$. With this new grid, we can assume a higher regularity $n = 2$ and we obtain, after optimization, $m = 7.8543$. This leads to the power $\alpha = 4.42715$ for the asymptotic behavior of the volatility. Again, although this power appears to be high given the numerical values $(x_k, f(x_k))_{1 \leq k \leq 4}$, the coefficients $(c_i)_{1 \leq i \leq 4}$ and hence the constant of proportionality are adjusted to exactly match the input points.

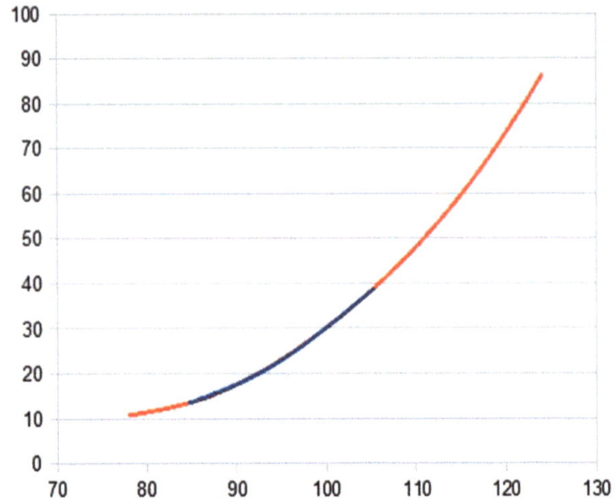

Fig. 23 RKHS based Extrapolation of σ^b

The extrapolated function obtained is the most consistent (within all the functions, in any $H_{2,m}$, that exactly match the input data) in terms of extending "naturally" the behavior of σ^b to $\mathbb{R}^+$. Again, we can conclude that there is a stock price bubble.

The Gold Bubble: Or Not?

Our final example is for the recent increase in gold prices (see [33]). Again, we obtained gold price tick data from Bloomberg[28] for the period August 25, 2011 to September 1, 2011. We used per second prices giving 73,695 data points. A graph of the spot price of gold for this period is given in Fig. 24.

We graph our estimated local volatility function for gold prices with its 95 % confidence interval in Fig. 25. As seen in Fig. 25, the volatility function is in fact decreasing as gold prices tend to ∞. This shows that speculative trading is not causing an increase in gold prices. Hence there is no gold price bubble.

Of course, our test only formally applies to the time period we have investigated, and there could be a regime change before or after this period giving a new function σ which might change whether or not a bubble is occurring. If a price bubble existed before our testing period, it may not be captured by our procedure. But, this is only true to the extent that the estimated volatility function's shape changes across the different time periods considered. Recall that our testing procedure determines the shape of the estimated volatility function for the *observed asset price range*.

[28]We thank Arun Verma of Bloomberg, again, for providing us with data.

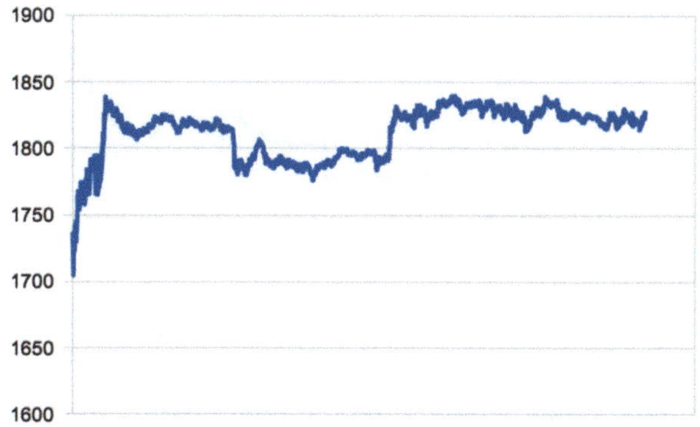

Fig. 24 Time series of gold spot prices

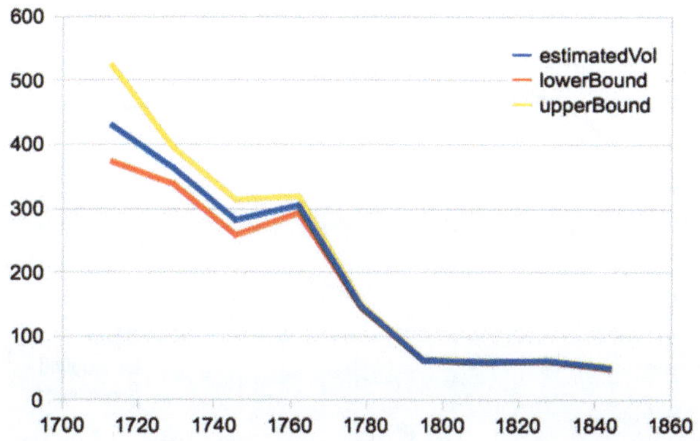

Fig. 25 Non-parametric gold price volatility estimate with 95 % confidence intervals

This volatility function's shape is then extrapolated to where the price becomes unbounded.

For gold, there is no reason to believe that the shape of the volatility function would change if we looked either backwards in time or used more current price observations. To verify this hypothesis, we studied two additional time intervals: July 4, 2011 to July 12, 2011 and September 26, 2011 to October 4, 2011. For each of these time periods we repeated the same bubble detection tests. The spot prices for gold are graphed in Figs. 26 and 27, and the estimated volatility functions are contained in Figs. 28 and 29, respectively. In both cases, the functions appear to be nicely bounded, so there is no gold price bubble in either period.

Despite the speculation that gold prices are a bubble (see for example [33]), our method shows that in fact there was not one, and that the bubbly fluctuations fall within the normal bounds of trading, rather than being indicative of excessive speculation. *That our method can distinguish this bubbly appearance from the reality (or lack thereof) of a bubble is precisely the point of our methodology.*

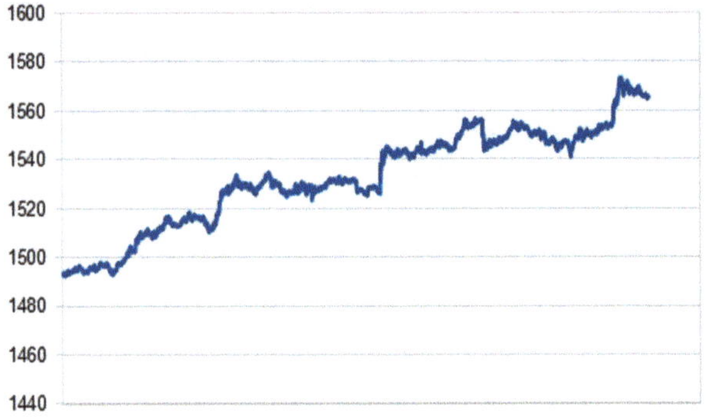

Fig. 26 Time series of gold spot prices—July 4, 2011 to July 12, 2011

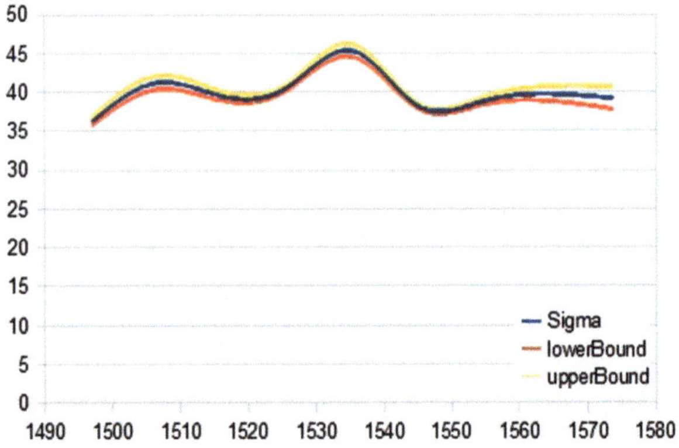

Fig. 27 Time series of gold spot prices—September 26, 2011 to October 4, 2011

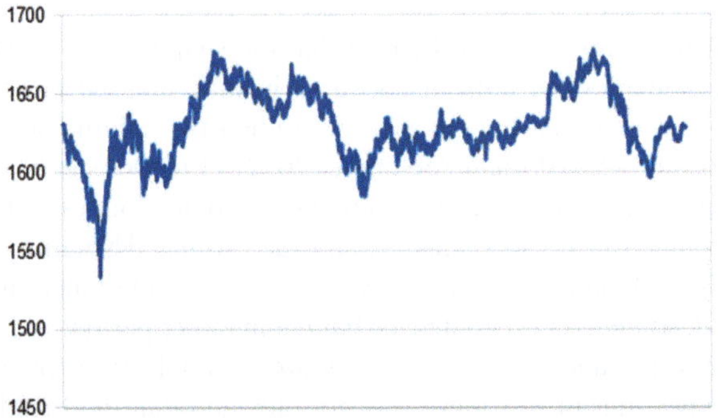

Fig. 28 Non-parametric gold price volatility estimate with 95 % confidence intervals—July 4, 2011 to July 12, 2011

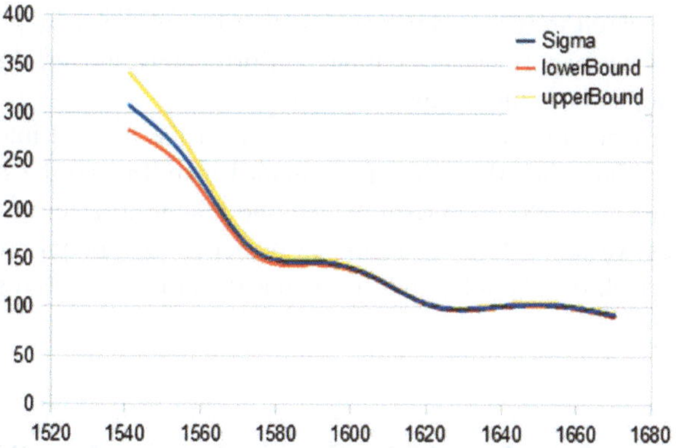

Fig. 29 Non-parametric gold price volatility estimate with 95 % confidence intervals—September 26, 2011 to October 4, 2011

But, going forward in time, the shape of an asset's volatility function can certainly change. Speculative trading can spontaneously increase due to a changing economic environment. And, bubbles that exist at any one time, can certainly burst and disappear. Whether or not a given time period's trading activity applies to other time periods, as discussed above, is beyond the capacity of our statistical procedure. But fortunately for us, the stability of speculative trading activity can be determined by an independent analysis of the economic environment. And, as long as the speculative trading activity is stable and unchanging which reasoned economic analysis should be able to determine, our method applies across time periods as well.

Summary of the Examples

Given the price process of a risky asset that follows a stochastic differential equation under the risk neutral measure of the form

$$dX_t = \sigma(X_t)dW_t$$

where W is a standard one dimensional Brownian motion, we provide methods for estimating the volatility coefficient $\sigma(x)$ at the values where it is observed. If the behavior of $\sigma(x)$ is reasonable, we extend this estimator to all of $\mathbb{R}_+$ via the technology of Reproducing Kernel Hilbert Spaces. Having done this, we are then able to decide on the convergence or the divergence of the integral

$$\int_{\epsilon}^{\infty} \frac{x}{\sigma(x)^2}dx,$$

for any $\epsilon > 0$, which in turn determines whether or not the risky price process is experiencing, or has experienced, a bubble. Unfortunately, the test does not always work, since it depends on the behavior of $\sigma(x)$.

We illustrated our methodology using data from the alleged internet dot com bubble of 1998–2001, the 2011 IPO of the stock LinkedIn, and the suspected gold bubble of August, 2011. Not surprisingly, we find that all three eventualities occur: in several cases we are able to confirm the presence of a bubble; in other cases we confirm the lack of a bubble, and in one particular case we find that the test is inconclusive.

11 The Issue of the Local Martingale Approach to Bubbles

There have been three rubrics of criticism to this approach. While the three are intertwined, nevertheless they should be separated into two types: the first is a criticism of the entire approach, and the second is a criticism of our bubble detection methodology. Of course, our bubble detection technique is pointless if one does not stipulate the validity (or at least the plausibility) of our mathematical approach, so let us first address the criticisms of the entire idea of modeling bubbles with this mathematical approach.

Discrete Time and Strict Local Martingales

There are two basic criticisms of the model. The first is based on an old controversy: modeling in discrete time versus modeling in continuous time. There is a consistent attitude, especially among economists, that continuous time is rather pointless, and needlessly complicates and obscures ideas that are relatively straightforward in discrete time; and besides, for implementation of continuous time models, often at some point one needs to discretize in any event. This idea is derived from the common belief that in economic theory both discrete and continuous time models are equivalent in the sense that one can always be used to approximate the other, or equivalently, any economic phenomena present in one is also present in the other.

One can indeed model stock prices, for example, as a discrete time series by looking (for example) at close of day data, but if one wants to model tick data, the data does not arrive in uniformly spaced time increments, and it seems more natural to view tick data as a frequently sampled collection of observations from an underlying continuous process. The sampling times are then stopping times in such a model. Of course a really fine analysis shows that even this interpretation might be naïve due to the presence of microstructure noise, and/or rounding errors (see for example [1, 77, 155], three of many recent papers on the subject). But while this approach might be naïve in this broader context, the noise does in invalidate it, but rather adds new layers of complexity.

Nevertheless some scholars have an issue with bubbles being the nuanced difference between a strict local martingale and a true martingale. In discrete time, it is widely believed that there are no strict local martingales; that all local martingales are actually true martingales. Technically this is not true (see [75,95]) but "morally" it is in fact true, because the standard definition of conditional expectation requires an L^1 condition, and as a consequence local martingales in the traditional sense are actually true martingales. This was shown by P.A. Meyer in 1973 [115]. To clarify, we have (from the textbook of Shiryaev [143]) the following theorem:

Theorem 47. *Let* $X = (X_n)_{n=0,1,\dots}$ *be an adapted process with* $X_0 = 0$. *Then the following conditions are equivalent:*

- *X is a local martingale;*
- *X is a generalized martingale, i.e.* $E(|X_{n+1}||\mathcal{F}_n) < \infty, E(X_{n+1}|\mathcal{F}_n) = X_n$ *for all* $n = 0, 1, 2, \dots$

The condition that $X_0 = 0$ in Theorem 47 is important: the theorem is no longer true if $X_0 \notin L^1$. However this characterization of discrete-time local martingales holds in the case when X is nonnegative and X_0 is integrable. We have then the corollary:

Corollary 7. *A local martingale* $X = (X_n)_{n=0,1,\dots,T}$ *with* $X_0 \in L^1$ *and* $X_T \geq 0$ *is a martingale.*

Since we are usually dealing with price processes that are nonnegative (certainly the case for stocks), and typically X_0 is assumed to be non-random and hence trivially in L^1, Corollary 7 does indeed give an equivalence between local martingales and true martingales. And since we are mostly concerned with bubbles on compact time intervals $[0, T]$ which must be strict local martingales to exist, the critics are correct that such a subtle distinction is meaningless in discrete time. Where this author disagrees with the critics is with the logical leap they make that this matters. Even in a subject as mundane as differential calculus, there are no continuous functions,[29] let alone differentiable ones, in discrete time; and try to teach the ideas of calculus using finite sums instead of integrals. So shall we discard calculus by the same reasoning? Another example of such reasoning would have us discard the normal distribution, since it cannot possibly exist in a finite, discretized world. Nevertheless, the normal distribution, and continuous functions and integrals can all be approximated as limits of discrete sums. But so can strict local martingales be approximated by discrete time processes; it is just that the discrete time processes will be true martingales, the strict local nature only occurring in the limit, just as the property of being a continuous function only occurs in the limit when approximating by discretized functions. For a more detailed discussion of this question we refer the interested reader to the recent article [86].

[29] That is, there are no continuous functions except for trivialities such as using the discrete topology and thereby making all functions continuous.

The Critique of Fragility of the Model

This critique comes principally from one paper of P. Guasoni and M. Rásonyi [61] which addresses the "fragility" of both the mathematical concept of No Free Lunch With Vanishing Risk, and that of the theory of mathematical bubbles presented in this paper. The basic premise is that if one has a sophisticated model of an economic phenomenon, it is by necessity of the subject only an approximation, and thus any model should be "robust" in some appropriate sense. For economics models, this makes sense at first blush, but one can stumble in the concept of robustness. The authors of [61] use the idea of the paths of an alternative model could be only ϵ close (on a logarithmic scale) to the originally proposed model, and yet have very different properties. The flaw in this logic (in this author's opinion) is that the reasoning has the reverse order of one that is appropriate. Indeed, their implicit assumption is that mathematical models of economic phenomena arise simply from fitting curves to graphs of data. We would contend they are anything but that: one comes up with a model through economic and probabilistic reasoning, and then one checks later to see if it is reasonable by testing if it matches data well, for example by a goodness of fit procedure. If it does not, one tries to improve the reasoning, or call into question the hypotheses that led to the model and change them appropriately, in order to arrive at a better model. Indeed, in analogy with physics, the motion of a baseball is based on calculations involving models that include major forces (initial velocity, gravity, Newton's laws of motion, friction with air resistance, etc.), usually ignoring minor forces such as the gravitational pull of the moon on the baseball. One then checks to see if predictions are valid and if observation is consistent with the model. One does not then invalidate the model if one can come up with another essentially arbitrary model that is "ϵ close" to the same trajectories, but without the physics reasoning and without some of the key properties of the original model.

The authors of [61] do have a point, however: robustness of a model is an appropriate question to ask. A more reasonable way to frame the problem would be, perhaps, that if one has a model given by an SDE of the form:

$$\frac{dS_t}{S_t} = \sigma(S_t)dB_t + \mu(S_t)dt; \quad S_0 = x \tag{155}$$

then one could approximate the coefficients with a sequence of functions σ_n, μ_n, $n = 1, 2, \ldots$, such that σ_n and μ_n converge to σ and μ respectively (and in an appropriate sense), and consider the sequence of SDEs

$$\frac{dS_t^n}{S_t^n} = \sigma_n(S_t^n)dB_t + \mu_n(S_t^n)dt; \quad S_0^n = x \tag{156}$$

and then as is well known, S^n converges to S, so one can ask if, for some N large enough and $n \geq N$, do the processes S^n possess the desired properties? For NFLVR, it is clear that they do indeed possess NFLVR if S of (155) does, and if the functions

σ_n, μ_n are reasonably (and not maliciously) chosen; for example one would want the solutions of (155) and (156) all to have unique, strong solutions. As far as bubbles are concerned, in the framework of (155), if the function σ satisfies the condition (142), then it is reasonable to choose approximating functions σ_n such that, at least for $n \geq N$ for some large N, the approximations σ_n have similar asymptotic behavior and also satisfy (142) for $n \geq N$, and therefore establish the robustness of the bubble property of the model.

Of course, there are other possible interpretations of robustness. For example, one could approximate the differentials in an appropriate way, such as with Emery's semimartingale topology, or (better) using the techniques of Kurtz and Protter [103], or alternatively those of Mémin and Slominski [113].[30] In this case one would have equations of the form

$$\frac{dS_t^k}{S_t^k} = \sigma(S_t^k)dB_t^k + \mu(S_t^k)dA_t^k; \quad S_0 = x \tag{157}$$

or even combine the two approximations to arrive at equations of the form

$$\frac{dS_t^{n,k}}{S_t^{n,k}} = \sigma_n(S_t^{n,k})dB_t^k + \mu_n(S_t^{n,k})dA_t^k; \quad S_0 = x \tag{158}$$

and again, if one were reasonable with the approximations (for example they should satisfy the condition UCV of [103]), one could preserve NFLVR (for example, one would have to choose $dA_t^k \ll d[B^k, B^k]_t$ a.s. for large enough k, in the case where B^k is a continuous local martingale), and also preserve the bubble property. We do not provide details here, because this is only tangential to the purpose of this paper.

"No Empirical Test Can Reliably Distinguish a Strict Local Martingale from a Martingale"

The title above is an actual quotation of a written report. It is true that any statistical based procedure can never produce truth, but at best only a good likelihood of a result. We assume this explains the presence of the word "reliably" in the quote above. However for the stochastic differential equations presented in Theorems 7 and 8 it certainly seems possible *a priori*, via the strong law of large numbers and the martingale central limit theorem, that one can identify (for example to the 95 % level) when the parameter is within the range where the solution is a true martingale, and within the range where the solution is a strict local martingale, even if nobody (to our knowledge) has yet tried to do so. (Indeed one should be

[30] It is shown in [104] that the two methods are equivalent.

able to use modern semimartingale estimation techniques such as those presented in [56] or more generally [72], especially for the cases of Theorems 7 and 8. For a treatise on more advanced techniques see [74].) Probably however the quotation above refers to the technique presented in Sect. 10. Of this technique, it would seem that the method of the estimation of the diffusion coefficient is beyond reproach; therefore the criticism is probably addressed to the extrapolation technique (and the idea of such) presented in the subsection titled "Bubble Detection." The idea is to use the time honored method of Reproducing Kernel Hilbert Spaces (RKHS) made famous within Statistics circles by E. Parzen and more recently by G. Wahba (see for example [123, 156, 157]); however we use RKHS techniques in a new way here, in order to extrapolate the diffusion coefficient function $x \mapsto \sigma(x)$ to an interval of the form $(\hat{x}_{\max}, \infty)$ where $\hat{x}_{\max}$ denotes the largest observed value $\hat{x}$. On the semi infinite interval $(\hat{x}_{\max}, \infty)$ observation data does not exist. Since our conclusion is based on this extrapolation, it is indeed beyond the usual domain of statistics, where procedures are consistent, in the sense that they converge to a limit as the procedure gets arbitrarily accurate. However since a consistent estimator is not possible here, the method proposed is at least an attempt to resolve the issue, and it is further enhanced by the fact that it seems to work, and to work well, when tested against data. We do not claim it is a definitive answer to this problem, but we do think it is an advance and represents the best possible method currently available. We eagerly await the work of others who hopefully will improve on this method, or propose alternative methods for the important problem of bubble detection.

A Brief Discussion of Some Alternative Methods

The literature, particularly the economics literature, concerning financial bubbles is vast. We make no attempt to give a survey here, although we have provided references to some key papers [19, 48, 49, 54, 55, 63, 68, 108, 112, 138, 139, 148, 154, 158], each of which in turn provides more references. Instead we limit ourselves to a discussion of proposed alternative methods for bubbles detection.

We know of four alternative methods that propose a methodology to detect financial bubbles.

The first method is that of "charges," and is proposed in the papers of Jarrow and Madan [81], Gilles [57], and Gilles and Leroy [58]. To explain this we need the technical concept of a "price operator." We let ν represent some fixed and constant (future) time. Let $\phi = (\Delta, \Xi^\nu)$ denote a payoff of an asset (or admissible trading strategy) where: (a) $\Delta = (\Delta_t)_{0 \le t \le \nu}$ is an arbitrary càdlàg nonnegative and nondecreasing semimartingale adapted to $\mathbb{F}$ which represents the asset's cumulative dividend process, and (b) $\Xi^\nu \in \mathcal{F}_\nu$ is a nonnegative random variable which represents the asset's terminal payoff at time ν. V^π denotes the wealth process corresponding to the trading strategy π. This recalls our original framework for defining the fundamental price of an asset.

Definition 8 (Set of Super-replicated Cash Flows).

$$\text{Let } \Phi := \{\phi \in \Phi_0 : \exists \pi \text{admissible}, a \in \mathbb{R}_+ \text{ such that } \Delta_\nu + \Xi^\nu \le a + V_\nu^\pi\}. \quad (159)$$

The set Φ represents those asset cash flows that can be super-replicated by trading in the risky asset and money market account. As seen below, it is the relevant set of cash flows for our no dominance assumption. We first show that this subset of asset cash flows is a convex cone.

We start with a price function $\Lambda_t : \Phi \to \mathbb{R}_+$ that gives for each $\phi \in \Phi$, its time t price $\Lambda_t(\phi)$. Let $\Phi_m \subset \Phi$ represent the set of traded assets. Take as our economy $\Phi_m = \{1, S\}$. The no dominance assumption implies the following:

Theorem 48. *(Positivity and Linearity on Φ) Let "$\ge_t$" denote dominance at time t.*

1. *Let $\phi', \phi \in \Phi$. If $\phi' \ge_t \phi$ for all t, then $\Lambda_t(\phi') > \Lambda_t(\phi)$ for all t almost surely.*
2. *Let $a, b \in \mathbb{R}_+$ and $\phi', \phi \in \Phi$. Then, $a\Lambda_t(\phi') + b\Lambda_t(\phi) = \Lambda_t(a\phi' + b\phi)$ for all t almost surely.*

The next theorem is established in [89] and shows that the local martingale characterization of market prices has a finitely additive market price operator if and only if bubbles exist.

Theorem 49. *Fix $t \in \mathbb{R}_+$. The market price operator Λ_t is countably additive if and only if bubbles do not exist.*

The second approach is that of Caballero et al. [18]. As described by Phillips et al. [125], they use a "simple general equilibrium model without monetary factors, but with goods that may be partially securitized. Date-stamping the timeline of the origination and collapse of the various bubbles is a critical element in the validity of this sequential hypothesis." They "put forward a sequential hypothesis concerning bubble creation and collapse that accounts for the course of the financial turmoil in the U.S. economy."

The third approach builds on the above approach of [18], Phillips et al. [125, 126] study bubbles more in the spirit that is presented in this paper. They posit the existence of a dividend process D_t that is a martingale under certain conditions and such that it is "reflecting market conditions that generate cash flows." They then define a fundamental process F_t by the relationship

$$F_t = \int_0^\infty \exp(-s(r_{t+s}))E\{D_{t+s}|\mathcal{F}_t\}ds \quad (160)$$

Here $(r_t)_{t\ge0}$ is the spot interest rate process, and r_D is an (assumed) constant growth rate for the interest rates such that one has the relationship

$$E\{D_{t+s}|\mathcal{F}_t\} = \exp(r_D s)D_t \quad (161)$$

and then if $r_D = 0$ one has that D is a martingale. Combining (160) and (161) one gets

$$F_t = \int_0^\infty \exp(-s(r_{t+s} - r_D))D_t ds$$

Phillips et al. then make a key assumption on the exact structure of the spot discount rate r_t, and under this assumption F satisfies an SDE of the form

$$dF_t = (1 - e^{-\gamma})c_a F_t dt + \sigma_t dD_t \tag{162}$$

where γ is such that $(1 - e^{-\gamma})c_a > 0$. The solution of (162) can have an explosive drift under certain assumptions on the structure of the interest rates, as it approaches a special time t_b. When the drift explodes this way, they claim one has a bubble. They observe that "the discrete time path of $F_t \ldots$ is therefore propagated by an explosive autoregressive process with coefficient $\rho > 1$." They explain their reasoning as follows:

> The heuristic explanation of this behavior is as follows. As $t \nearrow t_b$ there is growing anticipation that the discount factor will soon increase. Under such conditions, investors anticipate the present to become more important in valuing assets. This anticipation in turn leads to an inflation of current valuations and price fundamentals F_t become explosive as this process continues.

The fourth and last approach we shall mention is that of D. Sornette and co-authors (they have written many papers on financial bubbles; here is a sample selection of a rather large armamentarium: [10, 66, 92, 132, 147]). We are concerned with their model known as the "Johansen–Ledoit–Sornette Bubble Model," which we find to be the most mathematical, and closest to the spirit of this paper (see [132] for an exposition and discussion of this model). All quotations below are from the paper [132].

Sornette, together with his many co-authors over a long series of papers, propose that the dynamics of the price process satisfies a simple stochastic differential equation with drift and jump:

$$\frac{dp_t}{p_t} = \mu_t dt + dW_t - dj_t \tag{163}$$

where p is the stock market price, and W is a standard Wiener process, and j is a point process with hazard rate $h(t)$. The point process has one jump only, and it represents a market crash, and they introduce a random variable κ to denote the size of the crash. They assume that the aggregate effect of noise traders leads to a "crash hazard rate" of the form, with t_c denoting the time of the crash:

$$h(t) = B'(t - t_c)^{m-1} + C'(t - t_c)^{m-1}\cos(\gamma \ln(t - t_c) - \phi') \tag{164}$$

The authors interpret (164) by stating, "the cosine part of the second term in (164) takes into account the existence of possible hierarchical cascades of accelerating panic punctuating the growth of the bubble, resulting from a preexisting hierarchy in noise trader sizes and/or the interplay between market price impact inertia and nonlinear fundamental value investing." And assuming p is a martingale under the risk neutral measure (no mention is made of local martingales) and conditional that the crash has not yet occurred, the authors obtain the relation $\mu(t) = \kappa h(t)$, from which (using (164)) one derives a log periodic power law (LPPL):

$$\ln E(p_t) = A + B(t - t_c)^m + C(t - t_c)^m \cos(\gamma \ln(t - t_c) - \phi) \qquad (165)$$

where $B = \kappa B'/m$ and $C = -\kappa C'/\sqrt{m^2 + \gamma^2}$. This model, known as the JLS model, assumes that the parameter m is in between 0 and 1. Then a bubble exists when the crash hazard rate accelerates with time.

The JLS model claims that the price follows a "faster-than-exponential" growth rate during a bubble. For detection, the authors contend that financial crashes are preceded by bubbles with fluctuations. This leads to the claim that "both the bubble and the crash can be captured by the LPPL when specific bounds are imposed on the critical parameters m and γ." This is elaborated upon in [10].

In a very recent paper of Hüsler, Sornette, and Hommes [70] the three authors dismiss the bubble detection technique of [78] presented in this paper, by claiming that an earlier paper by Andersen and Sornette [4] has "shown that some (and perhaps most) bubbles are not associated with an increase in volatility." However an examination of their model (which is again a version of the JLS model) shows that the assumed extreme simplicity of their model of the evolution of a risky asset price, seems to make erroneous conclusions easy to reach.

Remark 50. The primary difference between the two alternative methods presented above (those of Phillips et al. and of Sornette et al.), and the one presented in this article, is that both alternative approaches make assumptions (albeit very different ones) on the drifts in their models that lead to bubbles (under their [different] understandings of what constitutes a bubble), whereas in our presentation the key assumptions related to bubbles revolve around the diffusive part of the model. One sees this in (162) for Phillips et al., and for the Sornette et al. model one sees it with the inclusion of a hazard rate implicit in (163), as seen in (165). In addition, the Sornette et al. alternative model above is inextricably tied to a relatively simple and specialized Brownian paradigm. The Phillips model includes dividends in the fundamental model as well as interest rates, but excludes what we have called X_τ, a final payoff in the event of bankruptcy or dissolution for some reason, such as a merger or a payout.

Acknowledgements The author wishes to thank Vicky Henderson and Ronnie Sircar for their kind invitation to do this project. He also wishes to thank Celso Brunetti, Roy DeMeo, John Hall, Jean Jacod, Robert Jarrow, Alexander Lipton, Wesley Phoa, and for helpful comments and criticisms, and his co-authors on the subject of bubbles, Younes Kchia, Soumik Pal, Sergio Pulido, Alexandre Roch, Kazuhiro Shimbo, and especially Robert Jarrow. He has also benefited from discussions with Sophia (Xiaofei) Liu, Johannes Ruf, Etienne Tanre, and especially Denis Talay. An anonymous referee gave us many useful suggestions and we thank this referee for his or her careful reading of this work. The author is grateful to the NSF for its financial support. He is also grateful for a sabbatical leave from Columbia University, and the author thanks the hospitality of the Courant Institute of NYU, and INRIA, Sophia–Antipolis, where he spent parts of his sabbatical leave.

References

1. Y. Aït-Sahalia, J. Yu, High frequency market microstructure noise estimates and liquidity measures. Ann. Appl. Stat. **3**, 422–457 (2009)
2. K. Amin, R. Jarrow, Pricing Foreign currency options under stochastic interest rates. J. Int. Money Financ. **10**(3), 310–329 (1991)
3. L.B.G. Andersen, V. Piterbarg, Moment explosions in stochastic volatility models. Financ. Stoch. **11**, 29–50 (2007)
4. J.V. Andersen, D. Sornette, Fearless versus fearful speculative financial bubbles. Phys. A **337**, 565–585 (2004)
5. N. Aronszajin, La théorie générale des noyaux reproduisants et ses applications, Première Partie. Proc. Camb. Philos. Soc. **39**, 133–153 (1943)
6. N. Aronszajn, Theory of reproducing kernels. Trans. Am. Math. Soc. **68**, 337–404 (1951)
7. Y. Balasko, D. Cass, K. Shell, Market participation and sunspot equilibria. Rev. Econ. Stud. **62**, 491–512 (1995)
8. C. Baldwin, LinkedIn shares were a bubble: Academic model (2 June 2011). Available (for example) at the URL http://www.easybourse.com/bourse/international/news/918606/linkedin-shares-were-a-bubble-academic-model.html,
9. E. Bayraktar, C. Kardaras, H. Xing, Strict local martingale deflators and valuing American call-type options. Financ. Stoch. **16**, 275–291 (2011)
10. K. Bastiaensen, P. Cauwels, Z.-Q. Jiang, D. Sornette, R. Woodard, W.-X. Zhou, Bubble diagnosis and prediction of the 2005–2007 and 2008–2009 Chinese stock market bubbles. J. Econ. Behav. Organ. **74**, 149–162 (2010)
11. A. Bentata, M. Yor, Ten notes on three lectures: From Black–Scholes and Dupire formulae to last passage times of local martingales, in *Notes from a Course at the Bachelier Seminar*, 2008
12. B. Bernanke, Senate Confirmation Hearing, December 2009. This is quoted in many places; one example is *Dealbook, edited by Andrew Ross Sorkin* (January 6, 2010)
13. J. Berman, Apple not bringing overseas cash back home, Blames U.S. tax policy. The Huffington Post (March 19, 2012), online at the URL: http://www.huffingtonpost.com/2012/03/19/apple-us-tax-law_n_1362934.html
14. F. Biagini, H. Föllmer, S. Nedelcu, Shifting martingale measures and the slow birth of a bubble, Working paper, 2012
15. P. Biane, M. Yor, Quelques Précisions sur le Méandre Brownien. Bull. de la Soc. Math. 2 Sér. **112**, 101–109 (1988)
16. M. Blais, P. Protter, An analysis of the supply curve for liquidity risk through book data. Int. J. Theor. Appl. Financ. **13**(6), 821–838 (2010)
17. A. Bris, W.N. Goetzmann, N. Zhu, Efficiency and the bear: Short sales and markets around the world. J. Financ. **62**(3), 1029–1079 (2007)
18. R. Caballero, E. Fahri, P.-O. Gourinchas, Financial crash, commodity prices and global imbalances. Brookings Pap. Econ. Act. Fall, 1–55 (2008)
19. C. Camerer, Bubbles and fads in asset prices. J. Econ. Sur. **3**(1), 3–41 (1989)
20. R. Carmona (ed.), *Indifference Pricing: Theory and Applications* (Princeton University Press, Princeton, 2008)

21. P. Carr, T. Fisher, J. Ruf, On the hedging of options On exploding exchange rates, preprint (2012)
22. D. Cass, K. Shell, Do sunspots matter? J. Polit. Econ. **91**(2), 193–227 (1983)
23. U. Çetin, R. Jarrow, P. Protter, Liquidity risk and arbitrage pricing theory. Financ. Stoch. **8**, 311–341 (2004)
24. U. Çetin, M. Soner, N. Touzi, Option hedging for small investors under liquidity costs. Financ. Stoch. **14**, 317–341 (2010)
25. A. Charoenrook, H. Daouk, A study of market-wide short-selling restrictions (February 2005). Available at SSRN: http://ssrn.com/abstract=687562orhttp://dx.doi.org/10.2139/ssrn.687562
26. J. Chen, H. Hong, J. Stein, Breadth of ownership and stock returns. J. Financ. Econ. **66**, 171–205 (2002)
27. P. Cheridito, D. Filipovic, M. Yor, Equivalent and absolutely continuous measure changes for jump-diffusion processes. Ann. Probab. **15**, 1713–1732 (2005)
28. R. Chernow, *The House of Morgan* (Atlantic Monthly Press, New York, 1990)
29. K.L Chung, R.J.Williams, *Introduction to Stochastic Integration*, 2nd edn. (Birkhäuser, Boston, 1990)
30. A. Cox, D. Hobson, Local martingales, bubbles and option prices. Financ. Stoc. **9**, 477–492 (2005)
31. J. Cox, J. Ingersoll, S. Ross, The relationship between forward prices and futures prices. J. Financ. Econ. **9**(4), 321–346 (1981)
32. J. Creswell, Analysts are Wary of LinkedIn's stock Surge. New York Times Digest 4 (Monday, May 23, 2011)
33. S.M. Davidoff, How to deflate a gold bubble (that might not even exist). Dealbook; The New York Times (August 30, 2011)
34. F. Delbaen, W. Schachermayer, A general version of the fundamental theorem of asset pricing. Math. Ann. **300**(3), 463–520 (1994)
35. F. Delbaen, W. Schachermayer, The fundamental theorem of asset pricing for unbounded stochastic processes. Math. Ann. **312**(2), 215–250 (1998)
36. F. Delbaen, W. Schachermayer, A simple counter-example to several problems in the theory of asset pricing. Math. Financ. **8**, 1–12 (1998)
37. F. Delbaen, W. Schachermayer, in *The Mathematics of Arbitrage*. Springer Finance (Springer, Heidelberg, 2005)
38. F. Delbaen, H. Shirakawa, No arbitrage condition for positive diffusion price processes. Asia Pacific Financ. Mark. **9**, 159–168 (2002)
39. C. Dellacherie, *Capacités et Processus Stochastiques* (Springer, Berlin, 1972)
40. H. Dengler, R.A. Jarrow, Option pricing using a binomial model with random time steps (a formal model of gamma hedging). Rev. Deriv. Res. **1**, 107–138 (1997)
41. K. Diether, C. Malloy, A. Scherbina, Differences of opinion and the cross section of stock returns. J. Financ. **52**, 2113–2141 (2002)
42. D. Duffie, *Dynamic Asset Pricing Theory*, 3rd edn. (Princeton University Press, Princeton, 2001)
43. B. Dupire, Pricing and hedging with smiles, in *Mathematics of Derivative Securities* (Cambridge University Press, Cambridge, 1997), pp. 103–112
44. E. Ekström, J. Tysk, Convexity and the Black–Scholes equation. Ann. Appl. Probab. **19**, 1369–1384 (2009)
45. E. Ekström, P. Lötstedt, L. Von Sydow, J. Tysk, Numerical option pricing in the presence of bubbles. Quant. Financ. **11**(8), 1125–1128 (2011)
46. K.D. Elworthy, X.M. Li, M. Yor, The importance of strictly local martingales; applications to radial Ornstein–Uhlenbeck processes. Probab. Theory Relat. Fields **115**, 325–355 (1999)
47. H.J. Engelbert, W. Schmidt, Strong Markov continuous local martingales and solutions of one-dimensional stochastic differential equations, Parts I, II, and III. Math. Nachr. **143**, 167–184; **144**, 241–281; **151**, 149–197 (1989/1991)

48. G. Evans, A test for speculative bubbles in the sterling-dollar exchange rate: 1981–1984. Am. Econ. Rev. **76**(4), 621–636 (1986)
49. J.D. Farmer, The economy needs agent-based modelling. Nature **460**, 680–681 (2009)
50. R. Fernholz, I. Karatzas, Relative arbitrage in volatility stabilized markets. Ann. Financ. **1**, 149–177 (2005)
51. Finfacts web site: http://www.finfacts.com/Private/curency/djones.htm. Accessed 2012
52. D. Florens-Zmirou, On estimating the diffusion coefficient from discrete observations. J. Appl. Probab. **30**, 790–804 (1993)
53. H. Föllmer, M. Schweizer, Hedging of contingent claims under incomplete information, in *Applied Stochastic Analysis*, ed. by M.H.A. Davis, R.J. Elliott (Gordon and Breach, New York, 1991), pp. 389–414
54. K. Froot, M. Obstfeld, Intrinsic bubbles: The case of stock prices. Am. Econ. Rev. **81**(5), 1189–1214 (1991)
55. J.K. Galbraith, *A Short History of Financial Euphoria* (Penguin Books, New York, 1993)
56. V. Genon-Catalot, J. Jacod, On the estimation of the diffusion coefficient for multi dimensional diffusion processes. Ann. de l'I.H.P B **29**, 119–151 (1993)
57. C. Gilles, Charges as equilibrium prices and asset bubbles. J. Math. Econ. **18**, 155–167 (1988)
58. C. Gilles, S.F. LeRoy, Bubbles and charges. Int. Econ. Rev. **33**(2), 323–339 (1992)
59. J. Goldstein, Interview with William Dudley, the President of the New York Federal Reserve. Planet Money (April 9, 2010)
60. P. Grandits, T. Rheinlander, On the minimal entropy martingale measure. Ann. Probab. **30**, 1003–1038 (2002)
61. P. Guasoni, M. Rasonyi, Fragility of arbitrage and bubbles in diffusion models, preprint (2011). Available at SSRN: http://ssrn.com/abstract=1856223orhttp://dx.doi.org/10.2139/ssrn.1856223
62. M. Guidolina, A. Timmermann, Asset allocation under multivariate regime switching. J. Econ. Dyn. Control **31**, 3503–3544 (2007)
63. S. Heston, M. Loewenstein, G.A. Willard, Options and bubbles. Rev. Financ. Stud. **20**(2), 359–390 (2007)
64. M. Hoffmann, L_p Estimation of the diffusion coefficient. Bernoulli **5**, 447–481 (1999)
65. T. Hollebeek, T.S. Ho, H. Rabitz, Constructing multidimensional molecular potential energy surfaces from AB initio data. Annu. Rev. Phys. Chem. **50**, 537–570 (1999)
66. C.H. Hommes, A. Huesler, D. Sornette, Super-exponential bubbles in lab experiments: evidence for anchoring over-optimistic expectations on price, preprint (2012). Available at http://arxiv.org/abs/1205.0635
67. H. Hong, J. Scheinkman, W. Xiong, Advisors and asset prices: A model of the origins of bubbles. J. Financ. Econ. **89**, 268–287 (2008)
68. H. Hulley, The economic plausibility of strict local martingales in financial modeling, in *Contemporary Quantitative Finance*, ed. by C. Chiarella, A. Novikov (Springer, Berlin, 2010)
69. H. Hulley, E. Platen, A visual criterion for identifying Itô diffusions as martingales or strict local martingales, in *Seminar on Stochastic Analysis, Random Fields and Applications VI*. Progress in Probability, vol. 63, Part 1 (2011), pp. 147–157
70. A. Hüsler, D. Sornette, C.H. Hommes, Super-exponential bubbles in lab experiments: evidence for anchoring over-optimistic expectations on price, Swiss Finance Institute Research Paper No. 12–20 (2012). Available at SSRN: http://ssrn.com/abstract=2060978orhttp://dx.doi.org/10.2139/ssrn.2060978
71. J. Jacod, Rates of convergence to the local time of a diffusion. Ann. de l'Institut Henri Poincaré Sect. B **34**, 505–544 (1998)
72. J. Jacod, Non-parametric kernel estimation of the coefficient of a diffusion. Scand. J. Stat. **27**, 83–96 (2000)
73. J. Jacod, P. Protter, Risk neutral compatibility with option prices. Financ. Stoch. **14**, 285–315 (2010)
74. J. Jacod, P. Protter, *Discretization of Processes* (Springer, Heidelberg, 2012)

75. J. Jacod, A. Shiryaev, Local martingales and the fundamental asset pricing theorems in the discrete-time case. Financ. Stoch. **2**, 259–273 (1998)
76. J. Jacod, A. Shiryaev, *Limit Theorems for Stochastic Processes*, 2nd edn. (Springer, Heidelberg, 2003)
77. J. Jacod, Y. Li, P. Mykland, M. Podolskij, M. Vetter, Microstructure noise in the continuous case: The pre-averaging approach. Stoch. Process. Their Appl. **119**, 2249–2276 (2009)
78. R. Jarrow, Y. Kchia, P. Protter, How to detect an asset bubble. SIAM J. Financ. Math. **2**, 839–865 (2011)
79. R. Jarrow, Y. Kchia, P. Protter, Is there a bubble in LinkedIn's stock price? J. Portf. Manag. **38**, 125–130 (2011)
80. R. Jarrow, Y. Kchia, P. Protter, Is gold in a bubble? Bloomberg's Risk Newsletter 8–9 (October 26, 2011)
81. R. Jarrow, D. Madan, Arbitrage, martingales, and private monetary value. J. Risk **3**(1), 73–90 (2000)
82. R. Jarrow, G. Oldfield, Forward contracts and futures contracts. J. Financ. Econ. **9**(4), 373–382 (1981)
83. R. Jarrow, P. Protter, An introduction to financial asset pricing theory, in *Handbook in Operation Research and Management Science: Financial Engineering*, vol. 15, ed. by J. Birge, V. Linetsky. (North Holland, New York, 2007), pp. 13–69
84. R. Jarrow, P. Protter, Forward and futures prices with bubbles. Int. J. Theor. Appl. Financ. **12**(7), 901–924 (2009)
85. R. Jarrow, P. Protter, Foreign currency bubbles. Rev. Deriv. Res. **14**(1), 67–83 (2011)
86. R. Jarrow, P. Protter, Discrete versus continuous time models: Local martingales and singular processes in asset pricing theory. Financ. Res. Lett. **9**, 58–62 (2012)
87. R. Jarrow, Y. Yildirim, Pricing treasury inflation protected securities and related derivatives using an HJM model. J. Financ. Quant. Anal. **38**(2), 337–358 (2003)
88. R. Jarrow, P. Protter, K. Shimbo, Asset price bubbles in a complete market, in *Adv. Math. Financ. [in Honor of Dilip B. Madan]*, ed. by M.C. Fu, D. Madan (2006), pp. 105–130
89. R. Jarrow, P. Protter, K. Shimbo, Asset price bubbles in incomplete markets. Math. Financ. **20**, 145–185 (2010)
90. R. Jarrow, P. Protter, A. Roch, A liquidity-based model for asset price bubbles. Quant. Financ. (2011). doi:10.1080/14697688.2011.620976
91. R. Jarrow, P. Protter, S. Pulido, The effect of trading futures on short sales constraints, preprint (2012)
92. A. Johansen, D. Sornette, Modeling the stock market prior to large crashes. Eur. Phys. J. B **9**, 167–174 (1999)
93. G. Johnson, L.L. Helms, Class D supermartingales. Bull. Am. Math. Soc. **14**, 59–61 (1963)
94. B. Jourdain, Loss of martingality in asset price models with lognormal stochastic volatility, Preprint CERMICS 2004-267 (2004). http://cermics.enpc.fr/reports/CERMICS-2004/CERMICS-2004-267.pdf
95. Y. Kabanov, In discrete time a local martingale is a martingale under an equivalent probability measure. Financ. Stoch. **12**, 293–297 (2008)
96. I. Karatzas, C. Kardaras, The numéraire portfolio in semimartingale financial models. Financ. Stoch. **11**, 447–493 (2007)
97. I. Karatzas, S. Shreve, *Brownian Motion and Stochastic Calculus*, 2nd edn. (Springer, New York, 1991)
98. I. Karatzas, S. Shreve, *Methods of Mathematical Finance* (Springer, New York, 2010)
99. C. Kardaras, Valuation and parity formulas for exchange options, preprint (2012). Available at arXiv: arXiv:1206.3220v1 [q-fin.PR]
100. C. Kardaras, D. Kreher, A. Nikeghbali, Strict local martingales and bubbles, preprint (2012). Available at arXiv:1108.4177v1 [math.PR]
101. S. Kotani, On a condition that one dimensional diffusion processes are martingales, in *In Memoriam Paul-André Meyer*. Séminaire de Probabilitiés XXXIX. Lecture Notes in Mathematics, vol. 1874 (2006), pp. 149–156

102. D. Kramkov, Discussion on how to detect an asset bubble, by R. Jarrow, Y. Kchia, and P. Protter; Remarks given after the presentation of P. Protter at the meeting, *Contemporary Issues and New Directions in Quantitative Finance*, Oxford, England, 10 July 2010
103. T.G. Kurtz, P. Protter, Weak limit theorems for stochastic integrals and stochastic differential equations. Ann. Probab. **19**s, 1035–1070 (1991)
104. T.G. Kurtz, P. Protter, Characterizing the weak convergence of stochastic integrals, in *Stochastic Analysis*, ed. by M. Barlow, N. Bingham (Cambridge University Press, Cambridge, 1991), pp. 255–259
105. O.A. Lamont, R.H. Thaler, Can the market add and subtract? Mispricing in tech stock Carve-outs. J. Polit. Econ. **111**, 227–268 (2003)
106. X. Li, M. Lipkin, R. Sowers, Dynamics of Bankrupt stocks, preprint (2012). Available at SSRN: http://ssrn.com/abstract=2043631
107. P.L. Lions, M. Musiela, Correlations and bounds for stochastic volatility models. Ann. Inst. Henri Poincaré, (C) Nonlinear Anal. **24**(1), 1–16 (2007)
108. M. Loewenstein, G.A. Willard, Rational equilibrium asset-pricing bubbles in continuous trading models. J. Econ. Theory **91**, 17–58 (2000)
109. H.P. McKean, Jr., *Stochastic Integrals* (AMS Chelsea Publishing, 2005) [Originally published in 1969 by Academic Press, New York]
110. D. Madan, M. Yor, Itôs integrated formula for strict local martingales, in *In Memoriam Paul-André Meyer, Séminaire de Probabilités XXXIX*, ed. by M. Emery, M. Yor. Lecture Notes in Mathematics, vol. 1874 (Springer, Berlin, 2006), pp. 157–170
111. J. Markham, *A Financial History of Modern U.S. Corporate Scandals: From Enron to Reform* (M.E. Sharpe, Armonk, 2005)
112. R. Meese, Testing for bubbles in exchange markets: A case of sparkling rates? J. Polit. Econ. **94**(2), 345–373 (1986)
113. J. Mémin, L. Slominski, Condition UT et Stabilité en Loi des Solutions dEquations Différentielles Stochastiques, in *Sém. de Proba. XXV*. Lecture Notes in Mathematics, vol. 1485 (1991), pp. 162–177
114. R. Merton, Theory of rational option pricing. Bell J. Econ. **4**(1), 141–183 (1973)
115. P.A. Meyer, in *Martingales and Stochastic Integrals*. Lecture Notes in Mathematics, vol. 284 (Springer, Berlin, 1972/1973)
116. A. Mijatovic, M. Urusov, On the martingale property of certain local martingales. Probab. Theory Relat. Fields **152**, 1–30 (2012)
117. A. Mijatovic, M. Urusov, Convergence of integral functionals of one-dimensional diffusions. Probab. Theory Relat. Fields **152**, 1–30 (2012)
118. E. Miller, Risk, uncertainty and divergence of opinion. J. Financ. **32**, 1151–1168 (1977)
119. P. Monat, C. Stricker, Föllmer-Schweizer decomposition and mean-variance hedging for general claims. Ann. Probab. **23**, 605–628 (1995)
120. A. Nikeghbali, An essay on the general theory of stochastic processes. Probab. Sur. **3**, 345–412 (2006)
121. E. Ofek, M. Richardson, R.F. Whitelaw, Limited arbitrage and short sales restrictions: Evidence from the options markets. J. Financ. Econ. **74**(2), 305–342 (2004)
122. S. Pal, P. Protter, Analysis of continuous strict local martingales via h-transforms. Stoch. Process. Their Appl. **120**, 1424–1443 (2010)
123. E. Parzen, Statistical inference on time series by RKHS methods, in *Proceedings 12th Biennial Seminar*, ed. by R. Pyke. (Canadian Mathematical Congress, Montreal, 1970), pp. 1–37
124. C. Profeta, B. Roynette, M. Yor, *Option Prices as Probabilities: A New Look at Generalized Black-Scholes Formulae* (Springer, Heidelberg, 2010)
125. P.C.B. Phillips, J. Yu, Dating the timeline of financial bubbles during the subprime crisis. Quant. Econ. **2**, 455–491 (2011)
126. P.C.B. Phillips, Y. Wu, J. Yu, Explosive behavior in the 1990s Nasdaq: When did exuberance escalate asset values? Int. Econ. Rev. **52**, 201–226 (2011)

127. P. Protter, A partial introduction to financial asset pricing theory. Stoch. Process. Their Appl. **91**, 169–203 (2001)
128. P. Protter, *Stochastic Integration and Differential Equations*, version 2.1, 2nd edn. (Springer, Heidelberg, 2005)
129. P. Protter, The financial meltdown. Gazette de la Soc. des Math. de Fr. **119**, 76–82 (2009)
130. P. Protter, K. Shimbo, No arbitrage and general semimartingales, in *Markov Processes and Related Topics: A Festschrift for Thomas G. Kurtz*. IMS Lecture Notes–Monograph Series, vol. 4 (2008), pp. 267–283
131. S. Pulido, The fundamental theorem of asset pricing, the hedging problem and maximal claims in financial markets with short sales prohibitions. Ann. Appl. Probab., preprint (2011). Available at http://arxiv.org/abs/1012.3102 (to appear)
132. R. Rebib, D. Sornette, R. Woodard, W. Yan, Detection of crashes and rebounds in major equity markets. Int. J. Portf. Anal. Manag. **1**(1), 59–79 (2012)
133. A. Roch, Liquidity risk, volatility, and financial bubbles, Ph.D. Thesis, Applied Mathematics, Cornell University, 2009. Available at the URL dspace.library.cornell.edu/bitstream/1813/.../Roch,%20Alexandre.pdf
134. A. Roch, Liquidity risk, price impacts and the replication problem. Financ. Stoch. **15**(3), 399–419 (2011)
135. A. Roch, M. Soner, Resilient price impact of trading and the cost of illiquidity (2011). Available at SSRN: http://ssrn.com/abstract=1923840orhttp://dx.doi.org/10.2139/ssrn.1923840.
136. L.C.G. Rogers, S. Singh, The costs of illiquidity and its effect on hedging. Math. Financ. **20**(4), 597–615 (2010)
137. J. Ruf, Hedging under arbitrage. Math. Financ. **23**, 297–317 (2013)
138. J. Scheinkman, W. Xiong, Overconfidence and speculative bubbles. J. Polit. Econ. **111**(6), 1183–1219 (2003)
139. J. Scheinkman, W. Xiong, Heterogeneous beliefs, speculation and trading in financial markets, in *Paris-Princeton Lecture Notes on Mathematical Finance 2003*. Lecture Notes in Mathematics, vol. 1847 (2004), pp. 217–250
140. P. Schönbucher, A market model for stochastic implied volatility. R. Soc. Lond. Philos. Trans. Ser. A Math. Phys. Engr. Sci. **357**, 2071–2092 (1999)
141. M. Schweizer, J. Wissel, Term structures of implied volatilities: Absence of arbitrage and existence results. Math. Financ. **18**, 77–114 (2008)
142. M. Schweizer, J. Wissel, Arbitrage-free market models for option prices: The multi-strike case. Financ. Stoch. **12**, 469–505 (2008)
143. A.N. Shiryaev, *Probability* (Springer, Heidelberg, 1984)
144. S. Shreve, *Stochastic Calculus for Finance II: Continuous Time Models* (Springer, New York, 2004)
145. C. Sin, Complications with stochastic volatility models. Adv. Appl. Probab. **30**, 256–268 (1998)
146. D. Sondermann, in *Introduction to Stochastic Calculus for Finance: A New Didactic Approach*. Lecture Notes in Economic and Mathematical Systems (Springer, Berlin, 2006)
147. D. Sornette, R. Woodard, W. Yan, W-X. Zhou, Clarifications to questions and criticisms on the Johansen-Ledoit-Sornette bubble model, preprint (2011). Available at arXiv:1107.3171v1
148. M. Swayne, How to detect a gold bubble, in *eHow Money* (November 28, 2010), at the URL http://www.ehow.com/how_7415180_detect-gold-bubble.html
149. W.H. Taft, *Present Day Problems: A Collection of Addresses Delivered on Various Occasions*. (Books for Libraries Press, Freeport, 1967); Originally published in 1908
150. M. Taylor, Purchasing power parity. Rev. Int. Econ. **11**(3), 436–452 (2003)
151. A. Taylor, M. Taylor, The purchasing power parity debate. J. Econ. Perspect. **18**(4), fall, 135–158 (2004)
152. C. Thomas-Agnan, Computing a family of reproducing kernels for statistical applications. Numer. Algorithms **13**, 21–32 (1996)
153. J. Tirole, On the possibility of speculation under rational expectations. Econometrica **50**(5), 1163–1182 (1982)

154. J. Tirole, Asset bubbles and overlapping generations. Econometrica **53**(5), 1071–1100 (1985)
155. V. Todorov, Estimation of continuous-time stochastic volatility models with jumps using high-frequency data. J. Econom. **148**, 131–148 (2009)
156. G. Wahba, Spline models for observational data, in *CBMS-NSF Regional Conference Series in Applied Mathematics*, vol. 59 (SIAM, Philadelphia, 1990)
157. G. Wahba, An introduction to model building With reproducing kernel Hilbert spaces (2000). Available for free download at http://www.stat.wisc.edu/~wahba/ftp1/interf/index.html
158. P. Weil, On the possibility of price decreasing bubbles. Econometrica **58**(6), 1467–1474 (1990)
159. Wikipedia page http://en.wikipedia.org/wiki/Dot-com_bubble. Last updated 29 May 2013
160. N. Wingfield, Flush with cash, apple plans buyback and dividend. New York Times (March 19, 2012)
161. WRDS, Lastminute.com, *bid prices, from 14 March 2000 to 30 July 2004*; WRDS, eToys, *bid prices, from 20 May 1999 to 26 February 2001*; WRDS, Infospace, *bid prices, from 15 December 1998 to 19 September 2002*; WRDS, Geocities, *bid prices, from 11 August 1998 to 28 May 1999*

Stochastic Volatility and Dependency in Energy Markets: Multi-Factor Modelling*

Fred Espen Benth

Abstract We give a short introduction to energy markets, describing how they function and what products are traded. Next we survey some of the popular models that have been proposed in the literature. We extend the analysis of one of these models to include for stochastic volatility effects. In particular, we analyse a mean reverting stochastic spot price dynamics with a stochastic mean level modelled as an Ornstein–Uhlenbeck process. We include in this dynamics a stochastic volatility model of the Barndorff-Nielsen and Shephard type. Some properties of the dynamics are derived and discussed in relation to energy markets. Moreover, we derive a semi-analytical expression for the forward price based on such a spot dynamics. In the last part of these lecture notes we consider a cross-commodity spot price model including jumps. A Margrabe formula for options on the spread is derived, along with an analysis of the dependency risk under an Esscher measure transform. An empirical example demonstrates that the Esscher transform may increase the tail dependency in the bivariate jump part of the spot model.

*Financial support from the project "Energy Markets: Modeling, Optimization and Simulation" (EMMOS), funded by the Norwegian Research Council under grant 205328/v30, is greatly acknowledged. An anonymous referee is thanked for careful reading of the paper and the several suggestions which improved the presentation.

F.E. Benth (✉)
Center of Mathematics for Applications (CMA), University of Oslo, PO Box 1053 Blindern, N-0316 Oslo, Norway
e-mail: fredb@math.uio.no; http://folk.uio.no/fredb/

F.E. Benth et al., *Paris-Princeton Lectures on Mathematical Finance 2013*, Lecture Notes in Mathematics 2081, DOI 10.1007/978-3-319-00413-6_2,

1 An Introduction to Energy Markets

There exist many markets for trade in power and related products. In Europe, Australia and the US, the markets for selling and purchasing electricity have been liberalized the last decades. For example, one has the NordPool market covering generation and distribution of power in the Nordic countries, and the German power exchange EEX. In the US, there are several markets, as well as new markets emerging in Eastern Europe and Asia.

Typical fuels for power generation are oil, gas and coal. Oil in different qualities has been traded for a long time at different exchanges, including for example NYMEX and ICE in London. In recent years, gas and coal have been traded at the ICE and EEX markets, opening for more competitive prices than in the more traditional bilateral markets. Usually, when talking of *energy markets*, one is thinking of the markets for power, gas, coal and oil.

With the recent decade's attempt to regulate climate gas pollution, a market for emission allowances have emerged. In Europe, one can for example trade allowances for the emission of carbon dioxide at the European Climate Exchange and EEX. A gas or coal fired power plant must match its emission of carbon dioxide over a year with allowances, which can be bought at the exchange. This introduces an additional variable cost to the production.

Most of the energy markets, including the emission markets, offer platforms for trading in futures and forward contracts as well as call and put options on these. This creates opportunities for the market participants to manage their risk exposure towards fuel costs and power prices. However, a major risk factor in the energy markets is weather. For example, in the Nordic region, the demand for power is very dependent on the temperature. Cold winter temperature leads to an increase in demand due to household heating. In the US, one has the similar effect of warm summer temperature, where the demand for power goes up due to air-conditioning cooling. On the other hand, rainfall fills up reservoirs for hydro power production, while wind gives rise to wind power generation. In many markets, both hydro and wind generation of electricity are major sources of power.

To manage weather risk, Chicago Mercantile Exchange (CME) organizes trading in temperature futures contracts written on weather indices measured in various cities world-wide. The typical temperature indices are the cumulative amount of heating or cooling degree days. In addition to temperature futures, one can trade in options on these futures at the CME. In 2007, the US Futures Exchange announced that they would start a market for wind index futures contracts written on seven regions in the US where there are wind farms in 2008, however, the exchange closed down before this market came to be.

The power markets are interconnected through transmission lines. For example, one can send power produced in the Nordic countries to the German market, and vice versa, through air or sea-bed cables. This creates a dependency between the EEX and NordPool power prices, since a big price difference can be exploited by the producers and retailers in these two markets. Such opportunities are of course limited by the capacity in the transmission lines. However, an interesting effect of these inter-market dependencies were observed in the fall of 2007. Power prices at NordPool were higher than normal, attributed to the fact that from January 2008 carbon emission fees were to be introduced in the European Union, and power prices would increase in the German market. In the NordPool market, power producers could hold back production by storing water in the reservoirs and wait for the expected higher prices. This led to less supply in the autumn, and hence higher prices. The power spot prices in Germany, on the other hand, were not influenced before the emission fees became effective in January. The autumn prices for clean hydro power included emission fees in the Nordic market *before* these fees were introduced, while "dirty" power in the EEX area remained unaffected. The EEX is largely supplied by nuclear, coal and gas, and very little hydro power relative to the Nordic market. In fact, the NordPool and the emission market got connected via the transmission lines to the EEX market, although hydro power production does not emit carbon dioxide. We refer to Benth and Meyer-Brandis [5] for a discussion and mathematical modeling of this situation.

Transmission line capacity may also create price differences within a power market due to congestion. In the NordPool market, area prices are settled for each hour throughout the day to balance out the loss by transporting electricity through the network. For example, Norway may have up to five different prices for a given hour due to congestions between the different areas. Of course, the different area prices are highly dependent, but there exist also periods of high price differences even in neighboring areas.

The gas markets are also connected, via a network of pipelines for distribution of gas from the hubs to the various regions of Europe. LNG, liquefied natural gas, creates possibilities to transport gas from one continent to another by large tankers. Hence, producers and retailers have the opportunity to play on price differences, and ship the gas to the best markets. There is a market for freight called IMAREX, providing access to risk management tools for transportation. As coal is shipped from mines in Australia and Asia to continental Europe, the freight market will also play an important role here.

As we have already mentioned, weather impacts both demand and production of power, and therefore the power prices. A typical feature of electricity markets is the spiky behaviour of the prices. Occasionally, prices may rise by several hundreds of percent, and rapidly decay back to "normal levels". On the other hand, the seasonality of temperature creates a seasonally varying average price level. Both spikes and seasonality of power prices are clearly visible in Fig. 1, where we have depicted the spot prices at NordPool in the period April 1 1997 until July 14 2000.[1] The spikes are due to a sudden increase in demand due to a drop in temperature, say, or a fall-out of a major nuclear power plant in Sweden. If a nuclear power plant unexpectedly stops producing, prices will rise to compensate for the drop in supply.

[1]We have selected this rather old period of data for illustration only, since it was a period where prices had a very apparent seasonality and spike pattern.

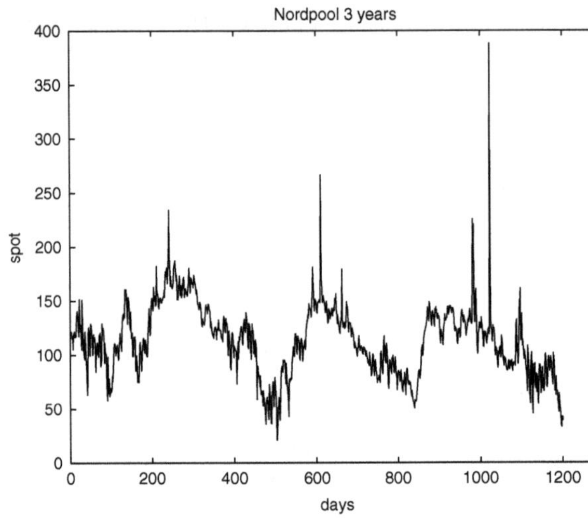

Fig. 1 The system price at NordPool from April 1 1997 untill July 14 2000. The prices are denominated in NOK per MWh

This will happen as a rather sharp increase since the supply is inflexible. However, prices will rapidly revert back since more expensive coal and gas fired power plants in Denmark will start operating and demand will decrease.

In the EEX market one has observed *negative* prices in the spot markets. The reason for this is the priority given to wind generated power in the network. If there is an unexpected increase in wind power generation, other producers may be forced to dump their production. In fact, since it is rather costly to shut down and next ramp up a coal fired power plant, say, it may be better financially to pay someone to consume the electricity production rather than adjust the power generation.

Coming back to the NordPool market, a decisive factor for the price level is reservoir filling. The amount of rain in the autumn and the snow levels during the winter, determines the production capacity for cheap hydro power, and therefore the prices during winter. A low reservoir filling, for example resulting from a very dry autumn, may lead to excessive prices for power during winter times. This is usually the period when spikes are observed in the NordPool market.

Let us move our attention to the specifics of the electricity market, where we use the NordPool market as the case of discussion. At NordPool, one can trade in spot electricity, forwards and futures contracts as well as plain vanilla call and put options on these. This division into three markets, a spot, forward and option market, is typical for most power markets, as well as other energy markets. The weather market is an obvious exception here, since there is no natural possibility to trade in "spot temperature". The spot market of power is a physical market where one must be either a producer or a consumer/retailer to participate. The two other segments are financially settled, and thus attracts speculators like investment banks, say.

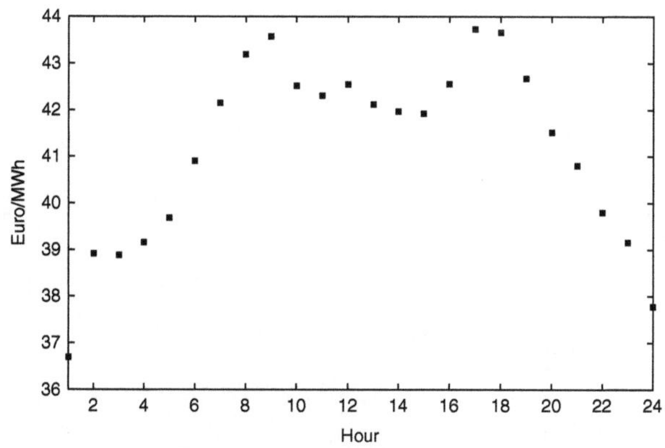

Fig. 2 The hourly system price at NordPool on 23 November 2011

The spot market is auction-based, where producers and consumers are handing in a limited amount of bids the day before. They can bid on buying or selling a certain amount (measured in MW=mega Watt) of power for a given price transmitted at a given hour the next day. Hence, the bids are for one or more of the 24 h the next day, giving the volume to buy or sell and the price at which the transaction can take place. The NordPool market is next feeding in all the bids and creating a demand and supply curve for each of the 24 h the coming day, and by noon the *system price* is settled for the next day. The system price is the spot price for delivery of 1 MW of electricity in a specific hour. In Fig. 2 we show the hourly system price in Euro per MWh at NordPool on November 23 2011. As is clearly seen, the power is most expensive in the peak hours around 8 in the morning and 6 in the afternoon. The evening and night prices are cheapest. This intra-day price pattern is rather typical. It is noticeable from a modeling point of view that the spot price is a time series, discretized at an hourly level. Furthermore, every day at noon 24 new prices for the next day will be revealed, very much in a similar fashion as a forward market is revealing prices for delivery at different times. In some markets, like the UK power market, one trades in half-hourly spot prices.

The hourly system price at NordPool will, however, not be the actual trading price for electricity in the market. Due to congestion, as discussed above, there will be different prices for different areas balancing production, demand and transmission capacity between these areas. These area prices will also be given for each hour, and constitute the actual price for power in that area. In Fig. 3 we have taken a screen shot from the web-page of the NordPool Spot market (see www.nordpoolspot.com), showing the average system price in the different areas in NordPool on 23 November 2011.

The forward and futures market at NordPool delivers power over a specific period of time. Unlike most other commodity markets that delivers the underlying asset at a specific delivery time, power has to be delivered over a *delivery period* by its physical nature. Hence, buying a future or forward contract will provide you

Elspot volumes

	Buy	Sell
NO1	103 994,2	77 712,8
NO2	64 390,3	111 192,1
NO3	31 813,8	27 909,6
NO4	26 262,0	65 587,4
NO5	18 196,7	30 196,7
DK1	71 118,5	58 722 ,2
DK2	50 701,7	26 171,9
SE1	27 704,5	59 430,4
SE2	38 202,9	123 979,9
SE3	241 970,8	188 972,7
SE4	90 755,7	19 513,6
FI	151 869,5	126 274,0
EE	18 836,2	20 153,5

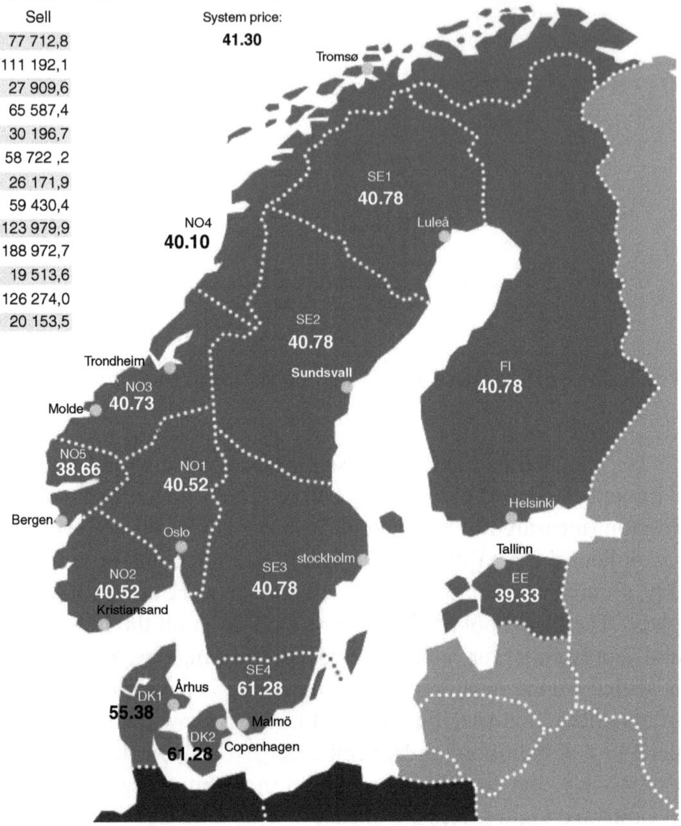

Fig. 3 The average area prices at NordPool on 23 November 2011. Prices are denominated in Euros per MWh

with power delivered over an agreed period. The NordPool exchange, similar to most other exchanges, is organizing the settlement in financial terms rather than actual physical delivery of power. Thus, buying a forward with delivery over the next month, say, will entitle you to receiving a stream of money equivalent to the spot price at each hour in the next month. If we denote by $S(t)$ the spot price at time t, you receive

$$\sum_{i=1}^{31\times 24} S(t_i) \tag{1}$$

where t_1 is the first hour in the next month, t_2 the second hour and so on throughout the month (assuming 31 days). The system price (and not the area prices) are used in settling the contract. In return, the owner of the forward will pay a fixed price,

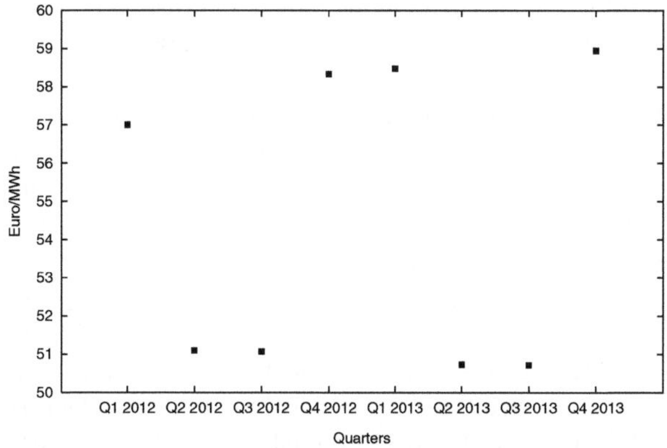

Fig. 4 Futures prices for contracts delivering over a quarter traded at EEX on 23 November 2011

called the *forward price*. It is the market convention to denote this price in terms of Euro per MWh, which is the same denomination as the system price for a given hour. Letting $F(t, T_1, T_2)$ be this forward price, with T_1 and T_2 being the first and last hour of the coming month, and $t \leq T_1$ being the time of entry into the forward contract, we must pay $24 \times 31 \times F(t, T_1, T_2)$ in return for receiving (1). Hence, we see that a forward contract is essentially a swap of a floating spot price with a fixed price. Often the forwards are called swaps in this market.

Since the forwards and futures contracts are financially settled, one does not need to have any physical capacity for producing or receiving electricity in order to participate in this market segment. Since the NordPool forward and futures are economically equivalent to physically settled contracts, producers and consumers may use them for hedging. But these contracts may also be traded by speculators, for example investment banks and funds, providing more liquidity to the market. This is also what has happened to some extent in the NordPool and EEX markets. In Fig. 4 we have plotted the futures prices for quarterly contracts at the EEX on 23 November 2011. As we clearly see, the prices for financial delivery of power in the first and fourth quarters are more expensive than in the second and third. This matches with the winter and summer periods, where one expects cold and warm weather, respectively. The EEX market trades in quarterly futures for 2 years ahead. The NordPool quarterly forward prices would follow a similar pattern.

NordPool offers both forward and futures contracts. The delivery periods offered in the market ranges from short term daily and weekly delivery, to longer term monthly, quarterly and yearly delivery. Yearly delivery contracts are offered for up to 3 years ahead, meaning that you can fix prices for delivery over 2014 today (which is November 2011, at the time of writing), for example. The short term contracts are futures-style, whereas the long-term are forward contracts. The market also distinguishes between peak and base load contracts. Peak load contracts are settled on the system price in peak hours, which are defined from 8 to 20 every working day. These hours are the times when demand is highest. Base load contracts, on the other hand, are settled on all hours in the delivery period. On some of the forward contracts, one can trade in European call and put options. The market for these options have been rather thin.

Although the option trading on the power market is not so active, there exists an abundance of various exotic option contracts traded OTC. So-called swing options, where the holder has the right to buy power at favourable prices, at the same time deciding the amount or volume to be traded, are very popular and appear in many different kinds. For example, the flexible load contracts gives the owner the right to buy electricity at a fixed price in a number of hours over a year. This is an American-style option, where the holder can decide when to exercise within a year, however, having multiple exercise rights. Each time the holder exercises, she also determines the volume of power she wants to buy. Hence, in mathematical terms, the owner of the flexible load contract must find an optimal strategy for the exercise times, *and* an optimal volume control at each time of exercise. Naturally, there are constraints on the volume to take out from the contract at each exercise, as well as a cumulative volume constraint. Such contracts have been analysed in Benth et al. [11] using stochastic control theory.

Spread options are very popular tools for managing cross-commodity risk. In the energy markets they appear as spark and dark spread options, say, being options written on the difference between power and gas/coal prices (or, rather the energy equivalent). They may also be combined into swing option like instruments. A typical example is the so-called *tolling agreement*, which is in effect a virtual power plant. For example, one may get a contract which gives the holder a stream of money every time she decides to produce power, in a virtual power plant using gas as fuel. She decides when to produce, and how much within certain boundaries. When producing, she will receive the spot price of power, in return to paying the spot price of gas. Such options can also be used to value projects of building a gas-fired power plant. Although most spread options are traded OTC, there exists a market for spreads between different refined oil products at NYMEX, as well as contracts on the difference between area prices in the NordPool market. The latter is called Contracts-for-Difference (CfD), and are futures written on the price spread between two areas. We refer to Carmona and Durrleman [14] for an extensive discussion and analysis of spread options in energy markets, and Eydeland and Wolyniec [15] for other exotic options.

There exist many other exotic options in the energy markets, like for example various average-type options and so-called quanto options. The latter are options written on an energy like gas or power, paying out in a call or put fashion. However, the payout is triggered by a weather index, say. For example, one may have a call option on the gas price, which is nulled if a temperature index falls within certain bounds. The latter may provide a control of risk towards demand, while the call structure provides a hedge against high prices of gas. Such products provide then a hedge towards volume risk for the participants in the energy market.

2 Stochastic Modeling of Energy Markets

We now outline some approaches to stochastic modelling of energy prices, without any intention to be exhaustive in our presentation. The purpose of this subsection is more to create a starting point to the topics presented in these lecture notes, namely cross-commodity and stochastic volatility modelling.

As the basic model for stock prices is the geometric Brownian motion, the Schwartz dynamics (see Schwartz [25]) is the canonical model in energy (and commodity) markets. Suppose that the spot price follows the stochastic process

$$S(t) = \Lambda(t)\exp(X(t)), \tag{2}$$

for $t \geq 0$, where $\Lambda(t)$ is a positive deterministic function modeling the average spot price, also called the seasonality, and $X(t)$ follows the Ornstein–Uhlenbeck (OU) process

$$dX(t) = -\alpha X(t)\,dt + \sigma\,dB(t). \tag{3}$$

Here, α and σ are positive constants and $B(t)$ is a Brownian motion defined on a complete filtered probability space $(\Omega, \mathcal{F}, \{\mathcal{F}_t\}_{t\geq 0}, P)$. We note that $\ln S(t)$, the logarithmic spot price, will mean revert towards its mean $\ln \Lambda(t)$, at a speed α and volatility given by σ. A straightforward application of Itô's Formula, assuming $\Lambda(t)$ being differentiable, yields the dynamics

$$dS(t) = \left(\frac{\Lambda'(t)}{\Lambda(t)} + \alpha \ln \Lambda(t) + \frac{1}{2}\sigma^2 - \alpha \ln S(t)\right) S(t)\,dt + \sigma S(t)\,dB(t)\,. \tag{4}$$

Thus, the Schwartz dynamics is a "geometric Brownian motion" with a state-dependent drift which is mean-reverting to a seasonal level.

In our survey on energy markets, we saw that the spot prices of electricity were discrete, quoting only prices for hourly delivery. A continuous-time model may seem inappropriate in this context. However, it is rather standard to make such an approximation. One may view the continuous price dynamics as an unobserved price for immediate delivery of electricity, and the actual hourly spot price are simply observations of this. There is a small problem with the filtration using such a view, since the 24 hourly prices are settled the day before, and not according to a stream of information arriving continuously over the day. On the other hand, it is very practical to have a continuous-time model for the spot price dynamics, as one is interested to derive forward prices based on this. The forward prices evolve in a continuous-time market, and not at discrete hourly times.

The model in (2)–(3) has been suggested for NordPool spot prices by Lucia and Schwartz [21]. In [21], they calibrated this model to spot price data, but also extended it by introducing a second factor driving the spot price. They considered the model

$$S(t) = \Lambda(t)\exp(X(t) + Y(t)) \tag{5}$$

for a non-stationary term $Y(t) = \mu t + \eta W(t)$, $W(t)$ being a correlated Brownian motion (see Schwartz and Smith [26] for a first application of this model to oil markets). In their paper, Lucia and Schwartz [21] made an extensive study of both spot and forward price modeling in the NordPool market.

The models suggested by Schwartz and co-authors above are driven by Brownian motion, and unlikely to create the large price spikes that one observes in power markets. Also in the gas market one may observe large price increases that may be attributed as spikes, being a result of increase of demand and low storage. A natural way to model spikes is to apply Lévy processes, which may produce a sudden increase of the price from a large upward jump. For example, we may substitute the Brownian motion $B(t)$ in (3) by a Lévy process $L(t)$. The speed of mean reversion α will then make sure that a large positive jump in $L(t)$, is followed by an exponentially fast decrease in prices. Letting the speed of mean reversion be sufficiently fast, we can obtain price paths with spikes, a feature which is typically observed in real power price data. Since Lévy processes are rather flexible, one may combine big jumps with many small, and even include a Brownian motion. Alternatively, one may separate the spike behaviour with the "normal variations" of prices, using $Y(t)$ as an OU process driven by Brownian motion rather than a drifted Brownian motion. The choices are many, and the model must be selected by properties of the data which vary considerably between markets. We refer to Benth et al. [7] for a thorough analysis of multi factor models.

There is a debate whether power spot prices are stationary or not. The two-factor model suggested by Lucia and Schwartz [21] above is clearly non-stationary, but letting $Y(t)$ be an OU-process creates a stationary model. For example, in Barndorff-Nielsen et al. [2], one finds a very good fit for a stationary one-factor dynamics using the general class of Lévy semistationary (LSS) models. LSS models generalize OU processes, and offer a great deal of flexibility to capture the probabilistic properties of spot price data. An unfortunate effect of using stationary spot price models is that the forward prices (theoretically) become constant in the long end of the market. This is not a property one observes for actual forward prices. This could suggest that non-stationary models are more appropriate, or that the connection between spot and forward prices are far more complex in power markets than traditional modelling and pricing suggests (see Benth et al. [8] for a discussion and an equilibrium approach to power forward pricing).

In the EEX market we may have negative prices for the electricity spot. In fact, the NordPool market also allows negative prices. The Schwartz model (2) is on an exponential form, yielding positive prices at all times. An alternative specification could be to state the spot prices directly as a one or two factor model, like for example

$$S(t) = \Lambda(t) + X(t) + Y(t). \tag{6}$$

As it turns out, such a model may successfully calibrate the market data. Benth et al. [10] has estimated such a model to EEX spot prices, using a stable Lévy process driving a CARMA model $X(t)$, and a non-stationary Lévy processes to model $Y(t)$. A CARMA model is a continuous-time autoregressive moving average process, being a specific class of multi-dimensional OU-processes. Apart from allowing for negative prices, such arithmetic models are useful when pricing power forward contracts (which is done in Benth et al. [10]).

In these lecture notes we want to investigate stochastic volatility in a class of spot price models based on the Schwartz dynamics. Leaving aside the issue of jumps, we consider a two-factor model where the stochastic volatility (SV) is driven by a superposition of subordinators, called the Barndorff-Nielsen and Shephard (BNS) SV model. The dynamics generalizes the simpler one-factor stochastic volatility model proposed and analysed by Benth [3], which was fitted gas prices in the UK. In Hikspoors and Jaimungal [19], various stochastic volatility models are analysed in the context of energy.

We will also consider a cross-commodity spot model, where each commodity (or energy) is modelled as a two-factor process. The aim is to price spread options and to analyse some effects of the dependency structure between the two commodities. Although our models are rather simple and specific, we apply them as cases to illustrate some of the main issues and challenges in mathematical finance applied to energy markets.

In power markets, one cannot trade in the underlying spot price of electricity since it is non-storable, and thus one cannot perform a buy-and-hold strategy to hedge a forward position. The notion of convenience yield does not make sense either (see Geman [17]). Thus, the relationship between the spot and forwards in power markets is an open question with (yet) no clear answer. What we do know, however, is that the dynamics of the forward price has to be a (local) martingale under some pricing measure $Q \sim P$ in order to ensure an arbitrage-free dynamics of the forward price. Since the underlying spot is not tradeable in the financial sense of the word, such a pricing measure does not have to be a (local) martingale measure for the spot price dynamics. A forward contract with delivery of power at time T and *forward price* $f(t,T)$ agreed at time $0 \leq t \leq T$, will yield a payoff $f(t,T) - S(T)$ to the seller. From the arbitrage theory for pricing derivatives, we reach the definition of the forward price with respect to the measure Q as

$$f(t,T) = \mathbb{E}_Q[S(T) \mid \mathcal{F}_t], \tag{7}$$

since the investment costs are zero. A forward contract delivering over $[T_1, T_2]$ will have a forward price $F(t, T_1, T_2)$ naturally defined as

$$F(t, T_1, T_2) = \mathbb{E}_Q\left[\frac{1}{T_2 - T_1}\int_{T_1}^{T_2} S(T)\,dT \mid \mathcal{F}_t\right], \tag{8}$$

where we use the approximation that the forward delivers continuously over the delivery period rather than at the discrete hours. Of course, we need to impose certain integrability conditions on the spot price dynamics under Q in order to make these definitions well-posed.

Interchanging expectation and integration leads to the equation

$$F(t, T_1, T_2) = \frac{1}{T_2 - T_1} \int_{T_1}^{T_2} f(t, T)\, dT, \tag{9}$$

which lends itself to the obvious interpretation of a power forward as a stream of fixed-delivery forwards. In many cases, it is possible to derive the price $f(t, T)$ for a given pricing measure Q rather explicitly, but there is no analytical formula for the integral defining $F(t, T_1, T_2)$. This is the case for exponential models of the type in (2). Hence, one must perform numerical integration in order to obtain the prices of power forwards. Recalling that the electricity spot prices are settled on an hourly resolution, a natural numerical integration scheme would be a Riemann sum over $f(t, T_i)$, where T_i are the different hours of the delivery period. In fact, this would point back to the very definition of a forward contract with delivery period in the electricity market, where in practise the settlement is on the hourly spot prices in the delivery period.

Using arithmetic models of the type (6) will in many interesting situations give analytic power forward prices $F(t, T_1, T_2)$, at least up to knowledge of the characteristics of the driving noise processes of the factors. This is an attractive aspect of arithmetic models, paving the way for pricing of options on power forwards using Fourier methods which are far more efficient than Monte Carlo simulation, say.

Of course, to pin down the right forward price requires a specification of the 'pricing measure' Q. This is typically done by choosing a parametric class of measures using Girsanov or Esscher transform (see Benth et al. [7]). From this, one may be able to compute theoretical prices, which next can be fitted to observed ones in order to estimate the parameters in the measure transform. This procedure is targeted to explain the *risk premium* in the market, defined as the difference between forward prices and the predicted spot:

$$R(t, T_1, T_2) = F(t, T_1, T_2) - \mathbb{E}\left[\frac{1}{T_2 - T_1} \int_{T_1}^{T_2} S(T)\, dT \mid \mathcal{F}_t\right]. \tag{10}$$

Empirical studies of the risk premium in power markets reveal quite a complex behaviour (see Benth et al. [8] for a discussion and references). In commodity markets, one usually expects the risk premium to be negative, a result of hedging pressure from producers accepting to pay a premium to the speculators for locking in prices of their commodities. This is usually referred to as *normal backwardation* in the market. However, in power markets we may encounter *positive* risk premia. Such positive premia can be found in the short end of the forward market, that is, for contracts which are close to delivery and with a relatively short delivery period. The reason for this is that the *consumers* want to lock in prices to hedge the spike risk. The producers are of course not afraid of excessively high prices, but this may be harmful to retailers which may be engaged in fixed-price contracts with their clients. The producers want to hedge using longer term contracts, which shows up as a negative premium in the long end of the forward market. Hence, the picture we see in power markets is a risk premium which changes sign from positive to negative as a function of time to delivery. We refer to Benth et al. [10] for a confirmation of this in the EEX market based on a spot price modeling approach, and to Benth et al. [8] for a simple equilibrium model explaining this along with an empirical case at the EEX. Benth and Sgarra [6] show that the change in sign of the risk premium may possibly be explained by a seasonality in the occurrence of spikes.

The Heath–Jarrow–Morton (HJM) approach in interest rate theory suggests to model the forward rates directly rather than via the short rate of interest (see Heath, Jarrow and Morton [18]). This idea was adopted to power markets by Benth and Koekebakker [4], where forward prices are modelled directly rather than via spot prices. The main challenge with the HJM methodology in power markets is the inclusion of a delivery period in the dynamics.

As we have discussed several times already, forward contracts in power markets deliver over various periods. These may in fact be overlapping. In the NordPool market, one can trade in contracts for delivery in the months January, February and March, say, but at the same time one can enter a contract delivering over the first quarter of the year. Of course, the three monthly contracts overlap completely with the quarterly contract. If T_i, $i = 0, 1, 2, 3$ are the first of each of the months January, February, March and April, we must have that the forward prices satisfy the condition

$$F(t, T_0, T_3) = \sum_{i=0}^{2} \frac{T_{i+1} - T_i}{T_3 - T_0} F(t, T_i, T_{i+1}), \tag{11}$$

in order to avoid arbitrage opportunities between the four contracts. If, in an HJM approach, we insist on specifying a model of $F(t, T_s, T_e)$ for all possible delivery periods $[T_s, T_e]$, with $0 \leq t \leq T_s < T_e$, we are led to the no-arbitrage condition (see Benth and Koekebakker [4])

$$F(t, T_s, T_e) = \frac{1}{T_e - T_s} \int_{T_s}^{T_e} F(t, T, T) \, dT. \tag{12}$$

This condition comes in addition to the martingale restriction on the dynamics of $F(t, T_s, T_e)$ under the pricing measure Q. Since $F(t, T, T)$ is the forward price of a contract delivering *at* time T, we are in a situation where any model of $F(t, T_s, T_e)$ is brought back to a model for $f(t, T)$. Hence, it is natural to model fixed-delivery power forward prices $f(t, T)$, although these do not exist in the market (see Benth et al. [7] for a discussion on this, both analytically and empirically).

Alternatively, one may take a LIBOR modeling point of view (see for example Brigo and Mercurio [13]), and simply focus on the *traded delivery periods*, that is, to model only the forwards which are traded in the market. We first single out the smallest delivery periods, and model these exclusively. For example, going back to the case above, we model only the forward price dynamics of the three monthly contracts, and let the quarterly contract has a forward price given by (11). This program was proposed and studied in the context of the NordPool market in Benth and Koekebakker [4], and further extended in Benth et al. [7].

3 A Multi-factor Stochastic Volatility Model for Energy Prices

Let us consider the question of modelling the spot price dynamics of an energy commodity. We denote the price at time $t \geq 0$ by $S(t)$, and choose to model the dynamics in continuous time although it may be discrete in some markets like for example electricity as discussed above. We focus our attention on a single market, and propose a multi-factor model which accounts for many of the stylized facts observed in power and energy markets. Inspired by Hikspoors and Jaimungal [19], we analyse a spot price model of exponential type which incorporates stochastic volatility, and where prices are mean-reverting towards a stochastic level.

To be slightly more specific, we suppose that the spot price on logarithmic scale is given as an Ornstein–Uhlenbeck process reverting towards a stochastic level, and driven by a Brownian motion scaled by a stochastic volatility. The stochastic volatility follows the so-called Barndorff-Nielsen and Shephard (BNS) model, see [1]. The level towards which the log-spot prices are reverting will be assumed to be again an Ornstein–Uhlenbeck dynamics. Hikspoors and Jaimungal [19] assume a Brownian-based dynamics of the stochastic volatility, including for example the Heston model, and focus on an asymptotic analysis of derivatives. We now introduce our model rigourously, starting out by defining the stochastic volatility part.

We denote by L_j, $j = 1, \ldots, n$, n independent subordinator processes, that is, increasing Lévy processes. We choose to work with the RCLL version of the L_j's. Define for $j = 1, \ldots, n$ the Ornstein–Uhlenbeck process

$$dY_j(t) = -\lambda_j Y_j(t)\,dt + dL_j(t)\,, \tag{13}$$

where $\lambda_j > 0$ is constant. The Lévy measure of L_j is denoted ℓ_j. Let $w_j > 0$ and $w_1 + \ldots + w_n = 1$, and define a volatility process $\sigma(t)$ by

$$\sigma^2(t) = \sum_{j=1}^{n} w_j Y_j(t). \tag{14}$$

Note that since the L_j's are subordinators, it follows that $Y_j(t)$ are non-negative for all $j = 1, \ldots, n$, and thus $\sigma^2(t)$ is non-negative as well. Therefore, $\sigma(t)$, the square-root of $\sigma^2(t)$, is well defined. We shall assume that the subordinators are driftless, that is, that $L_j(1)$ have cumulant functions given by

$$\psi_j(\theta) := \ln \mathbb{E}\left[\exp(\mathrm{i}\theta L_j(1))\right] = \int_0^\infty \{\mathrm{e}^{\mathrm{i}\theta z} - 1\}\,\ell_j(dz), \tag{15}$$

for $j = 1, \ldots, n$.

Note that we have a constant volatility process $\sigma(t)$ whenever $\lambda_j = 0$ and $L_j = 0$ for all j. If only the latter holds, the volatility becomes deterministic, converging to zero with time. To account for possible seasonal effects one may allow for time-dependent coefficients w_j and λ_j. However, we shall not consider this case, but restrict our attention to constant coefficients.

In our spot price model we suppose that the seasonal level is modelled by a bounded and measurable function $\Lambda : [0, \infty) \mapsto \mathbb{R}_+$. In case there is no seasonality, $\Lambda(t)$ is simply a constant, usually put equal to 1.

Define the spot price as

$$S(t) = \Lambda(t) \exp(X(t)), \tag{16}$$

with

$$dX(t) = (Z(t) - \alpha X(t))\, dt + \sigma(t)\, dB_1(t) \tag{17}$$

$$dZ(t) = (\mu - \beta Z(t))\, dt + \eta\, dB_2(t). \tag{18}$$

The Brownian motions B_1 and B_2 are correlated by a factor ρ, and independent of the subordinators L_j, $j = 1, \ldots, n$. This means that the volatility process $\sigma(t)$ is independent of the stochastic drivers of the mean level and the log-spot price. The deseasonalized log-spot prices are mean-reverting like in the one-factor model, however, now towards a stochastic mean, which again is a mean reverting process. From an applied point of view, it is natural to imagine that the stochastic mean is slowly mean reverting, while the prices themselves mean revert at a higher speed. Geman [17] argues for such a dynamics for oil prices using a Cox–Ingersoll–Ross model for $\sigma^2(t)$.

In the literature on energy spot price models (see Chap. 3 in Benth et al. [7] and the references therein) factor models are usually stated directly as a sum of stochastic processes. A simple two-factor model for the spot price on exponential form is given as

$$S(t) = \Lambda(t) \exp(X_1(t) + X_2(t)),$$

with

$$dX_i(t) = -\alpha_i X_i(t)\, dt + \sigma_i\, dB_i(t)$$

for $i = 1, 2$. Letting $X(t) := X_1(t) + X_2(t)$, we find the dynamics of $X(t)$ to be

$$dX(t) = ((\alpha_1 - \alpha_2) X_2(t) - \alpha_1 X(t))\, dt + \sigma_1\, dB_1(t) + (\alpha_1 - \alpha_2)\sigma_2\, dB_2(t).$$

Hence, by identifying $Z(t) = (\alpha_1 - \alpha_2) X_2(t)$ we see that we recover our mean-reversion model with a stochastic level in (16). We remark in passing that the two-factor model of Lucia and Schwartz [21] assumes the factor $X_2(t)$ to be non-stationary. We may incorporate that case by supposing $X_2(t)$ having zero mean-reversion, $\alpha_2 = 0$. This would correspond to a stochastic mean-level being non-stationary, that is, equivalent to letting $\beta = 0$ in the process $Z(t)$ in the model (16). The Lucia and Schwartz model was first proposed by Schwartz and Smith [26] for spot oil prices.

In the analysis of our spot price model, we want to derive probabilistic properties like the induced covariance structure, as well as the implied forward prices for contracts written on the spot. In the sequel, we suppose that $\alpha \neq \beta$ to avoid a singular case. Our first result derives the explicit dynamics of $X(t)$, $Z(t)$ and $Y_j(t)$, $j = 1, \dots, n$:

Lemma 3.1. *Suppose that $u \mapsto \sigma(u)\exp(-\alpha(u-t))$ is Itô integrable on $u \in [s,t]$ for $s \geq t$. The explicit dynamics of $X(s)$, $Z(s)$ and $Y_j(s)$ given $X(t)$, $Z(t)$ and $Y_j(t)$ for $s \geq t$ are*

$$X(s) = X(t)\mathrm{e}^{-\alpha(s-t)} + \int_t^s Z(u)\mathrm{e}^{-\alpha(s-u)}\,du + \int_t^s \sigma(s)\mathrm{e}^{-\alpha(s-u)}\,dB_1(u)\,,$$

$$Z(s) = Z(t)\mathrm{e}^{-\beta(s-t)} + \frac{\mu}{\beta}(1-\mathrm{e}^{-\beta(s-t)}) + \int_t^s \eta\mathrm{e}^{-\beta(s-u)}\,dB_2(u)\,.$$

and

$$Y_j(s) = Y_j(t)\mathrm{e}^{-\lambda_j(s-t)} + \int_t^s \mathrm{e}^{-\lambda_j(s-u)}\,dL_j(u)\,.$$

Proof. We apply Itô's Formula on the process $\exp(\alpha u)X(u)$ to find

$$\begin{aligned} d(\mathrm{e}^{\alpha u}X(u)) &= \alpha\mathrm{e}^{\alpha u}X(u)\,dt + \mathrm{e}^{\alpha u}dX(u) \\ &= \mathrm{e}^{\alpha u}Z(u)\,du + \sigma(u)\mathrm{e}^{\alpha u}\,dB_1(u)\,. \end{aligned}$$

Integrating from t to s yields the result for $X(s)$. A similar computation shows the result for $Z(s)$. Finally, using the Itô Formula for jump processes (see Ikeda and Watanabe [20]) yields the result for $Y_j(s)$. □

To this end, define the function

$$\gamma(u;\alpha,\beta) = \frac{1}{\alpha-\beta}\left(\mathrm{e}^{-\beta u} - \mathrm{e}^{-\alpha u}\right)\,. \tag{19}$$

This function will appear in several places in connection with the analysis of our spot price model. It is simple to observe that $\gamma(u;\alpha,\beta)$ is a non-negative continuous function, with $\gamma(0;\alpha,\beta) = 0$ and $\lim_{u\to}\gamma(u;\alpha,\beta) = 0$. Furthermore, by a straightforward differentiation, it attains its maximum value at

$$u^* = \frac{\ln\alpha - \ln\beta}{\alpha-\beta}\,.$$

The maximal value can be computed to be

$$\gamma^*(\alpha,\beta) := \max_{u\geq 0}\gamma(u;\alpha,\beta) = \gamma(u^*;\alpha,\beta) = \frac{\beta^{\beta/\alpha-\beta}}{\alpha^{\alpha/\alpha-\beta}}\,. \tag{20}$$

We shall make use of the properties of the function γ throughout this section.

By applying the explicit form of $Z(s)$ in Lemma 3.1, we can derive the following explicit dynamics of $X(s)$:

Lemma 3.2. *The explicit dynamics of $X(s)$ given $X(t)$ for $s \geq t$ can be represented as*

$$X(s) = X(t)\mathrm{e}^{-\alpha(s-t)} + \left(Z(t) - \frac{\mu}{\beta}\right)\gamma(s-t;\alpha,\beta) + \frac{\mu}{\beta}\gamma(s-t;\alpha,0)$$
$$+ \int_t^s \sigma(u)\mathrm{e}^{-\alpha(s-u)}\,dB_1(u) + \int_t^s \eta\gamma(s-u;\alpha,\beta)\,dB_2(u)\,.$$

Proof. From the explicit dynamics of $Z(u)$ given $Z(t)$, $u \geq t$ in Lemma 3.1, we get

$$\int_t^s Z(u)\mathrm{e}^{\alpha u}\,du = Z(t)\int_t^s \mathrm{e}^{\alpha u-\beta(u-t)}\,du + \int_t^s \mathrm{e}^{\alpha u}\frac{\mu}{\beta}(1-\mathrm{e}^{-\beta(u-t)})\,du$$
$$+\,\eta\int_t^s \mathrm{e}^{\alpha u}\int_t^u \mathrm{e}^{-\beta(u-v)}\,dB_2(v)\,du\,.$$

Applying the stochastic Fubini theorem and the definition of the function $\gamma(u;\alpha,\beta)$ yield the result. □

We continue with analysing the covariance structure of the spot price dynamics. For this study, we must suppose that the log-spot prices have finite variance. By inspection of $X(s)$ in the lemma above, the log-spot price has finite variance as long as $X(t)$ is of finite variance. But $X(t)$ has finite variance if and only if the stochastic integral with respect to B_1 has finite variance, since the other stochastic integral is a simple Wiener integral of a deterministic function. By the Itô isometry, we find

$$\mathbb{E}\left[\left(\int_0^t \sigma(u)\mathrm{e}^{-\alpha(t-u)}\,dB_1(u)\right)^2\right] = \mathbb{E}\left[\int_0^t \sigma^2(u)\mathrm{e}^{-2\alpha(t-u)}\,du\right]$$
$$= \int_0^t \mathbb{E}\left[\sigma^2(u)\right]\mathrm{e}^{-2\alpha(t-u)}\,du\,.$$

But, from the definition of $\sigma^2(u)$, we have

$$\mathbb{E}[\sigma^2(u)] = \sum_{j=1}^n w_j\mathbb{E}[Y_j(u)]$$
$$= \sum_{j=1}^n w_jY_j(0)\mathrm{e}^{-\lambda_j t} + \sum_{j=1}^n w_j\mathbb{E}\left[\int_0^t \mathrm{e}^{-\lambda_j(t-u)}\,dL_j(u)\right]\,.$$

We know from Benth et al. [7] that

$$\mathbb{E}\left[\int_0^t e^{-\lambda_j(t-u)}\,dL_j(u)\right] = \mathbb{E}[L_j(1)]\int_0^t e^{-\lambda_j(t-u)}\,du\,.$$

Hence, if $L_j(1)$ has finite expectation for all $j = 1,\dots,n$, then $X(t)$ has finite variance. From the definition of the cumulant functions of $L_j(1)$ in (15), $L_j(1)$, $j = 1,\dots,n$ have finite expectation if and only if

$$\int_0^\infty z\,\ell_j(dz) < \infty\,, j = 1,\dots,n\,. \tag{21}$$

From now on, we suppose that (21) holds. Under this condition we have the required Itô integrability in Lemma 3.1.

In the next proposition we find the covariance between deseasonalized log-spot prices at different time instances.

Proposition 3.3. *For $t,\tau > 0$, it holds that*

$$\mathrm{Cov}(X(t+\tau), X(t)) = a(t)e^{-\alpha\tau} + b(t)e^{-\beta\tau}\,,$$

where

$$a(t) = \mathrm{Var}(X(t)) - b(t)\,, \qquad b(t) = \frac{\mathrm{Cov}(Z(t), X(t))}{\alpha-\beta}\,.$$

Proof. Since, by Lemma 3.1

$$X(t+\tau) = X(t)e^{-\alpha\tau} + \int_t^\tau Z(s)e^{-\alpha(t+\tau-s)}\,ds + \int_t^{t+\tau}\sigma(s)e^{-\alpha(t+\tau-s)}\,dB_1(s)\,,$$

we find

$$\mathrm{Cov}(X(t+\tau), X(t)) = e^{-\alpha\tau}\left(\mathrm{Var}(X(t)) + \int_t^{t+\tau} e^{-\alpha(t-s)}\mathrm{Cov}(Z(s), X(t))\,ds\right)$$

since $X(t)$ and $\int_t^{t+\tau}\sigma(s)\exp(-\alpha(t+\tau-s))\,dB_1(s)$ are zero correlated due to the independent increment property of Brownian motion. Recalling $Z(s)$ given $Z(t)$ for $s \geq t$ in Lemma 3.1, we have

$$\mathrm{Cov}(Z(s), X(t)) = e^{-\beta(s-t)}\mathrm{Cov}(Z(t), X(t))\,.$$

since $X(t)$ and $\int_t^s \eta\exp(-\beta(t-u))\,dB_2$ are independent. The proposition follows after a straightforward integration. □

In the next lemmas we calculate the variance of $X(t)$ and the covariance of $Z(t)$ and $X(t)$, and investigate their asymptotics when time goes to infinity.

Lemma 3.4. *It holds that*

$$\begin{aligned}\mathrm{Var}(X(t)) &= \int_0^t \mathbb{E}[\sigma^2(s)]\mathrm{e}^{-2\alpha(t-s)}\,ds \\ &+ 2\rho\frac{\eta}{\alpha-\beta}\int_0^t \mathbb{E}[\sigma(s)]\left(\mathrm{e}^{-(\alpha+\beta)(t-s)} - \mathrm{e}^{-2\alpha(t-s)}\right)ds \\ &+ \frac{\eta^2}{(\alpha-\beta)^2}\left\{\frac{1}{2\beta}(1-\mathrm{e}^{-2\beta t}) - \frac{2}{\alpha+\beta}(1-\mathrm{e}^{-(\alpha+\beta)t}) + \frac{1}{2\alpha}(1-\mathrm{e}^{-2\alpha t})\right\}.\end{aligned}$$

Proof. From Lemma 3.1 we compute using the Itô isometry,

$$\begin{aligned}\mathrm{Var}(X(t)) &= \mathrm{e}^{-2\alpha t}\mathrm{Var}\left(\int_0^t Z(s)\mathrm{e}^{\alpha s}\,ds\right) \\ &+ 2\mathrm{e}^{-2\alpha t}\mathrm{Cov}\left(\int_0^t \sigma(s)\mathrm{e}^{\alpha s}\,dB_1(s), \int_0^t Z(s)\mathrm{e}^{\alpha s}\,ds\right) \\ &+ \int_0^t \mathbb{E}[\sigma^2(s)]\mathrm{e}^{-2\alpha(t-s)}\,ds.\end{aligned}$$

We consider the first two terms. Applying the stochastic Fubini Theorem, it holds that

$$\begin{aligned}\int_0^t Z(s)\mathrm{e}^{\alpha s}\,ds &= \frac{Z(0)}{\alpha-\beta}(\mathrm{e}^{(\alpha-\beta)t}-1) + \frac{\mu}{\beta}\left\{\frac{1}{\alpha}(\mathrm{e}^{\alpha t}-1) - \frac{1}{\alpha-\beta}(\mathrm{e}^{(\alpha-\beta)t}-1)\right\} \\ &+ \frac{\eta}{\alpha-\beta}\int_0^t \mathrm{e}^{\alpha u}(\mathrm{e}^{(\alpha-\beta)(t-u)}-1)\,dB_2(u).\end{aligned}$$

Thus,

$$\mathrm{Var}\left(\int_0^t Z(s)\mathrm{e}^{\alpha s}\,ds\right) = \frac{\eta^2}{(\alpha-\beta)^2}\int_0^t \mathrm{e}^{2\alpha u}(\mathrm{e}^{(\alpha-\beta)(t-u)}-1)^2\,du,$$

which gives us the last term involving $\eta^2/(\alpha-\beta)^2$ in the expression of $\mathrm{Var}(X(t))$. Finally, by Itô's isometry and the correlation between B_1 and B_2, we find that

$$\begin{aligned}&\mathrm{Cov}\left(\int_0^t \sigma(s)\mathrm{e}^{\alpha s}\,dB_1(s), \int_0^t Z(s)\mathrm{e}^{\alpha s}ds\right) \\ &= \rho\frac{\eta}{\alpha-\beta}\int_0^t \mathbb{E}[\sigma(s)]\mathrm{e}^{2\alpha s}(\mathrm{e}^{(\alpha-\beta)(t-s)}-1)ds.\end{aligned}$$

Hence, the lemma follows. □

The covariance between $Z(t)$ and $X(t)$ is derived next:

Lemma 3.5. *It holds that*

$$\mathrm{Cov}(Z(t), X(t)) = \rho\eta \int_0^t \mathbb{E}[\sigma(s)]\mathrm{e}^{-(\alpha+\beta)(t-s)}\, ds + \frac{\eta^2}{\alpha-\beta}\left\{\frac{1}{2\beta}(1-\mathrm{e}^{-2\beta t}) - \frac{1}{\alpha+\beta}(1-\mathrm{e}^{-(\alpha+\beta)t})\right\}.$$

Proof. Using the expression for

$$\int_0^t Z(s)\mathrm{e}^{\alpha s}\, ds$$

calculated in the proof of Lemma 3.4, it follows that

$$\begin{aligned}
&\mathrm{Cov}(Z(t), X(t)) \\
&= \frac{\eta^2}{\alpha-\beta}\mathrm{e}^{-(\alpha+\beta)t}\mathrm{Cov}\left(\int_0^t \mathrm{e}^{\beta s}\, dB_2(s), \int_0^t \mathrm{e}^{\alpha s}(\mathrm{e}^{(\alpha-\beta)(t-s)}-1)\, dB_2(s)\right) \\
&\quad + \eta\mathrm{e}^{-(\alpha+\beta)t}\mathrm{Cov}\left(\int_0^t \mathrm{e}^{\beta s}\, dB_2(s), \int_0^t \sigma(s)\mathrm{e}^{\alpha s}\, dB_1(s)\right) \\
&= \frac{\eta^2}{\alpha-\beta}\mathrm{e}^{-(\alpha+\beta)t}\int_0^t \mathrm{e}^{(\alpha+\beta)s}(\mathrm{e}^{(\alpha-\beta)(t-s)}-1)\, ds \\
&\quad + \rho\eta\mathrm{e}^{-(\alpha+\beta)t}\int_0^t \mathbb{E}[\sigma(s)]\mathrm{e}^{(\alpha+\beta)s}\, ds,
\end{aligned}$$

where we have used the Itô isometry in the last equality. The lemma follows after a straightforward integration. □

Notice that when $\rho = 0$, that is, when the noises of X and Z are independent, the term involving $\mathbb{E}[\sigma(s)]$ in $\mathrm{Var}(X(t))$ and $\mathrm{Cov}(Z(t), X(t))$ disappears.

We want to show that $X(t)$ has a "stationary" autocorrelation function: observe from Sato [24], Theorem 17.5, that $\sigma^2(t)$ has a stationary distribution function with cumulant

$$\psi_\infty(\theta) = \sum_{j=1}^n \int_0^\infty \psi_j(\theta\mathrm{e}^{-\lambda_j u})\, du,$$

as long as the condition

$$\int_2^\infty \ln z\, \ell_j(dz) < \infty,$$

holds for each subordinator L_j, $j = 1, \ldots, n$. But for $z > 2$, $\ln z \leq z$, and by our standing assumption of finite expectation of the $L_j(1)$, this condition holds true. Hence, the stochastic volatility process $\sigma^2(t)$ has a stationary distribution when t tends to infinity, which will be supported on the positive real line. Hence, both $\mathbb{E}[\sigma^2(t)]$ and $\mathbb{E}[\sigma(t)] = \mathbb{E}[\sqrt{\sigma^2(t)}]$ will have limits being strictly positive when t tends to infinity. From the explicit expressions of Y_j in Lemma 3.1, it holds in particular that

$$\lim_{t\to\infty} \mathbb{E}[\sigma^2(t)] = \sum_{j=1}^{n} \frac{1}{\lambda_j} \int_0^\infty z \ell_j(dz) .$$

If $\xi > 0$, we therefore find by L'Hopitals rule that

$$\lim_{t\to\infty} e^{-\xi t} \int_0^t \mathbb{E}[\sigma^2(s)] e^{\xi s} \, ds = \lim_{t\to\infty} \frac{\mathbb{E}[\sigma^2(t)] e^{\xi t}}{\xi e^{\xi t}} = \sum_{j=1}^{n} \frac{1}{\xi \lambda_j} \int_0^\infty z \ell_j(dz) .$$

Similarly,

$$\lim_{t\to\infty} e^{-\xi t} \int_0^t \mathbb{E}[\sigma(s)] e^{\xi s} \, ds = c/\xi ,$$

where c is the limit of $\mathbb{E}[\sigma(t)]$. In conclusion, both $\mathrm{Var}(X(t))$ and $\mathrm{Cov}(Z(t), X(t))$ have limits when t goes to infinity. Therefore, we see from Proposition 3.3 that

$$\lim_{t\to\infty} \mathrm{Corr}(X(t+\tau), X(t)) = c_1 e^{-\alpha\tau} + c_2 e^{-\beta\tau} , \tag{22}$$

for two positive constants c_1, c_2 such that $c_1 + c_2 = 1$. The autocorrelation function of the deseasonalized log-spot prices is thus given as a sum of two exponentially decaying functions. This can be utilized in calibration of the spot model, since we can find the speeds of mean-reversion α and β by minimizing the distance between the theoretical and empirical autocorrelation functions. The characteristics of the stochastic volatility $\sigma(t)$ and its square enter in c_1 and c_2, that is, in the weighting of the two exponential functions.

We observe by Lemma 3.1 that for a small $\Delta > 0$, it approximately holds

$$e^{\alpha\Delta} X(t+\Delta) - X(t) \approx Z(t)\Delta + \sigma(t)\Delta B_1(t) ,$$

with $\Delta B_1(t) = B_1(t+\Delta) - B_1(t)$. If we suppose that we can observe the mean-level process $Z(t)$ (using for example filtering), we can find observations of the residual process $\sigma(t)\Delta B_1(t)$ by the relation

$$\sigma(t)\Delta B_1(t) \approx e^{\alpha\Delta} X(t+\Delta) - X(t) - Z(t)\Delta .$$

Note that $X(t) = \ln S(t) - \ln \Lambda(t)$, and therefore $X(t)$ is directly observable from the spot prices given that we know the seasonality function $\Lambda(t)$. We have, by independence between $\sigma(t)$ and $B_1(t)$, that $\sigma(t)\Delta B_1(t)$ is a variance-mixture model, where $\sigma(t)\Delta B_1(t)$ conditioned on $\sigma^2(t)$ will be normally distributed with zero mean and variance $\sigma^2(t)\Delta$. Taking $\sigma^2(t)$ stationary, we can obtain a rich class of heavy-tailed distributions for these residuals, including for example the normal inverse Gaussian distribution. We refer to Barndorff-Nielsen and Shephard [1] for an in-depth analysis of this, where methods for estimating the factors Y_j of the stochastic volatility model are presented and discussed.

3.1 Forward Prices

Recall from Sect. 2 that $f(t,T)$ denotes the forward price at time $t \geq 0$ of a contract delivering the energy spot at time T, where $0 \leq t \leq T$. As argued in Sect. 2, the arbitrage theory of mathematical finance tells us that the process $t \mapsto f(t,T)$ must be a martingale with respect to some equivalent probability Q. This led to the definition

$$f(t,T) = \mathbb{E}_Q[S(T) \mid \mathcal{F}_t]\,. \tag{23}$$

In the standard pricing theory, the underlying asset of the forward is also a tradeable asset, and Q is therefore a martingale measure for S. However, in energy markets, trading constraints like storage and transportation of the energy (in case of gas and oil), or no-storage possibilities at all (in the electricity case) create an incomplete market where the buy-and-hold hedging strategy in the spot cannot be applied. Hence, the measure Q does not need to be a martingale measure for S. One typically lets Q be part of the modelling, and chooses it in a parametric class of equivalent probability measures, using (23) as the definition of the forward price dynamics.

Before deriving expressions for the forward price based on our spot price model, we look at a parametric class of equivalent probability measures Q which changes the Brownian motions in the spot model by a Girsanov transform. To simplify our considerations, we have chosen not to consider any measure change of the subordinators driving the stochastic volatility process $\sigma(t)$.

Represent the Brownian motion B_2 as

$$B_2(t) = \rho B_1(t) + \sqrt{1-\rho^2}U(t)\,,$$

with U being a Brownian motion independent of B_1. Next, let θ_1, θ_2 be two constants which we will call the *market prices of risk*, and define the adapted stochastic processes

$$\tilde{\theta}_1(t) \triangleq \frac{\theta_1}{\sigma(t)}$$

$$\tilde{\theta}_2(t) \triangleq \frac{\theta_2 - \rho\eta\tilde{\theta}_1(t)}{\eta\sqrt{1-\rho^2}} .$$

Then, by the Girsanov Theorem, it is easy to see that there exists a probability Q such that W_1 and V defined as

$$dB_1(t) = \tilde{\theta}_1(t)\,dt + dW_1(t)$$

$$dU(t) = \tilde{\theta}_2(t)\,dt + dV(t) ,$$

are two independent Brownian motions on a finite time interval $[0, T_{\max}]$ for any $T_{\max} < \infty$. Indeed, the Novikov condition for Girsanov's Theorem is satisfied since $\sigma(t)$ is bounded from below by a weighted sum of exponential functions: by Lemma 3.1 we have $Y_j(t) \geq Y_j(0)\exp(-\lambda_j t)$ for $j = 1, \ldots, n$. Hence,

$$\sigma^2(t) \geq \sum_{j=1}^{n} w_j Y_j(0)\mathrm{e}^{-\lambda_j t} .$$

Note that the characteristics of Y_j remain unaltered under the probability Q. This means that we suppose the *market price of volatility risk* be zero, although there are empirical studies showing that such a risk is present in energy markets (see Trolle and Schwartz [27]). There exist many other risk neutral probabilities that can be used, which can account for such risk premia as well. We refer to the class of Esscher transformed measures that will be introduced later in Sect. 4 as a possible choice.

here we focus on the standard class of measure change frequently used in commodity analysis.

We find the Q-dynamics of X and Z to be,

$$dX(t) = (\theta_1 + Z(t) - \alpha X(t))\,dt + \sigma(t)\,dW_1(t) \tag{24}$$

$$dZ(t) = (\mu + \theta_2 - \beta Z(t))\,dt + \eta\,dW_2(t) , \tag{25}$$

where $W_2 \triangleq \rho W_1 + \sqrt{1-\rho^2}V$ is a Q-Brownian motion correlated with W_1 by the factor ρ. Computing as in Lemma 3.2, we find for $t \leq s \leq T_{\max}$

$$\begin{aligned} X(s) = {} & X(t)\mathrm{e}^{-\alpha(s-t)} + \left(Z(t) - \frac{\mu+\theta_2}{\beta}\right)\gamma(s-t;\alpha,\beta) \\ & + \left(\theta_1 + \frac{\mu+\theta_2}{\beta}\right)\gamma(s-t;\alpha,0) \\ & + \int_t^s \sigma(u)\mathrm{e}^{-\alpha(s-u)}\,dW_1(u) + \eta\int_t^s \gamma(s-u;\alpha,\beta)\,dW_2(u) , \end{aligned} \tag{26}$$

Before moving on to compute the forward price, we must ensure that $S(T)$ is integrable with respect to the probability Q. The integrability of $S(T)$ is equivalent to exponential integrability of $X(T)$. Inspecting the explicit relation for $X(s)$ in (26), we see that $X(T)$ is exponentially integrable as long as

$$\mathbb{E}_Q\left[\exp\left(\int_0^T \sigma(s)\mathrm{e}^{-\alpha(T-s)}\,dW_1(s)\right)\right] < \infty\,. \tag{27}$$

We claim that a sufficient condition for this to hold is that

$$\int_0^\infty (\mathrm{e}^{0.5\gamma^*(2\alpha,\lambda_j)z} - 1)\,\ell_j(dz) < \infty\,, \tag{28}$$

for $j = 1,\dots,n$, and γ^* defined in (20). To show this, first note that by double conditioning using the σ-algebra $\mathcal{G}_T$ generated by $\sigma^2(s)\,,s \leq T$, we find by independence between B_1 and σ^2 that

$$\begin{aligned}
&\mathbb{E}_Q\left[\exp\left(\int_0^T \sigma(s)\mathrm{e}^{-\alpha(T-s)}\,dW_1(s)\right)\right] \\
&= \mathbb{E}_Q\left[\mathbb{E}_Q\left[\exp\left(\int_0^T \sigma(s)\mathrm{e}^{-\alpha(T-s)}\,dW_1(s)\right)\mid \mathcal{G}_T\right]\right] \\
&= \mathbb{E}\left[\exp\left(\frac{1}{2}\int_0^T \sigma^2(s)\mathrm{e}^{-2\alpha(T-s)}\,ds\right)\right].
\end{aligned}$$

From $Y_j(s)$ in Lemma 3.1 we have

$$\begin{aligned}
\int_0^T \sigma^2(s)\mathrm{e}^{-2\alpha(T-s)}\,ds &= \sum_{j=1}^n w_j Y_j(0)\int_0^T \mathrm{e}^{-2\alpha(T-s)}\mathrm{e}^{-\lambda_j s}\,ds \\
&\quad + \sum_{j=1}^n w_j \int_0^T \mathrm{e}^{-2\alpha(T-s)}\int_0^s \mathrm{e}^{-\lambda_j(s-u)}\,dL_j(u)\,ds \\
&= \gamma(T;2\alpha,\lambda_j) + \int_0^T \gamma(T-u;2\alpha,\lambda_j)\,dL_j(u)\,,
\end{aligned}$$

after using the stochastic Fubini theorem. By independence of L_j, $j = 1,\dots,n$, we find that (27) holds whenever

$$\mathbb{E}\left[\exp\left(\frac{1}{2}\int_0^T \gamma(T-u;2\alpha,\lambda_j)\,dL_j(u)\right)\right] < \infty\,, j = 1,\dots,n\,.$$

But $\gamma(u;2\alpha,\lambda_j)$ is a non-negative function which has the maximum $\gamma^*(2\alpha,\lambda_j)$, and L_j being a subordinator implies the bound

$$\int_0^T \gamma(T-u;2\alpha,\lambda_j)\,dL_j(u) \leq \gamma^*(2\alpha,\lambda_j)L_j(T)\,.$$

Finally, under condition (28) we have that

$$\mathbb{E}\left[\exp\left(\frac{1}{2}\gamma^*(2\alpha,\lambda_j)L_j(T)\right)\right] = \exp\left(T\int_0^\infty \{e^{0.5\gamma^*(2\alpha,\lambda_j)z} - 1\}\,\ell_j(dz)\right),$$

from the definition of the moment generating function of a subordinator. This shows that (28) is a sufficient condition for the exponential integrability of $X(T)$ with respect to Q for all $T \leq T_{\max}$. Note that it also is a sufficient condition for exponential integrability of $X(T)$ with respect to the market probability P.

In the next proposition we state a semi-analytical expression for the forward price:

Proposition 3.6. *Assume condition* (28) *holds. Then the forward price $f(t,T)$ at time $t \geq 0$ of a contract maturing at time $t \leq T \leq T_{max}$ is*

$$f(t,T) = \Lambda_f(t,T)\Theta(T-t)\,S(t)^{\exp(-\alpha(T-t))}\exp\left(Z(t)\gamma(T-t;\alpha,\beta)\right)$$
$$\times H(t,T,Y_1(t),\ldots,Y_n(t))$$

where

$$H(t,T,y_1,\ldots,y_n)$$
$$= \mathbb{E}\left[\exp\left(\frac{1}{2}\int_t^T (\rho\eta\gamma(T-s;\alpha,\beta) + \sigma(s)e^{-\alpha(T-s)})^2\,ds\right)\,\Big|\,Y_j(t) = y_j\right],$$

and

$$\ln\Lambda_f(t,T) = \ln\Lambda(T) - e^{-\alpha(T-t)}\ln\Lambda(t)$$
$$\ln\Theta(u) = \left(\theta_1 + \frac{\mu+\theta_2}{\beta}\right)\gamma(u;\alpha,0) - \frac{\mu+\theta_2}{\beta}\gamma(u;\alpha,\beta)$$
$$+\frac{1}{2}\eta^2(1-\rho^2)\int_0^u \gamma^2(s;\alpha,\beta)\,ds.$$

Proof. The forward price is

$$f(t,T) = \mathbb{E}_Q[S(T)\,|\,\mathcal{F}_t] = \Lambda(T)\mathbb{E}_Q\left[e^{X(T)}\,|\,\mathcal{F}_t\right].$$

Apply the explicit expression for the Q-dynamics of $X(T)$ given $X(t)$ and $Z(t)$ in (26), to obtain

$$\mathbb{E}_Q\left[e^{X(T)}\,|\,\mathcal{F}_t\right] = \exp\left(X(t)e^{-\alpha(T-t)} + Z(t)\gamma(T-t;\alpha,\beta)\right)$$
$$\times\exp\left((\theta_1 + \frac{\mu+\theta_2}{\beta})\gamma(T-t;\alpha,0) - \frac{\mu+\theta_2}{\beta}\gamma(T-t;\alpha,\beta)\right)$$

$$\times \mathbb{E}_Q\left[\exp\left(\int_t^T (\rho\eta\gamma(T-s;\alpha,\beta)+\sigma(s)\mathrm{e}^{-\alpha(T-s)})\,dW_1(s)\right) \mid \mathcal{F}_t\right]$$
$$\times \mathbb{E}_Q\left[\exp\left(\eta\sqrt{1-\rho^2}\int_t^T \gamma(T-s;\alpha,\beta)\,dV(s)\right) \mid \mathcal{F}_t\right].$$

Here, we have made use of the adaptedness of $X(t)$ and $Z(t)$ and the independence of W_1 and V. By the independent increment property of the Brownian motion V, we easily compute the second conditional expectation to be

$$\mathbb{E}_Q\left[\exp\left(\eta\sqrt{1-\rho^2}\int_t^T \gamma(T-s;\alpha,\beta)\,dV(s)\right) \mid \mathcal{F}_t\right]$$
$$= \exp\left(\frac{1}{2}\eta^2(1-\rho^2)\int_t^T \gamma^2(T-s;\alpha,\beta)\,ds\right).$$

Let now $\mathcal{G}_{t,T}$ be the product σ-algebra generated by the paths $\sigma(s)$; $s \le t \le T$ and $\mathcal{F}_t$. By double condition, we have by independence of $\sigma(t)$ and $W_1(t)$ and the independent increment property of W_1,

$$\mathbb{E}_Q\left[\exp\left(\int_t^T \rho\eta\gamma(T-s;\alpha,\beta)+\sigma(s)\mathrm{e}^{-\alpha(T-s)}\,dW_1(s)\right) \Big| \mathcal{F}_t\right]$$
$$= \mathbb{E}_Q\left[\mathbb{E}_Q\left[\exp\left(\int_t^T \rho\eta\gamma(T-s;\alpha,\beta)+\sigma(s)\mathrm{e}^{-\alpha(T-s)}\,dW_1(s)\right) \Big| \mathcal{G}_{t,T}\right] \Big| \mathcal{F}_t\right]$$
$$= \mathbb{E}\left[\exp\left(\frac{1}{2}\int_t^T (\rho\eta\gamma(T-s;\alpha,\beta)+\sigma(s)\mathrm{e}^{-\alpha(T-s)})^2\,ds\right) \Big| \mathcal{F}_t\right].$$

By the Markov property of $Y_j(s)$ we find the function H. The proof is hence complete. □

The forward price is explicit as a function of both the spot price and the seasonal level $Z(t)$. Moreover, it also varies explicitly as a function of the volatility of $S(t)$ through H which depends on each of the volatility factors $Y_1(t),\dots,Y_n(t)$. All these factor dependencies add up to a complex forward price evolution. In addition, we have two deterministic terms contributing to the shape of $f(t,T)$. First, the seasonal term $\Lambda_f(t,T)$, which can be interpreted as the change in seasonal level between today and maturity, adjusted by the mean-reversion between the two dates. The final term $\Theta(T-t)$ is coming from a drift contribution of $Z(t)$ as well as the market prices of risk θ_1 and θ_2. It is the only place where we see the effect of the pricing measure Q explicitly in the forward price. The parameters θ_1 and θ_2 are unknown, since they cannot be estimated from observations of the spot price dynamics. The typical approach to estimate (or calibrate) the market prices of risk, is to minimize the distance between theoretical and observed forward prices. We remark that in practical situations, one would often prefer to use market prices of risk which are time dependent in order to facilitate for exact calibration to observed prices. At the expense of more technical computations and expressions, it is not a difficult task to extend our results to market prices of risk θ_1, θ_2 being functions of time.

In commodity markets, one is often interested in the *risk premium*, which is defined as the difference between the forward price and the predicted spot price:

$$R_Q(t,T) := f(t,T) - \mathbb{E}[S(T) \mid \mathcal{F}_t]. \tag{29}$$

Note that we indicate the dependency on the pricing measure Q in the notation for the risk premium, which comes from the fact that we can express it as

$$R_Q(t,T) = \mathbb{E}_Q[S(T) \mid \mathcal{F}_t] - \mathbb{E}[S(T) \mid \mathcal{F}_t].$$

We obtain the predicted spot price from Proposition 3.6 simply by choosing $\theta_1 = \theta_2 = 0$. Hence, the risk premium becomes

$$\begin{aligned} R_Q(t,T) = \Lambda_f(t,T)\tilde{\Theta}_0(T-t)S(t)^{\exp(-\alpha(T-t))} \\ \exp(Z(t)\gamma(T-t;\alpha,\beta))H(t,T;Y_1(t),\ldots,Y_n(t)), \end{aligned}$$

with

$$\begin{aligned} \tilde{\Theta}_0(u) = \exp\left(\frac{\mu}{\beta}(\gamma(u;\alpha,0) - \gamma(u;\alpha,\beta)) + \frac{1}{2}\eta^2(1-\rho^2)\int_0^u \gamma^2(s;\alpha,\beta)\,ds\right) \\ \times \left\{\exp\left((\theta_1 + \frac{\theta_2}{\beta})\gamma(u;\alpha,0) - \frac{\theta_2}{\beta}\gamma(u;\alpha,\beta)\right) - 1\right\}. \end{aligned}$$

Hence, we observe that all terms in the risk premium are positive, except possibly the expression inside the curly brackets of $\tilde{\Theta}_0$. The sign of this term is determined by θ_1 and θ_2. In the simple case of $\theta_2 = 0$, we obtain a *negative* risk premium R_Q if and only if $\theta_1 < 0$. A negative risk premium means that forward prices are lower than the predicted spot, which says that those *selling* the energy in the forward market accept a reduced price compared to what they predict to get if selling in the spot market instead. This can be a result of producers wishing to hedge their production using the forward market, and thereby accepting a discount in prices compared to the spot market. The risk premium can be interpreted as the insurance premium paid by those producers. A negative risk premium corresponds to a market in so-called *backwardation*.

The effect of θ_2 on the risk premium is similar: suppose again for simplicity that $\theta_1 = 0$, and we see that the sign of the risk premium is negative whenever

$$\theta_2(\gamma(u;\alpha,0) - \gamma(u;\alpha,\beta)) < 0.$$

Define for the moment the function

$$g(u) = \gamma(u;\alpha,0) - \gamma(u;\alpha,\beta),$$

for $u \geq 0$. Note that $g(0) = 0$ and that $g(u)$ tends to $1/\alpha > 0$ when u tends to infinity. Since $Z(t)$ is the mean-level of the spot price, it is natural to suppose that this mean-reverts slower than the actual spot price itself, yielding that $\alpha > \beta$ is a natural situation. We claim that $g(u)$ is strictly positive for $u > 0$ in this case: indeed, the derivative of $g(u)$ is

$$g'(u) = \mathrm{e}^{-\alpha u} - \frac{1}{\alpha - \beta}\left(\beta \mathrm{e}^{-\beta u} - \alpha \mathrm{e}^{-\alpha u}\right).$$

But, $g'(0) = 2$ and $g'(u) = 0$ in only one point, namely

$$u^* = \frac{\ln(2\alpha - \beta) - \ln \beta}{\alpha - \beta}.$$

Since $\alpha > \beta$, we have that $2\alpha - \beta > \beta$, which implies that $u^* > 0$. Hence, the continuous function $g(u)$ must be positive for $u > 0$ since it starts increasing at $u = 0$ from the origin, has only one extremal point at $u = u^* > 0$ and is asymptotically converging to the positive constant $1/\alpha$ at infinity. In conclusion, as long as $\alpha > \beta$, it holds that the risk premium is negative as long as $\theta_2 < 0$. We obtain the same result in the case $\alpha < \beta$, but with the additional condition $2\alpha > \beta$, i.e., $\alpha < \beta < 2\alpha$.

We return to the analysis of the forward price $f(t, T)$. First, we present the dynamics of F:

Proposition 3.7. *Assume condition* (28)*. The dynamics of $f(t, T)$ for $t \leq T$ is*

$$\frac{df(t,T)}{f(t-,T)} = \sigma(t)\mathrm{e}^{-\alpha(T-t)}\,dW_1(t) + \eta\gamma(T-t;\alpha,\beta)\,dW_2(t)$$
$$+\sum_{j=1}^{n}\int_0^\infty \left\{\frac{H(t,T,Y_1(t-),\ldots,Y_j(t-)+z,\ldots,Y_n(t-))}{H(t,T,Y_1(t-1),\ldots,Y_n(t-))} - 1\right\}\tilde{N}_j(dz,dt)\,.$$

Proof. This follows from an application of Itô's Formula for jump processes exploiting the simplifying fact that F is a Q-martingale. □

We observe that the forward price evolves as a geometric jump-diffusion model, with the Brownian evolutions driven by the stochastic volatility $\sigma(t)$ discounted by the mean-reversion. In addition, we have explicit jump terms coming from the volatility. Although the spot price dynamics has continuous paths, the forward price dynamics will have jumps. Every time there is a change in volatility of the spot resulting from a jump in one or more of the subordinators L_j, the forward price will jump accordingly as well as getting an increase in volatility. In fact, the forward price will include a leverage effect in its price dynamics, since the volatility affects directly its price level. For the sake of illustration, consider the case of $n = 1$, that is, only one factor Y in the stochastic volatility specification. If $\rho \geq 0$ we find that the function

$$y \mapsto (\rho\eta\gamma(T-s;\alpha,\beta) + \sigma(s)\mathrm{e}^{-\alpha(T-s)})^2$$

is increasing, and therefore $y \mapsto H(t, T, y)$ must be increasing as well. Thus, the forward dynamics includes an *inverse leverage* effect in the sense that the forward price increases with the volatility $\sigma(t)$. The case $\rho < 0$ is more involved, and here we potentially may have a "classical" leverage effect where forward prices are pulled *down* with higher volatility.

The price dynamics of the forward is semi-analytic in general, due to the function H which is hard to compute explicitly. The problem is that the expectation in H will involve a term

$$\rho\eta\int_t^T \gamma(T-s;\alpha,\beta)\mathrm{e}^{-\alpha(T-s)}\sigma(s)\,ds\,.$$

We know $\sigma^2(s)$ explicitly, however, $\sigma(s)$ is the square root of a sum of OU-processes and the integral above seems hard to compute. In the case of no correlation between B_1 and B_2, that is, $\rho = 0$, we can indeed obtain an explicit expression for H, and thus for the forward price. This is the content of the next proposition:

Proposition 3.8. *Assume condition* (28), *and suppose that* $\rho = 0$. *Then*

$$H(t,T,y_1,\ldots,y_n) = \exp\left(\frac{1}{2}\sum_{j=1}^n w_j\gamma(T-t;2\alpha,\lambda_j)y_j \right.$$
$$\left. +\sum_{j=1}^n\int_0^{T-t}\psi_j\left(-\mathrm{i}\frac{w_j}{2}\gamma(s;2\alpha,\lambda_j)\,ds\right)\right).$$

Proof. For $\rho = 0$, the function H defined in Proposition 3.6 reduces to

$$H(t,T,y_1,\ldots,y_n) = \mathbb{E}\left[\exp\left(\frac{1}{2}\int_t^T\sigma^2(s)\mathrm{e}^{-2\alpha(T-s)}\,ds\right)\middle|\,Y_j(t)=y_j\right].$$

From the explicit dynamics of $Y_j(s)$ given $Y_j(t) = y_j$ in Lemma 3.1, we find

$$\begin{aligned}
\int_t^T\sigma^2(s)\mathrm{e}^{-2\alpha(T-s)}\,ds &= \sum_{j=1}^n w_jy_j\int_t^T \mathrm{e}^{-\lambda_j(s-t)}\mathrm{e}^{-2\alpha(T-s)}\,ds\\
&\quad+\sum_{j=1}^n w_j\int_t^T\int_t^s \mathrm{e}^{-\lambda_j(s-u)}\,dL_j(u)\mathrm{e}^{-2\alpha(T-s)}\,ds\\
&=\sum_{j=1}^n w_jy_j\gamma(T-t;2\alpha,\lambda_j)\\
&\quad+\sum_{j=1}^n w_j\int_t^T\int_u^T \mathrm{e}^{-\lambda_j(s-u)}\mathrm{e}^{-2\alpha(T-s)}\,ds\,dL_j(u)\\
&=\sum_{j=1}^n w_jy_j\gamma(T-t;2\alpha,\lambda_j)+\sum_{j=1}^n w_j\int_t^T\gamma(T-u;2\alpha,\lambda_j)dL_j(u),
\end{aligned}$$

where we have invoked the stochastic Fubini theorem in the second step. The Corollary follows by using the definition of the cumulant function of L_j and condition (28). □

In the case of zero correlation between the driving Brownian motions, we also observe that the long-term influence of H is simply a constant value,

$$\lim_{T-t\to\infty} H(t,T,y_1,\ldots,y_n) = \exp\left(\sum_{j=1}^{n}\int_0^\infty \psi_j\left(-\mathrm{i}\frac{w_j}{2}\gamma(s;2\alpha,\lambda_j)\,ds\right)\right),$$

as long as the indefinite integrals exist.

Recall that the function $\gamma(u;\alpha,\beta)$ starts at the origin for $u = 0$, and tends to zero when $u \to \infty$. Moreover, it is non-negative and has a maximal value for $u^* = (\ln\alpha - \ln\beta)/(\alpha-\beta) > 0$. Let now $n = 1$ in Proposition 3.8 above. We see that the H function is depending on the states as

$$h(T-t) = \exp\left(\frac{1}{2}\gamma(T-t;2\alpha,\lambda)y\right).$$

Hence, for $T-t = 0$ we have $h(0) = 1$, and when time-to-maturity $T-t$ tends to infinity, $h(T-t)$ tends to one. But, for

$$T-t = \frac{\ln(2\alpha) - \ln\lambda}{2\alpha-\lambda}$$

we have a maximal value of h strictly bigger than one. In fact, this maximum will be the product of the maximal value of $\gamma(u;2\alpha,\lambda)$ and the state y. Thus, we find that $h(T-t)$ has a so-called *hump-shaped* structure, where the size of the hump will depend on the current state of the volatility, being $\sigma^2 = y$. Thus, if we are in a market which currently is in a very volatile period, the model predicts a significant hump in the forward prices implied from the function $H(t,T,y)$. The hump will be in the shorter or longer end of the market, depending on the relative size between the speed of mean reversion α of the base component $X(t)$ and the speed of mean reversion of the volatility λ.

Note from the expression of the forward price in Proposition 3.6 that it is also explicitly a function of the current state of $Z(t)$, given by the term

$$G(t,T) = \exp\left(Z(t)\gamma(T-t;\alpha,\beta)\right).$$

As for the stochastic volatility, this term will also contribute with a hump shape, where the location and size of the hump will be dependent on the parameters α and β, and on the state of the stochastic mean level. If the mean level is very high, then the hump will be very pronounced, whereas a low mean level in the market will lead to a relatively small hump shape. Notice that for a given speed of mean reversion

β of the mean level process $Z(t)$, we find that the maximal value of $\gamma(u;\alpha,\beta)$ will have the property that

$$\lim_{\alpha\to 0} u^* = \lim_{\alpha\to 0} \frac{\ln\alpha - \ln\beta}{\alpha - \beta} = +\infty\,.$$

Hence, the hump will be far out on the forward curve when the speed of mean reversion of the logarithmic price process is very slow. On the other hand, a big α relative to β will give a hump in the very short end of the market, as

$$\lim_{\alpha\to\infty} \frac{\ln\alpha - \ln\beta}{\alpha - \beta} = 0\,.$$

We recall from (20) the maximal value of the function $\gamma(u;\alpha,\beta)$ to be $\gamma^*(\alpha,\beta)$. Taking limits using L'Hopital's rule reveals that

$$\lim_{\alpha\to 0} \gamma^*(\alpha,\beta) = \beta^{-1}\,,$$

while

$$\lim_{\alpha\to\infty} \gamma^*(\alpha,\beta) = 0\,.$$

Hence, a hump in the short end of the forward curve (implied by α being very big), is hardly visible except if the mean level is dramatically high. If the hump is far out (implied by a very slow mean reversion α), we will see a hump basically given by $Z(t)/\beta$, which can become very large.

Remark that the terms $\Lambda_f(t,T)$ and $\Theta(T-t)$ in the forward price will scale the effects discussed above deterministically, as functions of the seasonality and market prices of risk. We might have humps arising from these terms as well, but such humps will occur at given times and of a given size. For example, a hump could occur every winter due to a seasonality effect in the market. The factor involving the current spot price $S(t)$ will yield a curve which decreases from the current spot in the short end to 1 in the long end (or the other way, if $S(t) < 1$).

In Fig. 5 we have plotted the forward curve of WTI crude oil monthly contracts from February 28 2011. There is a clear hump shape in the forward curve, which may be attributed to an increase on the mean level of crude oil prices. In this period, the spot prices increased from about 90 Dollars per barrel to around 105, which may be attributed to an increase in the mean level $Z(t)$ (and possibly the volatility $\sigma(t)$). Our model predicts in such a case a hump shape, which therefore may explain the forward prices observed for WTI crude oil. (see Geman [17] for a discussion of hump shaped forward curves for Brent oil).

We move on with our analysis of the forward prices with an investigation of the effect of the correlation ρ. In the case when $\rho \leq 0$, we have the trivial majorization

$$(\sigma(s)\mathrm{e}^{-\alpha(T-s)} + \rho\eta\gamma(T-s;\alpha,\beta))^2 \leq \sigma^2(s)\mathrm{e}^{-2\alpha(T-s)} + \rho^2\eta^2\gamma^2(T-s;\alpha,\beta)\,.$$

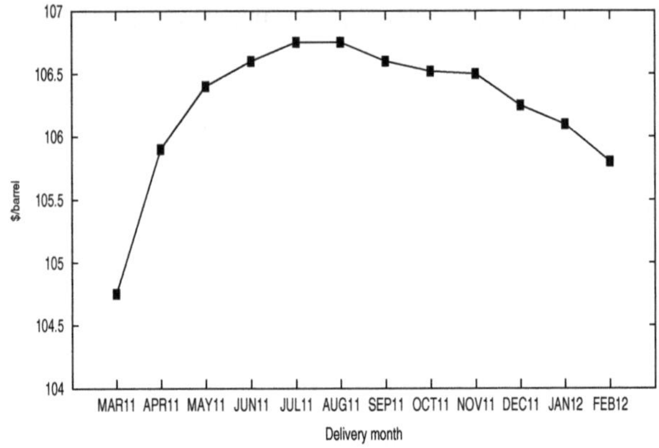

Fig. 5 The WTI crude oil forward curve on February 28 2011

From Propositions 3.6–3.8, it follows that

$$f(t,T) \leq f_0(t,T),$$

where f_0 denotes the forward price for $\rho = 0$. A negative correlation will lead to more concentrated spot prices compared to no or positive correlation. Less variation in spot price reduces the forward price. From the same arguments, the opposite holds when $\rho \geq 0$, that is,

$$f(t,T) \geq f_0(t,T).$$

A positive correlation creates a bigger variation in the spot prices, and we recognize the effect as higher forward prices compared to the benchmark at zero correlation.

For a negative correlation ρ, we have that the function $H(t,T,y_1,\ldots,y_n)$ is bounded by the expression given in Proposition 3.8 (being the function H with $\rho = 0$). This bound has a stationary limit under some mild hypothesis on the cumulant functions of the subordinators driving the volatility. Hence, as time to maturity goes to infinity, we find that the function H will be contained within the interval $(1,c)$, where c is the stationary limit of H for $\rho = 0$.

Let us consider the case with positive correlation $\rho > 0$. The lower limit for H will be c, the limit of H in the zero correlation case. However, we can also bound H from above. By elementary inequalities, it holds that

$$\frac{1}{2}\left(\rho\eta\gamma(T-s;\alpha,\beta)+\sigma(s)\mathrm{e}^{-\alpha(T-s)}\right)^2 \leq \rho^2\eta^2\gamma^2(T-s;\alpha,\beta)+\sigma^2(s)\mathrm{e}^{-2\alpha(T-s)}.$$

Thus,

$$H(t, T, y_1, \ldots, y_n) \le \exp\left(\rho^2\eta^2 \int_0^{T-t} \gamma^2(s; \alpha, \beta)\, ds\right) \times \mathbb{E}\left[\exp\left(\int_t^T \sigma^2(s) \mathrm{e}^{-2\alpha(T-s)}\, ds\right) \mid Y_j(t) = y_j\right].$$

As $\gamma(s; \alpha, \beta)$ is the difference of two exponentially decaying functions, the first term above has a limit. Appealing to the same arguments as in the proof of Proposition 3.8 reveals that the expectation operator also has a limit when $T - t$ tend to infinity. Hence, H is bounded from above when $T - t$ becomes large, and there will be an interval (c, d) within which H is contained. We leave the analysis of the asymptotic limit of H when $T - t$ tends to infinity as an open question.

We end this Section with an example of a stochastic volatility specification. Let $n = 1$, such that $\sigma^2(t) = Y(t)$, and assume that the subordinator L driving Y is a compound Poisson process with exponentially distributed jumps, that is,

$$L(t) = \sum_{k=1}^{N(t)} J_k$$

where $N(t)$ is a Poisson process with intensity $\delta > 0$ and $\{J_k\}$, are independent and distributed according to an exponential distribution with mean $1/a$, $a > 0$. We first compute the cumulant of J:

$$\psi_J(x) = \ln \mathbb{E}[\exp(\mathrm{i}xJ)] = \ln \int_0^\infty \mathrm{e}^{\mathrm{i}xy} a \mathrm{e}^{-ay}\, dy = \ln\left(\frac{a}{a - \mathrm{i}x}\right).$$

We observe that the moment generating function $\phi_J(y) = \psi_J(-\mathrm{i}y)$ exists for all $y \le a$. By conditioning, we can next compute the cumulant of $L(1)$:

$$\begin{aligned}
\psi(x) &= \ln \mathbb{E}[\exp(\mathrm{i}xL(1))] \\
&= \mathbb{E}\left[\exp(\mathrm{i}x \sum_{k=1}^{N(1)} J_k)\right] \\
&= \ln \mathrm{e}^{-\delta} \sum_{n=0}^{\infty} \frac{\delta^n}{n!} \mathbb{E}[\exp(\mathrm{i}xJ)]^n \\
&= \delta \left(\mathrm{e}^{\psi_J(x)} - 1\right) \\
&= \delta \frac{\mathrm{i}x}{a - \mathrm{i}x}.
\end{aligned}$$

Denoting by $\psi_{\sigma^2(s)}(t,x)$ the conditional cumulant of $\sigma^2(s)$ given $\mathcal{F}_t$, $s \geq t$, we find by a direct computation that

$$\begin{aligned}
\psi_{\sigma^2(s)}(t,x) &= \ln \mathbb{E}\left[\exp\left(\mathrm{i}x\sigma^2(s)\right) \mid \mathcal{F}_t\right] \\
&= \ln \mathbb{E}\left[\exp\left(\mathrm{i}x\sigma^2(t)\mathrm{e}^{-\lambda(s-t)} + \int_t^s \mathrm{e}^{-\lambda(s-u)}\,dL(u)\right) \mid \mathcal{F}_t\right] \\
&= \mathrm{i}x\sigma^2(t)\mathrm{e}^{-\lambda(s-t)} + \ln \mathbb{E}\left[\exp\left(\mathrm{i}x\int_t^s \mathrm{e}^{-\lambda(s-u)}\,dL(u)\right)\right] \\
&= \mathrm{i}x\sigma^2(t)\mathrm{e}^{-\lambda(s-t)} + \int_0^{s-t} \psi(x\mathrm{e}^{-\lambda u})\,du \\
&= \mathrm{i}x\sigma^2(t)\mathrm{e}^{-\lambda(s-t)} + \delta\int_0^{s-t} \frac{\mathrm{i}x\mathrm{e}^{-\lambda u}}{a - \mathrm{i}x\mathrm{e}^{-\lambda u}}\,du \\
&= \mathrm{i}x\sigma^2(t)\mathrm{e}^{-\lambda(s-t)} + \frac{\delta}{\lambda}\ln\left(\frac{a - \mathrm{i}x\mathrm{e}^{-\lambda(s-t)}}{a - \mathrm{i}x}\right).
\end{aligned}$$

Here we have used the $\mathcal{F}_t$-measurability of $\sigma^2(t)$ and the independent increment property of L. As $s - t \to \infty$, we find that

$$\lim_{s-t\to\infty} \psi_{\sigma^2(s)}(t,x) = \ln(1 - \mathrm{i}\frac{x}{a})^{-\delta/\lambda}.$$

Hence, in stationarity $\sigma^2(s)$ becomes Γ distributed with shape parameter δ/λ and scale $1/a$. The probability density function of this distribution is given as

$$p_\Gamma(x;k,a) = \frac{a^k}{\Gamma(k)} x^{k-1}\mathrm{e}^{-ax},$$

with $k = \delta/\lambda$.

Let us analyse $\sigma(t)$, the volatility, in this case. We find that the characteristic function of $\sigma(t)$ is

$$\begin{aligned}
\mathbb{E}\left[\mathrm{e}^{\mathrm{i}x\sigma(t)}\right] &= \mathbb{E}\left[\mathrm{e}^{\mathrm{i}x\sqrt{\sigma^2(t)}}\right] \\
&= \int_0^\infty \mathrm{e}^{\mathrm{i}x\sqrt{y}}\,P_{\sigma^2(t)}(dy),
\end{aligned}$$

with $P_{\sigma^2(t)}$ being the distribution function of $\sigma^2(t)$. However, as we know from above, $P_{\sigma^2(t)}(dy) \to p_\Gamma(y;k,a)\,dy$ as $t \to \infty$. Thus,

$$\lim_{t\to\infty} \mathbb{E}\left[\mathrm{e}^{\mathrm{i}x\sigma(t)}\right] = \int_0^\infty \mathrm{e}^{\mathrm{i}x\sqrt{y}}\,p_\Gamma(y;k,a)\,dy.$$

As this integral can be computed (yielding a very long expression consisting of Whittaker parabolic and trigonometric functions), we find an expression for the characteristic function of the stationary distribution of $\sigma(t)$. The mean value of the stationary distribution is, however, expressible in a rather compact form (using Maple):

$$\lim_{t\to\infty} \mathbb{E}[\sigma(t)] = -\frac{2\pi \sec(k\pi)}{(2k+1)\sqrt{a}\Gamma(k)\Gamma(-k-\frac{1}{2})}.$$

The stationary mean of the volatility is therefore proportional to the square-root of the mean jump size $1/a$ of L. For example, if $\delta = \lambda$ (implying that $k = 1$), we find

$$\lim_{t\to\infty} \mathbb{E}[\sigma(t)] = \frac{\sqrt{\pi}}{2\sqrt{a}}.$$

Recall the analysis of the autocorrelation structure of $X(t)$ leading to (22), where the stationary mean value of the volatility is appearing explicitly.

4 Cross-Commodity Derivatives

In this Section we focus on cross-commodity models of energy prices. We want to investigate pricing of simple spread options in a cross-commodity multi-factor model, as well as sensitivity measures and dependency risk.

4.1 A Margrabe Formula for Energy Markets

We want to derive a Margrabe formula for energy markets. In the energy markets, there exist a plethora of various spread options, and we focus on exchange-type options on spot, including spark and dark spreads. We recall that a stationary model is the natural dynamics for energy spot prices rather than geometric Brownian motions, calling for an extension of the classical Margrabe formula (see Margrabe [22]). Moreover, spikes call for non-Gaussian models, which further complicates the pricing of spread options for energy markets.

Letting $S_1(t)$ and $S_2(t)$ be the spot price dynamics of two energies, we are interested in deriving a price for an option on the spread between them, that is,

$$P(t) = \mathrm{e}^{-r(T-t)}\mathbb{E}\left[(S_1(T) - hS_2(T))^+ \mid \mathcal{F}_t\right] \tag{30}$$

where $h > 0$ is a constant, $(x)^+ = \max(x, 0)$ and $r > 0$ a constant risk-free interest rate. For simplicity, we suppose throughout this Section that the pricing measure Q is chosen to be the market probability P, $Q = P$, i.e., there is no market price of risk. We may view the situation alternatively as the spot being defined under the pricing measure directly, interpreting P as this one. Obviously, we assume that S_1 and S_2 are integrable in order to make the expectation well-defined.

In the classical cases of energy spreads, S_1 may be the price of electricity, and $S_2(t)$ the fuel. For example, we can have that S_2 is the price of gas, and in that case h is known as the *heat rate*, the factor converting the price of gas energy into the electricity equivalent.

Suppose that the price dynamics of S_i, $i = 1,2$ are defined as

$$S_i(t) = \Lambda_i(t) \exp(X_i(t) + Y_i(t)), \tag{31}$$

where

$$dX_i(t) = (\mu_i - \alpha_i X_i(t))\,dt + \sigma_i\,dB_i(t) \tag{32}$$

$$dY_i(t) = (\gamma_i - \beta_i Y_i(t))\,dt + dL_i(t). \tag{33}$$

Here, $L = (L_1, L_2)$ is a bivariate square-integrable Lévy process independent of B_1, B_2, which are two correlated Brownian motions with correlation coefficient ρ. Furthermore, $\mu_i, \alpha_i, \gamma_i, \beta_i$ and σ_i, for $i = 1,2$ are all constants, with α_i, β_i and σ_i assumed positive. In the next Lemma, we state the explicit dynamics of the OU-processes:

Lemma 4.1. *For $0 \leq t \leq s$, it holds that*

$$X_i(s) = X_i(t)\mathrm{e}^{-\alpha_i(s-t)} + \frac{\mu_i}{\alpha_i}(1 - \mathrm{e}^{-\alpha_i(s-t)}) + \int_t^s \sigma_i \mathrm{e}^{-\alpha_i(s-u)}\,dB_i(u),$$

and

$$Y_i(s) = Y_i(t)\mathrm{e}^{-\beta_i(s-t)} + \frac{\gamma_i}{\beta_i}(1 - \mathrm{e}^{-\beta_i(s-t)}) + \int_t^s \mathrm{e}^{-\beta_i(s-u)}\,dL_i(u),$$

for $i = 1,2$.

Proof. This is a straightforward application of Itô's Formula for jump processes. □

In order for the spread option price to be well-defined, we need that $(S_1(T) - S_2(T))^+$ has finite expectation, which is true if both $S_1(T)$ and $S_2(T)$ have finite expectation. A sufficient condition for this to hold is that $\int_0^t \exp(-\beta_i(t-s))\,dL_i(s)$, $i = 1,2$ have finite exponential moment. To this end, introduce the rectangle $R \subset \mathbb{R}^2$ including the origin defined as all pairs (a,b) such that

$$\int_{\mathbb{R}_0^2} \{\mathrm{e}^{ax+by} - 1\}\,\ell(dx,dy) < \infty \tag{34}$$

with ℓ being the Lévy measure of L. A sufficient condition for $\int_0^t \exp(-\beta_i(t - s))\,dL_i(s)$, $i = 1,2$ to have finite exponential moments is that $R = [0,1]^2$ in (34). We assume this is true from now on.

Hence, by appealing to $\mathcal{F}_t$-adaptedness, we find that the conditional expectation for the price can be expressed as

$$\begin{aligned}
&\mathbb{E}\left[(S_1(T) - hS_2(T))^+ | \mathcal{F}_t\right] \\
&= \mathbb{E}\left[\left(\Lambda_1(T)\mathrm{e}^{X_1(T)+Y_1(T)} - h\Lambda_2(T)\mathrm{e}^{X_2(T)+Y_2(T)}\right)^+ | \mathcal{F}_t\right] \\
&= C_1(t,T,X_1(t),Y_1(t)) \\
&\quad \times \mathbb{E}\Big[\mathrm{e}^{\Xi_2(t,T)+\Psi_2(t,T)}\Big(\mathrm{e}^{\Xi_1(t,T)-\Xi_2(t,T)+\Psi_1(t,T)-\Psi_2(t,T)} \\
&\quad -h\frac{C_2(t,T,X_2(t),Y_2(t))}{C_1(t,T,X_1(t),Y_1(t))}\Big)^+ | \mathcal{F}_t\Big]
\end{aligned}$$

where

$$\begin{aligned}
&C_i(t,T,x,y) \qquad (35)\\
&= \Lambda_i(T)\exp\left(\frac{\mu_i}{\alpha_i}(1-\mathrm{e}^{-\alpha_i(T-t)}) + \frac{\gamma_i}{\beta_i}(1-\mathrm{e}^{-\beta_i(T-t)}) + x\mathrm{e}^{-\alpha_i(T-t)} + y\mathrm{e}^{-\beta_i(T-t)}\right),
\end{aligned}$$

and

$$\Xi_i(t,T) = \int_t^T \sigma_i \mathrm{e}^{-\alpha_i(T-u)}\,dB_i(u) \tag{36}$$

$$\Psi_i(t,T) = \int_t^T \mathrm{e}^{-\beta_i(T-u)}\,dL_i(u), \tag{37}$$

for $i = 1,2$. Hence, due to $\mathcal{F}_t$-adaptedness and independent increment property of the Brownian motions and Lévy processes, the pricing $P(t)$ of the spread option entails in computing the expectation

$$p(t,T,K) = \mathbb{E}\left[\mathrm{e}^{\Xi_2(t,T)+\Psi_2(t,T)}\left(\mathrm{e}^{\Xi_1(t,T)-\Xi_2(t,T)+\Psi_1(t,T)-\Psi_2(t,T)} - K\right)^+\right], \tag{38}$$

for a deterministic strike price K depending on the current states of the factors in the spot prices as well as current time t and maturity T. We write K in the sequel for simplicity.

In the next Proposition, we compute $p(t,T,K)$ in (38) using the change of measure technique with respect to Brownian motion (see Carmona and Durrleman [14] for this idea, used to derive the classical Margrabe formula):

Proposition 4.2. *The price $p(t,T,K)$ defined in* (38) *is given by*

$$\begin{aligned}
p(t,T,K) &= \exp\left(\frac{\sigma_2^2}{4\alpha_2}(1-\mathrm{e}^{-2\alpha_2(T-t)})\right) \\
&\quad \times \mathbb{E}\left[\mathrm{e}^{\Psi_2(t,T)}F(a(t,T),\Sigma(t,T),K;\Psi_1(t,T)-\Psi_2(t,T))\right]
\end{aligned}$$

where $F(a,b,K;x)$ is defined as

$$F(a,b,K;x)=\exp(a+x+\frac{1}{2}b^2)N\left(b+\frac{a+x-\ln K}{b}\right)-KN\left(\frac{a+x-\ln K}{b}\right),$$

with N being the cumulative standard normal distribution. Furthermore,

$$a(t,T)=\rho\frac{\sigma_1\sigma_2}{\alpha_1+\alpha_2}(1-\mathrm{e}^{-(\alpha_1+\alpha_2)(T-t)})-\frac{\sigma_2^2}{2\alpha_2}(1-\mathrm{e}^{-2\alpha_2(T-t)}),$$

and variance

$$\Sigma^2(t,T)=\frac{\sigma_1^2}{2\alpha_1}(1-\mathrm{e}^{-2\alpha_1(T-t)})-2\rho\frac{\sigma_1\sigma_2}{\alpha_1+\alpha_2}(1-\mathrm{e}^{-(\alpha_1+\alpha_2)(T-t)})$$
$$+\frac{\sigma_2^2}{2\alpha_2}(1-\mathrm{e}^{-2\alpha_2(T-t)}).$$

Proof. Recall that (Ψ_1,Ψ_2) and (Ξ_1,Ξ_2) are independent, and hence by the tower property of conditional expectation we find

$$\mathbb{E}\left[\mathrm{e}^{\Xi_2(t,T)+\Psi_2(t,T)}\left(\mathrm{e}^{\Xi_1(t,T)-\Xi_2(t,T)+\Psi_1(t,T)-\Psi_2(t,T)}-K\right)^+\right]$$
$$=\mathbb{E}\left[\mathrm{e}^{\Psi_2(t,T)}\mathbb{E}\left[\mathrm{e}^{\Xi_2(t,T)}\left(\mathrm{e}^{\Xi_1(t,T)-\Xi_2(t,T)+\Psi_1(t,T)-\Psi_2(t,T)}-K\right)^+|\Psi_1(t,T),\Psi_2(t,T)\right]\right].$$

Thus, our first problem is to compute the inner conditional expectation, which amounts to calculating the expectation

$$\mathbb{E}\left[\mathrm{e}^{\Xi_2(t,T)}\left(\mathrm{e}^{\Xi_1(t,T)-\Xi_2(t,T)+m}-K\right)^+\right],$$

for a constant $m=\Psi_1(t,T)-\Psi_2(t,T)$.

To this end, introduce the martingale process $Z(s)$ on $t\leq s\leq T$ as

$$Z(s)=\exp\left(\int_t^s\sigma_2\mathrm{e}^{-\alpha_2(T-u)}\,dB_2(u)-\frac{\sigma_2^2}{4\alpha_2}(\mathrm{e}^{-2\alpha_2(T-s)}-\mathrm{e}^{-2\alpha_2(T-t)})\right),$$

which, by Girsanov's Theorem is the density process of an equivalent probability P^* and such that

$$dW(s)=dB_2(s)-\sigma_2\mathrm{e}^{-\alpha_2(T-s)}\,ds,$$

is a P^*-Brownian motion on $s\in[t,T]$. Thus,

$$\begin{aligned}
&\mathbb{E}\left[\mathrm{e}^{\Xi_2(t,T)}\left(\mathrm{e}^{\Xi_1(t,T)-\Xi_2(t,T)+m}-K\right)^+\right] \\
&= \mathrm{e}^{\frac{\sigma_2^2}{4\alpha_2}(1-\mathrm{e}^{-2\alpha_2(T-t)})}\mathbb{E}\left[Z(T)\left((\mathrm{e}^{\Xi_1(t,T)-\Xi_2(t,T)+m}-K\right)^+\right] \\
&= \mathrm{e}^{\frac{\sigma_2^2}{4\alpha_2}(1-\mathrm{e}^{-2\alpha_2(T-t)})}\mathbb{E}_*\left[\left(\mathrm{e}^{\Xi_1(t,T)-\Xi_2(t,T)+m}-K\right)^+\right],
\end{aligned}$$

where $\mathbb{E}_*$ is the expectation operator under P^*. Since B_1 and B_2 are correlated Brownian motions, we find, for an independent Brownian motion B, that

$$B_1(t) = \rho B_2(t) + \sqrt{1-\rho^2}B(t)\,.$$

Hence,

$$\begin{aligned}
\Xi_1(t,T) - \Xi_2(t,T) &= \int_t^T \sigma_1\mathrm{e}^{-\alpha_1(T-u)}\,dB_1(u) - \int_t^T \sigma_2\mathrm{e}^{-\alpha_2(T-u)}\,dB_2(u) \\
&= \int_t^T \rho\sigma_1\mathrm{e}^{-\alpha_1(T-u)} - \sigma_2\mathrm{e}^{-\alpha_2(T-u)}\,dB_2(u) \\
&\quad + \int_t^T \sqrt{1-\rho^2}\sigma_1\mathrm{e}^{-\alpha_1(T-u)}\,dB(u) \\
&= \int_t^T \rho\sigma_1\mathrm{e}^{-\alpha_1(T-u)} - \sigma_2\mathrm{e}^{-\alpha_2(T-u)}\,dW(u) \\
&\quad + \int_t^T \sqrt{1-\rho^2}\sigma_1\mathrm{e}^{-\alpha_1(T-u)}\,dB(u) \\
&\quad + \int_t^T \sigma_2\mathrm{e}^{-\alpha_2(T-u)}(\rho\sigma_1\mathrm{e}^{-\alpha_1(T-u)} - \sigma_2\mathrm{e}^{-\alpha_2(T-u)})\,du
\end{aligned}$$

Note that B is a Brownian motion under P^*, since it is independent of B_2. Thus, under P^*, we have that $\Xi_1(t,T) - \Xi_2(t,T) + m$ is a normally distributed random variable, with mean equal to $m + a(t,T)$ where

$$\begin{aligned}
a(t,T) &= \int_t^T \sigma_2\mathrm{e}^{-\alpha_2(T-u)}(\rho\sigma_1\mathrm{e}^{-\alpha_1(T-u)} - \sigma_2\mathrm{e}^{-\alpha_2(T-u)})\,du \\
&= \rho\frac{\sigma_1\sigma_2}{\alpha_1+\alpha_2}(1-\mathrm{e}^{-(\alpha_1+\alpha_2)(T-t)}) - \frac{\sigma_2^2}{2\alpha_2}(1-\mathrm{e}^{-2\alpha_2(T-t)}),
\end{aligned}$$

and variance

$$\Sigma^2(t,T) = \int_t^T (\rho\sigma_1 \mathrm{e}^{-\alpha_1(T-u)} - \sigma_2 \mathrm{e}^{-\alpha_2(T-u)})^2 + (1-\rho^2)\sigma_1^2 \mathrm{e}^{-2\alpha_1(T-u)}\,du$$
$$= \frac{\sigma_1^2}{2\alpha_1}(1-\mathrm{e}^{-2\alpha_1(T-t)}) - 2\rho\frac{\sigma_1\sigma_2}{\alpha_1+\alpha_2}(1-\mathrm{e}^{-(\alpha_1+\alpha_2)(T-t)})$$
$$+ \frac{\sigma_2^2}{2\alpha_2}(1-\mathrm{e}^{-2\alpha_2(T-t)}).$$

Using the same line of derivations as for a call option price, we find that

$$\mathbb{E}_*\left[(\mathrm{e}^{\Xi_1(t,T)-\Xi_2(t,T)+m} - K)^+\right]$$
$$= \mathrm{e}^{a(t,T)+m+\frac{1}{2}\Sigma^2(t,T)} N\left(\Sigma(t,T) + \frac{a(t,T)+m-\ln K}{\Sigma(t,T)}\right) - KN\left(\frac{a(t,T)+m-\ln K}{\Sigma(t,T)}\right).$$

Thus, by appealing to the definition of F, the proof is complete. □

In the Proposition above, we have reduced the problem of finding $p(t,T,K)$ to computing an expectation of a function of the difference of two Lévy integrals. Thus, we face the problem of pricing a spread option again, but now reduced to being a spread between the jump terms only. As it turns out, one may again appeal to a change of measure to express this expectation. Moreover, it is advantageous to apply the Fourier transform to obtain a "closed-form" expression for $p(t,T,K)$ that is possible to compute numerically by fast Fourier transform methods.

To prepare for this, we make a small excursion into the Esscher transform for bivariate Lévy processes. On $t \in [0,T]$, define the stochastic process

$$Z(t) = \exp\left(\int_0^t \theta_1(s)\,dL_1(s) + \int_0^t \theta_2(s)\,dL_2(s) - \int_0^t \phi(\theta_1(s),\theta_2(s))\,ds\right), \tag{39}$$

with θ_1, θ_2 being two bounded measurable functions, and $\phi(x,y)$ the log-moment generating function of $L = (L_1, L_2)$. Note that by our exponential integrability assumption on L, $\phi(x,y)$ is well-defined, and it follows that $Z(t)$ is a martingale process on $[0,T]$. We introduce the probability measure $\tilde{P}$ with density

$$\left.\frac{d\tilde{P}}{dP}\right|_{\mathcal{F}_t} = Z(t)\,, t \le T\,.$$

This is known as the Esscher transform of L. Define the conditional cumulant function of L under $\tilde{P}$ as

$$\psi_{\tilde{P}}(s,t,x,y) := \ln \mathbb{E}_{\tilde{P}}\left[\exp\left(\mathrm{i}x(L_1(t)-L_1(s)) + \mathrm{i}y(L_2(t)-L_2(s))\right) \mid \mathcal{F}_s\right] \tag{40}$$

for $T \ge t \ge s \ge 0$. It turns out that L becomes a bivariate independent increment process (see Benth et al. [7]) under $\tilde{P}$, with characteristics given in the next Lemma:

Lemma 4.3. *Under $\tilde{P}$, L has a conditional cumulant function*

$$\psi_{\tilde{P}}(s,t,x,y) = \int_s^t \psi(x - \mathrm{i}\theta_1(u), y - \mathrm{i}\theta_2(u))\,du - \int_s^t \psi(-\mathrm{i}\theta_1(u), -\mathrm{i}\theta_2(u))\,du\,,$$

where $\psi(x,y)$ is the cumulant of L under P.

Proof. With $T \geq t \geq s \geq 0$, the conditional characteristic function becomes, by using Bayes' Formula and the independent increment property of Lévy processes,

$$\begin{aligned}
&\mathbb{E}_{\tilde{P}}\left[\exp\left(\mathrm{i}x(L_1(t) - L_1(s)) + \mathrm{i}y(L_2(t) - L_2(s))\right) \mid \mathcal{F}_s\right] \\
&= \mathbb{E}\left[\frac{Z(t)}{Z(s)}\exp\left(\mathrm{i}x(L_1(t) - L_1(s)) + \mathrm{i}y(L_2(t) - L_2(s))\right) \mid \mathcal{F}_s\right] \\
&= \exp\left(-\int_s^t \phi(\theta_1(u), \theta_2(u))\,du\right) \\
&\times \mathbb{E}\left[\exp\left(\int_s^t \mathrm{i}x + \theta_1(s)\,dL_1(s) + \int_s^t \mathrm{i}y + \theta_2(u)\,dL_2(u)\right) \mid \mathcal{F}_s\right] \\
&= \exp\left(-\int_s^t \phi(\theta_1(u), \theta_2(u))\,du\right) \\
&\times \mathbb{E}\left[\exp\left(\int_s^t \mathrm{i}x + \theta_1(s)\,dL_1(s) + \int_s^t \mathrm{i}y + \theta_2(u)\,dL_2(u)\right)\right] \\
&= \exp\left(\int_s^t \psi(x - \mathrm{i}\theta_1(u), y - \mathrm{i}\theta_2(u))\,du - \int_s^t \psi(-\mathrm{i}\theta_1(u), -\mathrm{i}\theta_2(u))\,du\right)
\end{aligned}$$

Hence, the proof is complete. □

By the Lévy–Kintchine formula for L, we have that

$$\begin{aligned}
\psi_{\tilde{P}}(s,t,x,y) &= \mathrm{i}\left\langle \int_t^s (\xi + \int_{|z|<1} z\{\mathrm{e}^{\langle z,\theta(u)\rangle} - 1\}\,\ell(dz))\,du, w\right\rangle \\
&\quad + \int_t^s \int_{\mathbb{R}^2} \{\mathrm{e}^{\mathrm{i}\langle z,w\rangle} - 1 - \mathrm{i}\langle z,w\rangle 1_{|z|<1}\}\mathrm{e}^{\langle z,\theta(u)\rangle}\,\ell(dz)\,du\,,
\end{aligned}$$

where $w := (x,y)$, $\theta(u) = (\theta_1(u), \theta_2(u))$, $\xi \in \mathbb{R}^2$ is the drift of L and $\ell(dz)$ is the Lévy measure of L defined on $\mathbb{R}^2 \setminus \{0\}$. The drift will change from ξ to

$$\int_t^s (\xi + \int_{|z|<1} z\{\mathrm{e}^{\langle z,\theta(u)\rangle} - 1\}\,\ell(dz))\,du\,,$$

under $\tilde{P}$. Moreover, L will have a time-dependent jump measure under $\tilde{P}$ given by

$$\ell_{\tilde{P}}(dz, du) = \mathrm{e}^{\langle z,\theta(u)\rangle}\,\ell(dz)\,du\,.$$

One refers to this measure as the *compensator measure* of L. Since the compensator measure is time-dependent, but deterministic, L is an independent increment process under $\tilde{P}$. If θ_1, θ_2 are supposed to be constant, then L will have stationary increments under $\tilde{P}$, and therefore becomes a $\tilde{P}$-Lévy process in that case. We remark in passing that we could use the Esscher transform to change measure for the subordinators driving the stochastic volatility model discussed in Sect. 3, and thereby modelling the market price of volatility risk.

Before analysing $p(t,T,K)$ further, let us discuss the function $F(a,b,K;x)$ defined in Proposition 4.2. Recall from the proof of Proposition 4.2 that we can express F as

$$F(a,b,K;x) = \mathbb{E}\left[(\exp(a + bU + x) - K)^+\right],$$

where U is a standard normally distributed random variable. We note here that b is strictly positive. It is then simple to see that $\exp(-cx)F(a,b,K;x)$ is integrable on $\mathbb{R}$ for any $c > 1$, as the next Lemma proves:

Lemma 4.4. *For any constant* $c > 1$, *we have that* $\exp(-cx)F(a,b,K;x) \in L^1(\mathbb{R})$.

Proof. By Tonelli's theorem, it holds,

$$\begin{aligned}\int_{\mathbb{R}} e^{-cx} F(a,b,K;x)\,dx &= \mathbb{E}\left[\int_{\mathbb{R}} e^{-cx}\left(e^{a+bU+x} - K\right)^+ dx\right] \\ &= \mathbb{E}\left[\int_{\ln K - a - bU}^{\infty} e^{-cx}\left(e^{a+bU+x} - K\right) dx\right] \\ &\leq \mathbb{E}\left[e^{a+bU}\int_{\ln K - a - bU}^{\infty} e^{-(c-1)x}\,dx\right] \\ &= \frac{1}{c-1} e^{a-(c-1)(\ln K - a)}\mathbb{E}\left[e^{bcU}\right] < \infty,\end{aligned}$$

where in the last step we have used that a standard normal distributed random variable has finite exponential moments of all orders. Hence, the Lemma follows. □

In the next Lemma we find the Fourier transform of the function $\exp(-cx)F(a,b,K;x)$ for $c > 1$. For this purpose, we apply the definition of the Fourier transform of a function $g \in L^1(\mathbb{R})$ given in Folland [16];

$$\hat{g}(y) = \int_{\mathbb{R}} g(x) e^{-ixy}\,dx\,. \tag{41}$$

Notice the sign in the complex exponent. With this definition, it holds that the inverse Fourier transform can be expressed as

$$g(x) = \frac{1}{2\pi}\int_{\mathbb{R}} \hat{g}(y)e^{ixy}\,dy, \tag{42}$$

as long as $\hat{g} \in L^1(\mathbb{R})$.

Lemma 4.5. *The Fourier transform of $F_c(a,b,K;x) := \exp(-cx)F(a,b,K;x)$ is*

$$\hat{F}_c(a,b,K;y) = \frac{e^{a-(iy+(c-1))(\ln K-a)+\frac{1}{2}b^2(1+(iy+(c-1)))^2}}{(c-1)+iy} - K\frac{e^{-(iy+c)(\ln K-a)+\frac{1}{2}b^2(iy+c)^2}}{c+iy},$$

for every $c > 1$.

Proof. By the Fubini–Tonelli Theorem, we compute as follows:

$$\begin{aligned}\int_{\mathbb{R}} e^{-cx}F(a,b.K;x)e^{-ixy}\,dx &= \mathbb{E}\left[\int_{\mathbb{R}} e^{-cx}\left(e^{a+bU+x}-K\right)^+ e^{-ixy}\,dx\right] \\ &= \mathbb{E}\left[e^{a+bU}\int_{\ln K-a-bU}^{\infty} e^{-((c-1)+iy)x}\,dx\right] \\ &\quad - K\mathbb{E}\left[\int_{\ln K-a-bU}^{\infty} e^{-(c+iy)x}\,dx\right].\end{aligned}$$

Hence, the Lemma follows after a straightforward integration of exponentials. □

Note that the Fourier transformed function $\hat{F}_c(a,b,K;\cdot) \in L^1(\mathbb{R})$ since both terms will consist of expressions involving $\exp(-b^2y^2/2)$. Hence, using the inverse Fourier transform, we find

$$F(a,b,K;x) = \frac{1}{2\pi}\int_{\mathbb{R}} \hat{F}_c(a,b,K;y)e^{(iy+c)x}\,dy. \tag{43}$$

We are ready for our next proposition on the derivation of $p(t,T,K)$ in (38):

Proposition 4.6. *For a given $c > 1$, suppose that $(c, c+1) \in R$. Then it holds that*

$$\begin{aligned}p(t,T,K) &= \exp\left(\frac{\sigma_2^2}{4\alpha_2}\left(1-e^{-2\alpha_2(T-t)}\right)\right) \\ &\quad\times \frac{1}{2\pi}\int_{\mathbb{R}} \hat{F}_c(a(t,T),\Sigma(t,T),K;y)\exp\left(\Phi(T-t,y)\right)dy,\end{aligned}$$

where $a(t,T)$ and $\Sigma(t,T)$ are defined in Proposition 4.2, and $\Phi(\tau,y)$ is given by

$$\Phi(\tau,y) = \int_0^{\tau} \phi\left((iy+c)e^{-\beta_1 u}, (iy+c+1)e^{-\beta_2 u}\right)du$$

Proof. First, define the density process

$$Z(t) = \exp\left(\int_0^t e^{-\beta_2(T-u)}\,dL_2(u) - \int_0^t \phi(0, e^{-\beta_2(T-u)})\,du\right),$$

corresponding to letting $\theta_1(u) = 0$ and $\theta_2(u) = \exp(-\beta_2(T-u))$ in the Esscher transform defined above. According the Lemma 4.3, this is a measure transform giving rise to an equivalent probability P_β such that (L_1, L_2) becomes an independent increment process under P_β, with explicitly known conditional cumulant function. Observe that

$$\frac{Z(T)}{Z(t)} = \exp\left(\Psi_2(t,T) - \int_t^T \phi(0, \mathrm{e}^{-\beta_2(T-u)})\,du\right).$$

Denoting $\mathbb{E}_\beta$ the expectation operator under P_β, we find from the Fourier inversion formula and Fubini–Tonelli,

$$\begin{aligned}
p(t,T,K) &= \exp\left(\frac{\sigma_2^2}{4\alpha_2}(1-\mathrm{e}^{-2\alpha_2(T-t)}) + \int_0^{T-t} \phi(0,\mathrm{e}^{-\beta_2 u})\,du\right) \\
&\quad \times \mathbb{E}_\beta\left[F(a(t,T),\Sigma(t,T),K;\Psi_1(t,T)-\Psi_2(t,T))\right] \\
&= \exp\left(\frac{\sigma_2^2}{4\alpha_2}(1-\mathrm{e}^{-2\alpha_2(T-t)}) + \int_0^{T-t} \phi(0,\mathrm{e}^{-\beta_2 u})\,du\right) \\
&\quad \times \frac{1}{2\pi}\int_{\mathbb{R}} \hat{F}_c(a(t,T),\Sigma(t,T),K;y) \\
&\quad \times \mathbb{E}_\beta\left[\exp\left((\mathrm{i}y+c)(\Psi_1(t,T)-\Psi_2(t,T))\right)\right]\,dy,
\end{aligned}$$

with $a(t,T)$ and $\Sigma(t,T)$ defined in Proposition 4.2. But, appealing to the definition of the measure P_β, we find

$$\begin{aligned}
\ln\mathbb{E}_\beta\left[\exp\left((\mathrm{i}y+c)(\Psi_1(t,T)-\Psi_2(t,T))\right)\right] &= -\int_0^{T-t}\phi(0,\mathrm{e}^{-\beta_2 u})\,du \\
+\ln\mathbb{E}\left[\exp\left(\int_t^T(\mathrm{i}y+c)\mathrm{e}^{-\beta_1(T-u)}\,dL_1(u) + \int_t^T(\mathrm{i}y+c+1)\mathrm{e}^{-\beta_2(T-u)}\,dL_2(u)\right)\right] \\
= -\int_0^{T-t}\phi(0,\mathrm{e}^{-\beta_2 u})\,du + \int_t^T \phi\left((\mathrm{i}y+c)\mathrm{e}^{-\beta_1(T-u)},(\mathrm{i}y+c+1)\mathrm{e}^{-\beta_2(T-u)}\right)\,du,
\end{aligned}$$

where we used the independent increment property of the Lévy process L. But then the result follows. □

We collect everything together, to state the Margrabe formula for energy spread options:

Theorem 4.7. *Suppose for a $c > 1$ that $(c, c+1) \in R$ in condition* (34). *Then the price $P(t)$ for the spread option is given by*

$$P(t) = \mathrm{e}^{-r(T-t)}C_1(t,T,X_1(t),Y_1(t))p(t,T,K)$$

where $p(t,T,K)$ is given in Proposition 4.6 and

$$K = h\frac{C_2(t,T,X_2(t),Y_2(t))}{C_1(t,T,X_1(t),Y_1(t))}$$

with $C_i(t,T,x_i,y_i)$, $i = 1,2$ defined in (35).

Let us discuss the asymptotic properties of this generalization of Margrabe's formula. Indeed, from the definition of $C_i(t,T,x_i,y_i)$, $i = 1,2$ defined in (35), we find that

$$C_i(t,T,x,y) \sim \Lambda_i(T)\exp\left(\frac{\mu_i}{\alpha_i}+\frac{\gamma_i}{\beta_i}\right),$$

when $T-t$ tends to infinity. This means that the strike K behaves asymptotically for maturities far in the future as

$$K \sim h\frac{\Lambda_2(T)}{\Lambda_1(T)}\exp\left(\frac{\mu_2}{\alpha_2}-\frac{\mu_1}{\alpha_1}+\frac{\gamma_2}{\beta_2}-\frac{\gamma_1}{\beta_1}\right).$$

Furthermore, from Proposition 4.6 and supposing natural integrability hypotheses, we find after letting $T-t \to \infty$

$$p(t,T,K) \sim \frac{1}{2\pi}\mathrm{e}^{\frac{\sigma_2^2}{4\alpha_2}}\int_{\mathbb{R}}\hat{F}_c(\tilde{a},\tilde{\Sigma},K;y)\exp\left(\tilde{\Phi}(y)\right)dy,$$

with

$$\tilde{a} = \rho\frac{\sigma_1\sigma_2}{\alpha_1+\alpha_2}-\frac{\sigma_2^2}{2\alpha_2},$$

and

$$\tilde{\Sigma} = \frac{\sigma_1^2}{2\alpha_1}-2\rho\frac{\sigma_1\sigma_2}{\alpha_1+\alpha_2}+\frac{\sigma_2^2}{2\alpha_2}.$$

Moreover,

$$\tilde{\Phi}(y) = \int_0^\infty \phi((\mathrm{i}y+c)\mathrm{e}^{-\beta_1 u},(\mathrm{i}y+c+1)\mathrm{e}^{-\beta_2 u})\,du.$$

This integral is well-defined under logarithmic integrability hypothesis of the jump processes, see Sato [24]. In conclusion, we have that the option price behaves asymptotically as

$$P(t) \sim k_1\mathrm{e}^{-r(T-t)}\Lambda_1(T)\int_{\mathbb{R}}\hat{F}_c\left(\tilde{a},\tilde{\Sigma},k_2\frac{\Lambda_2(T)}{\Lambda_1(T)}\right)\exp\left(\tilde{\Phi}(y)\right)dy,$$

for constants k_1 and k_2 independent of t and T. The option prices will not be influenced by the current spot price levels when we are far from exercise. This is an effect of the stationary processes driving the spot dynamics.

4.2 Computing Sensitivity Measures of Cross-Commodity Options

With the (semi-)explicit price for the spread option in Theorem 4.7 at hand, one can start to derive the "Greeks" for risk management purposes. By inspecting the price, we see that to find the Greek of $P(t)$ with respect to $X_1(t)$, say, will involve differentiation of the function $C_1(t, T, X_1(t), Y_1(t))$, which appears both explicitly and inside the inverse Fourier transform in the expression of the strike price K. To differentiate inside the inverse Fourier transform would yield an expression analogous to differentiating first the payoff function of the derivative and then apply Fourier methods to compute the resulting expectation.

In this Subsection we want to investigate a different approach based on the so-called *density method* (see Glasserman [17]). The density method allows for differentiation of option prices for many particular models, where one does not need to differentiate the payoff function. Our analysis will be valid for a rather general class of cross-commodity options, that is, European-style options written on the underlying bivariate commodity prices (S_1, S_2). Noting that each of the two price processes has a Brownian motion driven factor, one can exploit this by a conditioning argument to obtain expressions for the derivatives with respect to all four factors $X_i(t), Y_j(t), i, j = 1, 2$. This approach, called the *conditional density method*, was suggested and analysed for options written on one underlying asset in Benth et al. [9]. It was later extended to multi-factor models and applied to energy markets in Benth et al. [12]. We apply it here in our particular multi-factor cross commodity model, but remark that its potential is much larger.

Consider a cross commodity option paying $\tilde{g}(S_1(T), S_2(T))$ at time T, for some (nice) function $\tilde{g} : \mathbb{R}_+^2 \mapsto \mathbb{R}_+$. By a simple reformulation, we can express this payoff function as

$$\tilde{g}(S_1(T), S_2(T)) = g(T, X_1(T) + Y_1(T), X_2(T) + Y_2(T)),$$

for a function g. From now on, we suppress the dependency on T in this function (it comes from the seasonality functions), and suppose that

$$\mathbb{E}[|g(X_1(T) + Y_1(T), X_2(T) + Y_2(T))|] < \infty.$$

Our problem now is to find the derivative of the price functionals

$$P(t) = e^{-r(T-t)}\mathbb{E}[g(X_1(T) + Y_1(T), X_2(T) + Y_2(T)) \,|\, \mathcal{F}_t]. \qquad (44)$$

For simplicity, we let $r = 0$ in the rest of this subsection.

By the Markov property of the factors, we find that

$$P(t) = P(t, X_1(t), Y_1(t), X_2(t), Y_2(t)),$$

where

$$P(t, x_1, y_1, x_2, y_2) = \mathbb{E}\big[g\big(x_1 e^{-\alpha_1(T-t)} + y_1 e^{-\beta_1(T-t)} + \Xi_1(t,T) + \Psi_1(t,T), \\ x_2 e^{-\alpha_2(T-t)} + y_2 e^{-\beta_2(T-t)} + \Xi_2(t,T) + \Psi_2(t,T)\big)\big], \tag{45}$$

after using Lemma 4.1 and (36)–(37). By conditioning, we have the following representation:

Proposition 4.8. *It holds that*

$$P(t, x_1, y_1, x_2, y_2) = \mathbb{E}\left[\int_{\mathbb{R}^2} g(z_1, z_2) p_{\Xi}(z_1(x_1, y_1), z_2(x_2, y_2))\, dz_1\, dz_2\right]$$

where

$$z_i(x_i, y_i) = z_i - x_i e^{-\alpha_i(T-t)} - y_i e^{-\beta_i(T-t)} - \Psi_i(t,T),$$

for $i = 1, 2$, *and* $p_{\Xi}(z_1, z_2)$ *is the density function of the bivariate normal random variable* $(\Xi_1(t,T), \Xi_2(t,T))$.

Proof. First observe that by assumption, $\Psi_i(t,T)$ are independent of $\Xi_i(t,T)$, $i = 1, 2$. By conditioning, we find from properties of the conditional expectation that

$$\begin{aligned} P(t, x_1, y_1, x_2, y_2) = \mathbb{E}\big[\mathbb{E}\big[g\big(&x_1 e^{-\alpha_1(T-t)} + y_1 e^{-\beta_1(T-t)} + \psi_1 + \Xi_1(t,T), \\ &x_2 e^{-\alpha_2(T-t)} + y_2 e^{-\beta_2(T-t)} + \psi_2 + \Xi_2(t,T)\big) \\ &\mid \psi_i = \Psi_i(t,T), i = 1, 2\big]\big]. \end{aligned}$$

We see that the inner expression is an expectation of a function of $(\Xi_1(t,T), \Xi_2(t,T))$, and the result follows. □

From the definition of $\Xi_i(t,T)$, $i = 1, 2$, in (36) we find that it has expected value zero and variance given by the Ito isometry as

$$v_i(T-t) \triangleq \mathbb{E}\left[\left(\int_t^T \sigma_i e^{-\alpha_i(T-s)}\, dB_i(s)\right)^2\right] = \frac{\sigma_i^2}{2\alpha_i}\left(1 - e^{-2\alpha_i(T-t)}\right). \tag{46}$$

Furthermore, since B_1 and B_2 are two correlated Brownian motion, the Ito isometry yields the covariance between $\Xi_1(t,T)$ and $\Xi_2(t,T)$ as

$$v_{12}(T-t) \triangleq \mathbb{E}[\Xi_1(t,T)\Xi_2(t,T)] = \rho\frac{\sigma_1\sigma_2}{\alpha_1+\alpha_2}\left(1 - e^{-(\alpha_1+\alpha_2)(T-t)}\right). \tag{47}$$

This gives a full specification of the variance–covariance matrix $V(T-t) \in \mathbb{R}^{2\times 2}$ of $(\Xi_1(t,T), \Xi_2(t,T))$, and its bivariate probability density becomes

$$p_\Xi(z_1, z_2) = \frac{1}{2\pi\sqrt{\det(V(T-t))}} \exp\left(-\frac{1}{2} z^* V^{-1}(T-t) z\right), \tag{48}$$

with $z = (z_1, z_2)^*$ and $*$ meaning the transpose. It is easily seen that the gradient of p_Ξ is

$$\nabla p_\Xi(z_1, z_2) = -p_\Xi(z_1, z_2)\left(\mathbf{e}_1^* V^{-1} z, \mathbf{e}_2^* V^{-1} z\right) \tag{49}$$

where $\mathbf{e}_i$, $i = 1, 2$, are the Euclidean basis vectors in $\mathbb{R}^2$. We are now ready to derive the sensitivity of the option price with respect to the various factors.

For the sake of illustration, suppose we want to find the derivative of the option price with respect to $X_1(t)$, the first factor of the first commodity (energy) in the option. This is given via the derivative $\partial P(t, x_1, y_1, x_2, y_2)/\partial x_1$. By Proposition 4.8 this is now straightforwardly calculated. We find

$$\begin{aligned} \frac{\partial P(t, x_1, y_1, x_2, y_2)}{\partial x_1} &= \mathbb{E}\Big[\int_{\mathbb{R}^2} g(z_1, z_2) p_\Xi(z_1(x_1, y_1), z_2(x_2, y_2)) \\ &\quad \times \mathbf{e}_1^* V^{-1}(z_1(x_1, y_1), z_2(x_2, y_2))^* \mathrm{e}^{-\alpha_1(T-t)}\, dz_1\, dz_2\Big] \end{aligned} \tag{50}$$

According to Folland [16], Theorem 2.27, we can commute differentiation and integration in the above derivation as long as the integrand in (50) has a majorization uniformly in x_1. But this holds at least when restricting x_1 to a bounded subset of $\mathbb{R}$. Tracing back, we get

$$\begin{aligned} &\frac{\partial P(t, x_1, y_1, x_2, y_2)}{\partial x_1} \\ &= \mathrm{e}^{-\alpha_1(T-t)} \mathbb{E}\Big[g\Big(x_1 \mathrm{e}^{-\alpha_1(T-t)} + y_1 \mathrm{e}^{-\beta_1(T-t)} + \Xi_1(t,T) + \Psi_1(t,T), \\ &\quad x_2 \mathrm{e}^{-\alpha_2(T-t)} + y_2 \mathrm{e}^{-\beta_2(T-t)} + \Xi_2(t,T) + \Psi_2(t,T)\Big) \mathbf{e}_1^* V^{-1}(\Xi_1(t,T), \Xi_2(t,T))^*\Big]. \end{aligned} \tag{51}$$

Observe that the above Greek does not involve any differentiation of the payoff function g, and lends itself easily to Monte Carlo pricing. In fact, we are almost back to pricing the option itself, except for the additional weight functional

$$\mathbf{e}_1^* V^{-1}(\Xi_1(t,T), \Xi_2(t,T))^*,$$

that enters the expectation operator, and a "discounting" term given by the speed of mean reversion α_1.

Of course, the derivatives with respect to the other factors are calculable in the exact same fashion. Indeed, we can also compute derivatives with respect to some of the parameters, like for example the speeds of mean reversion using this approach. This would, however, yield more technically complex expressions, and we refrain from analyzing this further here.

4.3 Cross-Commodity Dependency Risk and Copulas

In this Subsection we consider the dynamics (32)–(33) defined under the market probability P, and focus on how to do a measure change from P to Q for bivariate jump process. In particular, we are interested in the potential effects on the dependency structure.

The Esscher transform (as previously introduced) is the standard approach to produce a parametric class of measure changes for jump processes. For the Brownian motions, one naturally applies the Girsanov theorem to change measure. To this end, we suppose that $\boldsymbol{\theta} = (\theta_1, \theta_2) \in \mathbb{R}^2$ and $\boldsymbol{\eta} = (\eta_1, \eta_2) \in \mathbb{R}^2$ are two constant vectors. Define the martingale process $Z_B(t)$ for $t \leq T$

$$Z_B(t) = \exp\left(\theta_1 B_1(t) + \theta_2 B_2(t) - \frac{1}{2}(\theta_1^2 + \theta_2^2)\, t\right).$$

Then, by Girsanov's Theorem, we find that $Z(t)$ is the density process of an equivalent probability measure Q_B such that

$$dW_i(t) = dB_i(t) - \theta_i\, dt\,, i = 1, 2\,, \tag{52}$$

are Q_B-Brownian motions on $[0, T]$. Observe that the correlation between B_1 and B_2 is preserved under this measure change, so that W_1 and W_2 also become correlated by the same factor ρ. The characteristics of the Lévy processes L_1, L_2 remain unchanged under Q_B.

Recall the Esscher transform defining a measure via a density process $Z(t)$ as in (39). We shall here focus on a constant measure change, and define the process $Z_L(t)$ for $t \leq T$ as

$$Z_L(t) = \exp\left(\eta_1 L_1(t) + \eta_2 L_2(t) - \phi(\eta_1, \eta_2) t\right)\,, \tag{53}$$

where $\phi(x, y)$ is the logarithm of the moment generating function of the bivariate random variable $(L_1(1), L_2(1))$. Choosing $\eta = (\eta_1, \eta_2) \in R$ (see (34)) implies that $Z_L(t)$ is a martingale with expectation equal to one, and therefore the density process of an equivalent probability Q_L. This measure transform of $(L_1(t), L_2(t))$ preserves the Lévy property of $(L_1(t), L_2(t))$ (recall Lemma 4.3 above). The change of measure from P to Q_L does not affect the Brownian part since the jump process and the Brownian motions are supposed independent.

We characterize the Lévy process $(L_1(t), L_2(t))$ under Q_L. By a direct computation, we find the logarithm of the moment generating function to be

$$\begin{aligned}\phi_Q(x,y) &= \ln \mathbb{E}_{Q_L}\left[\exp\left(xL_1(1) + yL_2(1)\right)\right] \\ &= \ln \mathbb{E}\left[\exp\left((x+\eta_1)L_1(1) + (y+\eta_2)L_2(1)\right)\right] - \phi(\eta_1,\eta_2) \\ &= \phi(x+\eta_1, y+\eta_2) - \phi(\eta_1,\eta_2)\,.\end{aligned}$$

To make this derivation rigorous, we must assume that $(x+\eta_1, y+\eta_2) \in R$. If $L(t)$ has drift given by the vector $\xi \in \mathbb{R}^2$ and a Lévy measure denoted by $\ell(dz_1, dz_2)$, we find from the Lévy–Kintchine representation that

$$\begin{aligned}\phi_Q(x,y) = \xi_1 x + \xi_2 y &+ \int_{|z|<1} (xz_1 + yz_2)\left(\mathrm{e}^{\eta_1 z_1 + \eta_2 z_2} - 1\right)\ell(dz_1, dz_2) \\ &+ \int_{\mathbb{R}_0}\int_{\mathbb{R}_0}\left\{\mathrm{e}^{xz_1 + yz_2} - 1 - \mathbf{1}_{\{|z|<1\}}(xz_1 + yz_2)\right\}\ell(dz_1, dz_2)\end{aligned}$$

where $|\cdot|$ is the norm on $\mathbb{R}^2$. Hence, the drift of $L = (L_1, L_2)$ is

$$\left(\xi_1 + \int_{|z|<1} z_1\left\{\mathrm{e}^{\eta_1 z_1 + \eta_2 z_2} - 1\right\}\ell(dz_1, dz_2), \xi_2 + \int_{|z|<1} z_2\left\{\mathrm{e}^{\eta_1 z_1 + \eta_2 z_2} - 1\right\}\ell(dz_1, dz_2)\right)$$

under Q_L, whereas the Lévy measure is

$$\ell_Q(dz_1, dz_2) = \mathrm{e}^{\eta_1 z_1 + \eta_2 z_2}\,\ell(dz_1, dz_2)\,.$$

We see that the effect of the Esscher transform is a linear shift in the drift and an exponential tilting of the Lévy measure.

We define a pricing measure $Q \sim P$ as $Q = Q_B \times Q_L$, which then will have a Radon–Nikodym derivative with density

$$\left.\frac{dQ}{dP}\right|_{\mathcal{F}_t} = Z_B(t) Z_L(t)\,, \tag{54}$$

for $t \leq T$. We know the characteristics of L and B_1, B_2 under Q.

We next compute the forward price dynamics for a contract delivering the spot at time T. By definition, we set the forward price at time $t \leq T$ on commodity i, denoted $f_i(t,T)$, as

$$f_i(t,T) = \mathbb{E}_Q\left[S_i(T) \mid \mathcal{F}_t\right]\,. \tag{55}$$

Proposition 4.9. *Suppose that* $(\eta_1 + 1, \eta_2 + 1) \in R$. *Then, the forward prices* $f_i(t,T)$, $i = 1, 2$, *are given by*

$$f_i(t,T) = \Lambda_i(T)\exp\left(X_i(t)\mathrm{e}^{-\alpha_i(T-t)} + Y_i(t)\mathrm{e}^{-\beta_i(T-t)}\right)\Upsilon_i(T-t)\Theta_i(T-t)\,,$$

where

$$\ln \Upsilon_i(u) = \frac{\mu_i}{\alpha_i}(1 - \mathrm{e}^{-\alpha_i u}) + \frac{\gamma_i}{\beta_i}(1 - \mathrm{e}^{-\beta_i u}) + \frac{\sigma_i^2}{4\alpha_i}(1 - \mathrm{e}^{-2\alpha_i u})$$

and

$$\ln \Theta_i(u) = \frac{\sigma_i \theta_i}{\alpha_i}(1 - \mathrm{e}^{-\alpha_i u}) + \int_0^u \phi(\eta_1 + \mathbf{1}_{\{i=1\}}\mathrm{e}^{-\beta_1 v}, \eta_2 + \mathbf{1}_{\{i=2\}}\mathrm{e}^{-\beta_2 v}) - \phi(\eta_1, \eta_2) dv,$$

with $0 \le u \le T$.

Proof. From Lemma 4.1 and adaptedness of $X_i(t)$ and $Y_i(t)$ to $\mathcal{F}_t$ we find

$$\begin{aligned} f_i(t,T) &= \Lambda_i(T) \exp\left(X_i(t)\mathrm{e}^{-\alpha_i(T-t)} + Y_i(t)\mathrm{e}^{-\beta_i(T-t)}\right) \\ &\quad \times \exp\left(\frac{\mu_i + \sigma_i\theta_i}{\alpha_i}(1 - \mathrm{e}^{-\alpha_i u}) + \frac{\gamma_i}{\beta_i}(1 - \mathrm{e}^{-\beta_i u})\right) \\ &\quad \times \mathbb{E}_Q\left[\exp\left(\int_t^T \sigma_i \mathrm{e}^{-\alpha_i(T-v)}\, dW_i(v) + \int_t^T \mathrm{e}^{-\beta_i(T-v)}\, dL_i(v)\right) \mid \mathcal{F}_t\right]. \end{aligned}$$

Focusing on the conditional expectation, we first recall that W_i and L_i have independent increments under Q, and moreover are independent. Hence,

$$\begin{aligned} &\mathbb{E}_Q\left[\exp\left(\int_t^T \sigma_i \mathrm{e}^{-\alpha_i(T-v)}\, dW_i(v) + \int_t^T \mathrm{e}^{-\beta_i(T-v)}\, dL_i(v)\right) \mid \mathcal{F}_t\right] \\ &\quad = \mathbb{E}_{Q^B}\left[\exp\left(\int_t^T \sigma_i \mathrm{e}^{-\alpha_i(T-v)}\, dW_i(v)\right)\right] \times \mathbb{E}_{Q^L}\left[\exp\left(\int_t^T \mathrm{e}^{-\beta_i(T-v)}\, dL_i(v)\right)\right]. \end{aligned}$$

The first expectation is simple to compute after observing that the Wiener integral is normally distributed. Hence,

$$\mathbb{E}_{Q^B}\left[\exp\left(\int_t^T \sigma_i \mathrm{e}^{-\alpha_i(T-v)}\, dW_i(v)\right)\right] = \exp\left(\frac{\sigma_i^2}{4\alpha_i}(1 - \mathrm{e}^{-2\alpha_i(T-t)})\right).$$

The last expectation is computed by appealing to the measure change Q_L and the definition of the logarithm of the cumulant function

$$\begin{aligned} &\mathbb{E}_{Q^L}\left[\exp\left(\int_t^T \mathrm{e}^{-\beta_i(T-v)}\, dL_i(v)\right)\right] \\ &\quad = \mathbb{E}\left[\exp\left(\int_t^T \mathrm{e}^{-\beta_i(T-v)}\, dL_i(v) + \eta_1(L_1(T) - L_1(t)) + \eta_2(L_2(T) - L_2(t))\right)\right] \\ &\quad \times \exp\left(-\phi(\eta_1, \eta_2)(T-t)\right). \end{aligned}$$

Observe that we must require that $(\eta_1 + \exp(-\beta_1 u), \eta_2 + \exp(-\beta_2 u)) \in R$ for $0 \leq u \leq T$ for this to hold. However, the condition in the proposition is sufficient for this. Hence, the proof is complete. □

We observe that θ and η describe the risk premium in the sense of determining the market price of risk. For the sake of illustration, suppose that $\eta = (0,0)$, and we find that the risk premium becomes

$$R_i(t,T) = \Lambda_i(T) \exp\left(X_i(t)e^{-\alpha_i(T-t)} + Y_i(t)e^{-\beta_i(T-t)}\right) \Upsilon_i(T-t)$$
$$\times \left\{\exp\left(\frac{\sigma_i \theta_i}{\alpha_i}(1 - e^{-\alpha_i(T-t)})\right) - 1\right\}.$$

Hence, as σ_i, the volatility, is naturally positive, we find a positive risk premium whenever the market price of risk θ_i is positive, and vice versa. Opposite, by setting $\theta = (0,0)$, we can obtain similar conclusions on the risk premium as a function of η. We refer to Benth and Sgarra [6] for a detailed analysis of the jump market price of risk. In fact, they show that in the case of seasonally occurring spikes, one may obtain a change in the sign of the risk premium from positive to negative. We refer to [6] for details.

We specialize our discussion next to compound Poisson processes, and investigate their dependency structure in light of our measure transform. As we shall see, there is an effect on the dependency structure when using the Esscher transform, contrary to what we find for the Girsanov transform.

Let $N(t)$ be a Poisson processes with intensity $\lambda > 0$, and (J^1, J^2) a bivariate random variable with $(J_i^1, J_i^2), i = 1, \ldots$ being independent copies of it. Let $F_i(x)$ be the probability distribution function of J^i, $i = 1, 2$. Define the compound Poisson processes

$$L_1(t) = \sum_{i=1}^{N(t)} J_i^1, \tag{56}$$

$$L_2(t) = \sum_{i=1}^{N(t)} J_i^2. \tag{57}$$

Suppose that the distribution function of (J^1, J^2) is defined via a copula function C, that is,

$$F_{1,2}(x,y) = C(F_1(x), F_2(y)).$$

Simply put, a copula C is a bivariate uniform distribution function (see Nelsen [23] for an introduction to copulas). We assume furthermore that there exist densities p_1, p_2 and $c(x,y)$. As long as exponential moments exist for the jumps (J^1, J^2), we find

$$\mathbb{E}\left[e^{xJ^1+yJ^2}\right] = \int_{\mathbb{R}_0^2} e^{xz_1+yz_2} c(F_1(z_1), F_2(z_2)) p_1(z_1) p_2(z_2)\, dz_1\, dz_2\,.$$

Hence, the log-MGF of (J^1, J^2) is

$$\phi_{1,2}(x, y) \triangleq \ln \int_{\mathbb{R}^2} e^{xz_1+yz_2} c(F_1(z_1), F_2(z_2)) p_1(z_1) p_2(z_2)\, dz_1\, dz_2\,. \tag{58}$$

Since

$$\phi(x, y) = \ln \mathbb{E}\left[e^{xL_1(1)+yL_2(1)}\right] = \lambda\left(e^{\phi_{1,2}(x,y)} - 1\right),$$

we find that the Lévy measure of the bivariate compound Poisson process $(L_1(t), L_2(t))$ is

$$\ell(dz_1, dz_2) = \lambda c(F_1(z_1), F_2(z_2)) p_1(z_1) p_2(z_2)\, dz_1\, dz_2\,.$$

On the other hand, one finds that the log-MGF of $(L_1(t), L_2(t))$ under Q is

$$\phi_Q(x, y) = \lambda \int_{\mathbb{R}^2} \left(e^{xz_1+yz_2} - 1\right) e^{\eta_1 z_1+\eta_2 z_2} F_{1,2}(dz_1, dz_2)\,.$$

Now, introduce the probability densities

$$p_1^Q(x) \triangleq k_1^{-1} e^{\eta_1 x} p_1(x) \tag{59}$$

$$p_2^Q(y) \triangleq k_2^{-1} e^{\eta_2 y} p_2(y)\,, \tag{60}$$

with k_1 and k_2 being normalizing constants. Letting

$$\lambda_Q \triangleq \lambda k_1 k_2 \tag{61}$$

we find the Lévy measure of L under Q to be

$$\ell_Q(dz_1, dz_2) = \lambda_Q c(F_1(z_1), F_2(z_2)) p_1^Q(z_1) p_2^Q(z_2)\, dz_1\, dz_2\,.$$

Finally, introduce the copula density under Q as

$$c_Q(x, y) \triangleq c\left(F_1(\{F_1^Q\}^{-1}(x)), F_2(\{F_2^Q\}^{-1}(y))\right) \tag{62}$$

Then,

$$\ell_Q(dz_1, dz_2) = \lambda_Q c_Q(F_1^Q(z_1), F_2^Q(z_2)) p_1^Q(z_1) p_2^Q(z_2)\, dz_1\, dz_2\,. \tag{63}$$

From this we conclude that $(L_1(t), L_2(t))$ is a bivariate compound Poisson process under Q with jump intensity λ_Q, and (J^1, J^2) having a distribution given by

$$F_{1,2}^Q(x, y) = C_Q(F_1^Q(x), F_2^Q(y)),$$

where $C_Q(x, y)$ is the copula with density $c_Q(x, y)$.

Let us investigate a simple example: assume that J^1 and J^2 are marginally exponentially distributed with means $1/a$ and $1/b$, resp., $a, b > 0$. Then, from the analysis above we find

$$F_1^Q(x) = 1 - \mathrm{e}^{-(a-\eta_1)x}$$
$$F_2^Q(y) = 1 - \mathrm{e}^{-(b-\eta_2)y},$$

and

$$\begin{aligned} c_Q(x, y) &= c\left(1 - \exp\left(\frac{a}{a-\eta_1}\ln(1-x)\right), 1 - \exp\left(\frac{b}{b-\eta_2}\ln(1-y)\right)\right) \\ &= c\left(1-(1-x)^{a/(a-\eta_1)}, 1-(1-y)^{b/(b-\eta_2)}\right). \end{aligned}$$

Hence, the density c^Q is nothing but a nonlinear transformation of the coordinates of c. The effect the change of measure on the copula density is to move the mass of c from coordinate (x, y) to the coordinate $(1-(1-x)^{a/a-\eta_1}, 1-(1-y)^{b/b-\eta_2})$. The function $g(x) = 1 - (1 - x)^q$ for $q > 0$ is monotonely increasing with $g(0) = 0$ and $g(1) = 1$. It holds that $g'(0) = q$, while $g'(1) = 0$ as long as $q > 1$ and $g'(1) = +\infty$ for $q < 1$. Obviously, the case $q = 1$ yields the identity mapping $g(x) = x$. Since $g(x)$ is concave for $q > 1$, it holds that $q(x) \geq x$ for $x \in [0, 1]$. Hence, if $a/a - \eta_1 > 1$, then the first coordinate of the copula density c is pushed towards higher values after transforming to the pricing measure Q. For $0 < q < 1$ we find $g(x) \leq x$, and the first coordinate of the copula density c is pushed to lower values when $0 < a/a - \eta_1 < 1$. For example, if $a/a - \eta_1 > 1$ and $b/b - \eta_2 > 1$, we move the mass of c at a point (x, y) towards the point $(1, 1)$, meaning that we obtain *more* big jumps appearing together. By making η_1 and η_2 close to a and b, resp., we can obtain a concentration of extreme tail dependency in the copula density c_Q. Hence, under Q we get both an emphasize on bigger marginal jumps, but they will also appear more often together. Interpreted in a power market context, this means that positive market risk premia η_1 and η_2 will lead to more spikes, occurring more often in both markets at the same time, compared to the situation under the physical probability P. If, on the other hand, we choose η_1 and η_2 negative, we can reduce any tail dependency of c under Q, since in this case $a/a - \eta_1$ and $b/b - \eta_2$ will become less than 1, and the mass of c at (x, y) will be moved towards the origin $(0, 0)$. In a power market, this would mean that under the pricing measure, we get less occurrence of spikes, being reduced in size marginally, and at the same time the spikes in the two markets will become more decoupled.

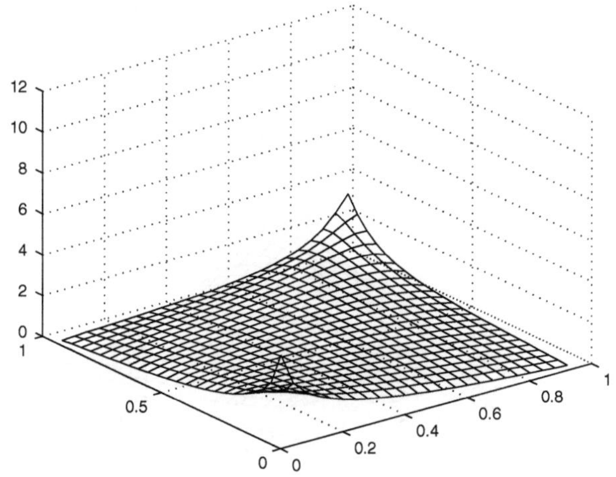

Fig. 6 The Gumbel copula density function for $\gamma = 1.5$

In order to gain further understanding, we include a numerical example using the Gumbel copula. The Gumbel copula is defined as a parametric class of copula functions given as

$$C_\gamma(x, y) = \exp\left(-\left((-\ln x)^\gamma + (-\ln y)^\gamma\right)^{1/\gamma}\right) \tag{64}$$

for $\gamma \geq 1$. The density is directly computable as

$$c_\gamma(x, y) = C_\gamma(x, y)\frac{((-\ln x)(-\ln y))^{\gamma-1}}{xy}((-\ln x)^\gamma + (-\ln y)^\gamma)^{\frac{1}{\gamma}-2} \\ \times \left\{((-\ln x)^\gamma + (-\ln y)^\gamma)^{\frac{1}{\gamma}} + (\gamma - 1)\right\} \tag{65}$$

In Fig. 6 we have plotted the Gumbel copula density for $\gamma = 1.5$. Next, we choose jump terms in S_1 and S_2 with $a = b = 0.9091$, which corresponds to an expected jump size of 1.1. Since $\exp(1.1) \approx 3.0$, we are looking at jumps which are on average scaling the price dynamics by 300 %, meaning spikes on average of the size of about 300 % price increase. For the sake of illustration, we use market prices of risk $\eta_1 = \eta_2 = 0.3$. As is clear from Fig. 7, the mass of c_Q has been transported towards $(1, 1)$, yielding more emphasis on common big jumps under Q than under P.

In the market place, there exist many swap contracts traded OTC, for example spark and dark spread swaps exchanging power with the energy equivalent of gas and coal, respectively. Furthermore, at NYMEX, one can trade in plain vanilla call and put options written on refined oil products. An analysis on the effect of Q on the dependency structure is valuable for pricing purposes.

Another interesting application of the above results is the valuation of a so-called *contracts-for-difference* (CfD) traded in the NordPool market. The CfD's are futures

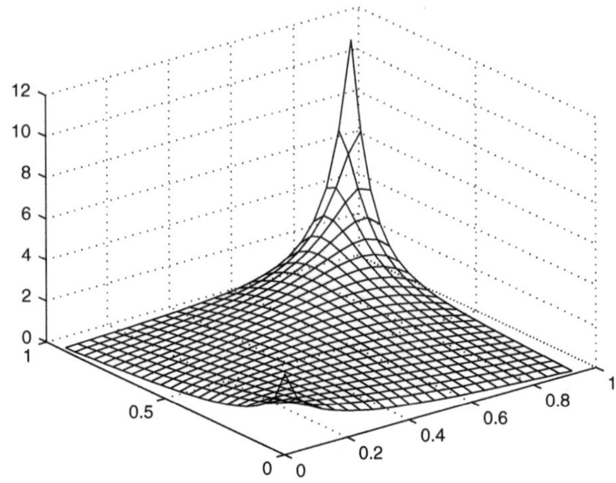

Fig. 7 The Gumbel copula density function for $\gamma = 1.5$ under the Esscher transform probability Q

contracts written on the spread between two area prices in the Nordic market. Different spot prices for different, pre-defined, areas are the result of transmission congestions in the market, where geographical separation of production and demand is resolved by differentiation in pricing. The CfD contract is a swap, where one area price is exchanged for another, yielding a payoff at time T given by $S_1(T) - S_2(T)$ for area spot prices S_1 and S_2. The CfD swap price $f_{\mathrm{CfD}}(t,T)$ at time $t \leq T$ is therefore given by

$$f_{\mathrm{CfD}}(t,T) = \mathbb{E}_Q\left[S_1(T) - S_2(T) \,|\, \mathcal{F}_t\right]$$

But from Proposition 4.9, we can easily derive an expression for this swap price dynamics. Recalling the discussion above with the Gumbel copula and exponentially distributed jumps, the effect of choosing an Esscher transform as the pricing measure Q will be more concentration of spot prices, thus reducing the swap price volatility compared with the choice $Q = P$. It is an interesting question to look at market data for CfD swap prices together with area spot prices to see how they interact.

5 Conclusions

In energy markets like gas and electricity, spot prices are typically mean reverting towards a seasonal mean. Further, spikes in prices occur as a result of sudden imbalances between supply and demand. Stochastic volatility effects like clustering are observed as well in price changes. These stylized facts on energy spot prices call for sophisticated stochastic models.

We have proposed an exponential two-factor model with stochastic volatility to model energy spot prices. The deseasonalized logarithmic spot prices are governed by an Ornstein–Uhlenbeck process reverting towards a stochastic mean level, again being an Ornstein–Uhlenbeck process. The mean level is slowly varying, whereas the short term factor can typically revert faster, and thus can potentially give bigger fluctuations in prices over short time. Both Ornstein–Uhlenbeck processes are governed by correlated Brownian motions, where the short term process is assumed to have stochastic volatility defined by the Barndorff-Nielsen and Shephard model. This will enable us to model leptokurtic residuals, which is a main characteristic in energy prices.

The proposed spot price model allow for semi-analytic forward prices. In the particular case of independent Brownian motions driving the mean level and short term dynamics, the forward prices are analytic. However, the general case requires the computation of an expectation functional in the derivation of the forward price dynamics. The main complication is the dependency on an integral of the linear combination of the volatility process and its square (the variance). However, we are able to analyse properties of the forward prices, showing among other things that the forward curves allow for humps which size depends on the state of the mean-level and/or the stochastic volatility. Humps in the forward market has been observed in energy markets, for example in oil, and we provide an explanation for this by stochastic volatility and randomness in the mean level of prices.

There are close dependencies between different energy prices. For example, the electricity markets are connected to gas and coal, as these are used as fuels in power production. Different electricity markets are connected via transmission lines. We study a simplified version of our proposed spot price model set in a bivariate market context. More specifically, the marginal demesnial logarithmic spot price dynamics is defined by a two-factor Ornstein–Uhlenbeck process driven by a Brownian motion and Lévy process. When considering two energy spot prices, we assume that both the Brownian motions and the Lévy processes are dependent. Using the Girsanov and Esscher transforms we are able to compute the price of spread options on the two energy spot prices. The price is derived based on the Fourier transform, and can be expressed in terms of the characteristics of the bivariate Lévy process. The trick in the derivation is a measure change, which reduces the bivariate option pricing problem into a problem of pricing a call option on a single underlying. Our pricing formula will extend the classical Margrabe formula for the price of a spread option on two correlated geometric Brownian motions. Sensitivity measures for the spread option is derived using a conditioning technique which exploits the independence between the Lévy processes and the Brownian motions. Spread options are traded on exchanges and bilaterally to a large extent, and their pricing and hedging is important in risk management.

The selection of the right pricing measure Q is a fundamental problem in energy markets. The challenge is to find a probability Q which are able to explain the risk premium. We focus on a specific issue related to this in the cross-commodity market setting, namely the question of the dependency risk premium. Looking at a bivariate compound Poisson process, we model the joint jump size distribution by a copula and analyse the change in characteristics of this particular Lévy process after performing an Esscher transform. It is known that marginally the Esscher transform will increase the spike intensity and size in the case of a positive market price of risk. We show that additionally, the jumps will happen more often in the two markets, that is, there will be a concentration of big jumps happening simultaneously in the two markets. In a numerical exercise, we illustrate this using the Gumbel copula, for which we observe an increase in the extremes for positive market prices of risk after doing an Esscher transform. These findings are of importance when pricing options on spreads and other cross-commodity derivatives.

The complexity of energy markets make them challenging to model. Cross-commodity models must account for both marginal price behaviour as well as dependencies between prices. The risk premium, in particular the dependency risk, is a delicate issue. Extensive data studies are called for to reveal the true nature of this risk premium. However, as models for the underlying spot are complex and the available data for spread options currently are rather limited, this remains a difficult task to solve.

References

1. O. Barndorff-Nielsen, N. Shephard, Non-Gaussian Ornstein-Uhlenbeck-based models and some of their uses in economics. J. R. Stat. Soc. B **63**(2), 167–241 (2001) (with discussion)
2. O.E. Barndorff-Nielsen, F.E. Benth, A. Veraart, Modelling energy spot prices by Lévy semistationary processes. Bernoulli (2010) (to appear)
3. F.E. Benth, The stochastic volatility model of Barndorff-Nielsen and Shephard in commodity markets. Math. Financ. **21**(4), 595–625 (2011)
4. F.E. Benth, S. Koekebakker, Stochastic modeling of financial electricity contracts. Energy Econ. **30**(3), 1116–1157 (2008)
5. F.E. Benth, T. Meyer-Brandis, The information premium for non-storable commodities. J. Energy Mark. **2**(3), 111–140 (2009)
6. F.E. Benth, C. Sgarra, The risk premium and the Esscher transform in power markets. Stoch. Anal. Appl. **30**, 20–43 (2012)
7. F.E. Benth, J. Šaltytė Benth, J. Koekebakker, *Stochastic Modelling of Electricity and Related Markets* (World Scientific, Singapore, 2008)
8. F.E. Benth, A. Cartea, R. Kiesel, Pricing forward contracts in power markets by the certainty equivalence principle: Explaining the sign of the market risk premium. J. Bank. Financ. **32**(10), 2006–2021 (2008)
9. F.E. Benth, G. Di Nunno, A. Khedher, Lévy model robustness and sensitivity, in *QP-PQ: Quantum Probability and White Noise Analysis*, vol. 25, ed. by H. Ouerdiane, A. Barhoumi. Proceedings of the 29th Conference in Hammamet, Tunisia, 13–18 October 2008 (World Scientific, Singapore, 2010), pp. 153–184
10. F.E. Benth, C. Klüppelberg, G. Müller, L. Vos, Futures pricing in electricity markets based on stable CARMA spot models. (2011) (submitted)
11. F.E. Benth, J. Lempa, T.K. Nilsen, On optimal exercise of swing options in electricity markets. J. Energy Mark. **4**(4), 3–28 (2012)

12. F.E. Benth, G. Di Nunno, A. Khedher, Computations of Greeks in multi-factor models with applications to power and commodity markets. J. Energy Mark. **5**(4), 3–31 (2013)
13. D. Brigo, F. Mercurio, *Interest Rate Models – Theory and Practice* (Springer, Berlin, 2001)
14. R. Carmona, V. Durrleman, Pricing and hedging spread options. SIAM Rev. **45**, 627–685 (2003)
15. A. Eydeland, K. Wolyniec, *Energy and Power Risk Management* (Wiley, New York, 2003)
16. G.B. Folland, *Real Analysis – Modern Techniques and their Applications* (Wiley, New York, 1984)
17. H. Geman, *Commodities and Commodity Derivatives* (Wiley-Finance, Chichester, 2005)
18. D. Heath, R. Jarrow, A. Morton, Bond pricing and the term structure of interest rates: A new methodology. Econometrica **60**, 77–105 (1992)
19. S. Hikspoors, S. Jaimungal, Asymptotic pricing of commodity derivatives for stochastic volatility spot models. Appl. Math. Financ. **15**(5, 6), 449–467 (2008)
20. N. Ikeda, S. Watanabe, *Stochastic Differential Equations and Diffusion Processes* (North-Holland/Kodansha, 1981)
21. J. Lucia, E.S. Schwartz, Electricity prices and power derivatives: Evidence from the Nordic power exchange. Rev. Deriv. Res. **5**(1), 5–50 (2002)
22. W. Margrabe, The value of an option to exchange one asset for another. J. Financ. **33**, 177–186 (1978)
23. R.B. Nelsen, *An Introduction to Copulas*, 2nd edn. (Springer, Berlin, 2010)
24. K. Sato, *Lévy Processes and Infinite Divisibility* (Cambridge University Press, Cambridge, 1999)
25. E.S. Schwartz, The stochastic behaviour of commodity prices: Implications for valuation and hedging. J. Financ. **LII**(3), 923–973 (1997)
26. E.S. Schwartz, J.E. Smith, Short-term variations and long-term dynamics in commodity prices. Manag. Sci. **46**(7), 893–911 (2000)
27. A.B. Trolle, E.S. Schwartz, Unspanned stochastic volatility and the pricing of commodity derivatives. Rev. Financ. Stud. **22**(11), 4423–4461 (2009)

Portfolio Choice with Transaction Costs: A User's Guide*

Paolo Guasoni and Johannes Muhle-Karbe

Abstract Recent progress in portfolio choice has made a wide class of problems involving transaction costs tractable. We review the basic approach to these problems, and outline some directions for future research.

1 Introduction

Transaction costs, originally considered one of many imperfections that are best neglected, have now become a very active and fast-growing theme in Mathematical

*The authors are grateful to Vicky Henderson, Ronnie Sircar, and Maxim Bichuch for useful comments, to Jens Jackwerth for additional references, and to Ren Liu for proofreading the manuscript.

P. Guasoni (✉)
Department of Mathematics and Statistics, Boston University, 111 Cummington Street Boston, MA 02215, USA

School of Mathematical Sciences, Dublin City University, Glasnevin, Dublin 9, Ireland
e-mail: guasoni@bu.edu

Partially supported by the ERC (278295), NSF (DMS-0807994, DMS-1109047), SFI (07/MI/008, 07/SK/M1189, 08/SRC/FMC1389), and FP7 (RG-248896)

J. Muhle-Karbe
ETH Zürich, Departement Mathematik, Rämistrasse 101, CH-8092, Zürich, Switzerland, and Swiss Finance Institute
e-mail: johannes.muhle-karbe@math.ethz.ch

Partially supported by the National Centre of Competence in Research "Financial Valuation and Risk Management" (NCCR FINRISK), Project D1 (Mathematical Methods in Financial Risk Management), of the Swiss National Science Foundation (SNF).

F.E. Benth et al., *Paris-Princeton Lectures on Mathematical Finance 2013*, Lecture Notes in Mathematics 2081, DOI 10.1007/978-3-319-00413-6_3,

Finance.[1] From the outset, such a growth may seem puzzling, since over the same period transaction costs have dramatically declined across financial markets, as stock exchanges have been fully automated, and paper trades replaced by electronic settlements. In fact, the interest for transaction costs reflects both the increased attention to robustness of financial models, and the growing role of high-frequency trading. The decline of bid-ask spreads has sparked a huge increase in trading volume, and high-volume strategies require a careful understanding of the effects of frictions on their returns.

At the same time, transaction costs help understand trading volume itself. In frictionless models, investors continuously rebalance their portfolios, as to hold a constant mix of assets over time. Since trading volume is proportional to the total variation of a portfolio, and prices follow diffusions that have infinite variation, such models lead to the absurd conclusion that trading volume is infinite over any time interval. With transaction costs, even small trading costs make it optimal for investors to trade infrequently, allowing wide oscillations in their portfolios.

This paper reviews a recent approach, which has made portfolio choice with transaction costs more tractable, and which appears to be applicable in more complex settings. This approach is not based on any new revolutionary concept, but it rather tries to combine several ideas that were previously used in isolation. Thus, we present a new toolbox that contains several used tools. In a nutshell, we argue that a natural approach to portfolio choice problems with transaction costs entails four steps: (a) heuristic control arguments to identify the long-run value function, (b) construction of a candidate shadow price using marginal rates of substitution, (c) verification and finite-horizon bounds using the myopic probability, and (d) asymptotic results from the implicit function theorem.

The advantages of this approach are threefold. First, it combines the dimension-reduction and higher tractability of the long-horizon problem with exact finite-horizon bounds, which keep a firm grip on the robustness of the solution. Second, we show that the free-boundaries arising with transaction costs can sometimes be identified explicitly in terms of a single parameter, the *equivalent safe rate*, which remains the only non-explicit part of the solution. This reduction is useful both for theoretical and for practical purposes, as it helps to simplify proofs as well as calculations. Third, this approach leads to the simultaneous computation of several related quantities, such as welfare, portfolios, liquidity premia, and trading volume.

The paper proceeds as follows: in the next section, we present a brief timeline of related research, which is far from exhaustive, and only aims at putting the paper in context. The following section introduces the main problem, discussing the relative advantages of the three main models with terminal wealth, consumption, and long-horizon. This section also discusses the typical heuristic arguments of stochastic control that lead to an educated guess for the value function, and the identification

[1]The Mathscinet database shows only nine publications with "transaction costs" in their title in the eighties (1980–1989). This figure rises to 52 in the nineties (1990–1999), and to 278 in the naughties (2000–2009).

of the corresponding free boundaries. For the long-run problem, the following section shows the passage from the heuristic calculations to a verification, which relies on two central ideas. The first one is the construction of a *shadow* market, an imaginary frictionless market, built to deliver the same optimal strategy as the original market with transaction cost. This shadow market harnesses transaction costs by hiding them inside a more complex model, without transaction costs, but in which investment opportunities are driven by a state variable that represents the portfolio composition of the investor. This insight—that transaction costs are essentially equivalent to state-dependent investment opportunities—in turn allows to exploit the approach to verification based on the change of measure to the myopic probability.

We conclude with a deliberately speculative section on three open problems: multiple assets, return predictability, and option spreads. We argue that with transaction costs, multivariate models present both a substantial technical challenge, and a potentially fertile ground for novel financial insights, which may alter the conventional wisdom on fund separation. Likewise, transaction costs may help reconcile statistical evidence on return predictability with the poor out-of-sample performance of market-timing strategies. Finally, the large bid-ask spreads observed in options on highly liquid assets still lack a theoretical basis, and transaction costs are a natural avenue to search for an explanation.

2 Literature Review

Portfolio choice with transaction cost starts with the seminal papers of [8,35], and [17], in the wake of the frictionless results of [36, 37]. From heuristic arguments, these early studies gleaned central insights that held up to subsequent formal proofs. First, optimal portfolios entail a no-trade region, in which it is optimal to keep existing holdings in all assets. Optimal portfolios always remain within this region, and hence trading should merely take place at its boundaries. The no-trade region is wide, even for small transaction costs, implying that investors should accept wide fluctuations around the frictionless target. Second, the large no-trade region has a small welfare impact [8], because the displacement loss is small near the frictionless optimum, and the wide no-trade region minimizes the effect of transaction costs.

On the mathematical side, Taksar et al. [44] reduce the maximization of logarithmic utility from terminal wealth at a long horizon to the solution of a nonlinear second-order ODE with free boundaries, to be determined numerically. Davis and Norman [15] accomplish this feat for power utility from consumption with infinite horizon. Shreve and Soner [40] extend their analysis with viscosity techniques, removing some parametric restrictions. Shreve and Soner [40] and Rogers [39] study the size of the no-trade interval and the utility loss due to transaction costs. They argue that these are of order $O(\varepsilon^{1/3})$ and $O(\varepsilon^{2/3})$, respectively, where ε is the proportional cost, in line with the numerical results of [8] alluded to above. Building on earlier heuristic results of [24, 46] explicitly determine the coefficients of the leading-order corrections around the frictionless case $\varepsilon = 0$. In a general Markovian setting and for arbitrary utility functions, Soner and Touzi [42] characterize the corresponding quantities in terms of an ergodic control problem. All these papers employ stochastic control as their main tool.

A new strand of literature, which finds its roots in the seminal work of [14, 25], seeks to bring martingale methods, now well-understood in frictionless markets, to bear on transaction costs. This idea has already shown its promise in the context of superreplication: Guasoni et al. [22] prove the face-lifting theorem of [43] for general continuous processes, using an argument based on shadow prices. Kallsen and Muhle-Karbe [26] explore this approach for optimal consumption from logarithmic utility, showing how shadow prices simplify verification theorems. Gerhold et al. [19] exploit this idea to obtain the expansions of [24] for logarithmic utility, but with an arbitrary number of terms. The present study reviews the approach put forward by [18], who prove a verification theorem and derive full asymptotics for the optimal policy, welfare, and implied trading volume in the long-run model of [17]. The duality-based verification is based on applying the frictionless long-run machinery of [21] to a fictitious shadow price, traded without transaction costs. Compared to [18], finding a candidate shadow price is greatly simplified by applying a observation originally made by [34]: Given a smooth candidate value function, it can simply be obtained via the marginal rate of substitution of risky for safe assets for the frictional investor. (Also cf. [23, 33] for applications of this idea to related problems.)

3 The Basic Model

3.1 Objectives

Let X_t^π denote the wealth of an investor who follows the portfolio π_t, and let c_t his consumption rate, both at time t. The three typical objectives for portfolio choice with power utility $U(x) = x^{1-\gamma}/(1-\gamma)$ are:

$$\max_\pi E\left[\frac{(X_T^\pi)^{1-\gamma}}{1-\gamma}\right], \qquad \text{(terminal wealth)} \qquad (1)$$

$$\max_{\pi,c} E\left[\int_0^\infty e^{-\delta t}\frac{c_t^{1-\gamma}}{1-\gamma}dt\right], \qquad \text{(consumption)} \qquad (2)$$

$$\max_\pi \liminf_{T\to\infty}\frac{1}{T}\log E\left[(X_T^\pi)^{1-\gamma}\right]^{\frac{1}{1-\gamma}}. \qquad \text{(long run)} \qquad (3)$$

Expected utility from terminal wealth (1) has attracted the attention of most of the semimartingale literature (see, for example, [28] and the references therein). This objective is the simplest for abstract questions, such as existence, uniqueness, well-posedness, and stability, which are by now largely understood. It is also relevant for problems such as retirement planning, which entail a known horizon and no intermediate consumption. Expected utility from intertemporal consumption (2) is more appealing for applications to macroeconomics, because it yields an endogenous consumption process c_t, and therefore has testable implications for consumption data.

The long run objective is probably the least intuitive, in view of the limit in (3). To understand its economic interpretation, note first that, for a fixed horizon T, the quantity $E\left[(X_T^\pi)^{1-\gamma}\right]^{\frac{1}{1-\gamma}}$ coincides with the *certainty equivalent* $U^{-1}(E\left[U(X_T^\pi)\right])$ of the payoff X_T^π. If we match this certainty equivalent with $xe^{\rho_T T}$, that is, the investor's initial capital x compounded at some constant rate ρ_T for the same horizon T, we recognize that:

$$\rho_T = \frac{1}{T}\log E\left[(X_T^\pi)^{1-\gamma}\right]^{\frac{1}{1-\gamma}}.$$

Thus, the limit in (3) has the interpretation of an *equivalent safe rate*, that is, the hypothetical safe rate that would make the investor indifferent between investing optimally in the market, and leaving all wealth invested at this hypothetical rate.

Both the consumption and long-run problems are *stationary* objectives, in that they lead to time-independent solutions (as long as investment opportunities are also stationary). Of course, the advantage of stationary problems is that the resulting optimization problems have one less dimension than similar nonstationary problems, such as utility maximization from terminal wealth. Both objectives model an investor with an infinite horizon, but with some important differences. First, the consumption objective involves the additional time-preference rate δ, which does not appear in the long-run objective. Second, in typical models (even in a Black–Scholes market, compare [6]), the consumption objective may not be well posed if risk aversion γ is less than the logarithmic value of one, and investment opportunities are sufficiently attractive. By contrast, the long-run objective is typically well-posed under more general conditions.

The irony of portfolio choice is that its most natural objectives are also the least tractable: the terminal-wealth problem admits closed-form solutions only in rare cases (cf., e.g., [31]). Even when such solutions exist, they are often too clumsy to yield clear insights on the role of preference and market parameters. Unfortunately, the consumption objective admits explicit solutions primarily in complete markets, or with investment opportunities independent of asset prices, a fact that severely limits our understanding of the effects of partial return predictability on consumption.

The good news is that the long-run problem admits explicit solutions in many situations in which the other two problems do not, its optimal portfolio is almost optimal even for the other objectives, and bounds on the resulting utility loss are available. This general insight is crucial in markets with frictions, such as transaction costs.

3.2 Control Heuristics

We now examine the differences in the Hamilton–Jacobi–Bellman equations arising from the three objectives (1)–(3), in the basic model with one safe asset growing at the riskless rate $r \geq 0$, and a risky asset with ask (buying) price S_t following geometric Brownian Motion:

$$\frac{dS_t}{S_t} = (\mu + r)dt + \sigma dW_t, \quad \mu, \sigma > 0. \tag{4}$$

The bid (selling) price is $(1-\varepsilon)S_t$, where $\varepsilon \in (0,1)$ is the relative bid-ask spread.

Denote the number of units of the safe asset by φ_t^0 and write the number of units of the risky asset $\varphi_t = \varphi_t^{\uparrow} - \varphi_t^{\downarrow}$ as the difference between cumulative purchases and sales. The values of the safe position X_t and of the risky position Y_t (quoted at the ask price) evolve as:

$$dX_t = rX_t dt - S_t d\varphi_t^{\uparrow} + (1-\varepsilon)S_t d\varphi_t^{\downarrow}, \tag{5}$$

$$dY_t = (\mu + r)Y_t dt + \sigma Y_t dW_t + S_t d\varphi_t^{\uparrow} - S_t d\varphi_t^{\downarrow}. \tag{6}$$

The second equation prescribes that risky wealth earns the return on the risky asset, plus units purchased, and minus units sold. In the first equation the safe position earns the safe rate, minus the units used for purchases (at the ask price S_t), and plus the units used for sales (at the bid price $(1-\varepsilon)S_t$).

For the maximization of utility from terminal wealth, denote the value function as $V(t, x, y)$, which depends on time t, on the safe position x, and on the risky position y. Itô's formula yields:

$$dV(t, X_t, Y_t) = V_t dt + V_x dX_t + V_y dY_t + \frac{1}{2} V_{yy} d\langle Y, Y\rangle_t \tag{7}$$

$$= \left(V_t + rX_t V_x + (\mu + r)Y_t V_y + \frac{\sigma^2}{2} Y_t^2 V_{yy}\right) dt \tag{8}$$

$$+ S_t(V_y - V_x)d\varphi_t^{\uparrow} + S_t((1-\varepsilon)V_x - V_y)d\varphi_t^{\downarrow} + \sigma X_t V_y dW_t, \tag{9}$$

By the martingale optimality principle of stochastic control, the value function $V(t, X_t, Y_t)$ must be a supermartingale for any choice of purchases and sales $\varphi_t^{\uparrow}, \varphi_t^{\downarrow}$. Since these are increasing processes, this implies $V_y - V_x \leq 0$ and $(1-\varepsilon)V_x - V_y \leq 0$, which means that

$$1 \leq \frac{V_x}{V_y} \leq \frac{1}{1-\varepsilon}. \tag{10}$$

In the interior of this "no-trade region", where the number $\varphi_t = \varphi_t^{\uparrow} - \varphi_t^{\downarrow}$ of risky shares remains constant, the drift of $V(t, X_t, Y_t)$ cannot be positive, and must become zero for the optimal policy. This leads to the HJB equation:

$$V_t + rX_t V_x + (\mu + r)Y_t V_y + \frac{\sigma^2}{2} Y_t^2 V_{yy} = 0 \quad \text{if} \quad 1 < \frac{V_x}{V_y} < \frac{1}{1-\varepsilon}. \tag{11}$$

Next, the value function is homogeneous in wealth, i.e. $V(t, X_t, Y_t) = (X_t)^{1-\gamma} v(t, Y_t/X_t)$, whence setting $z = y/x$:

$$\frac{\sigma^2}{2} z^2 v_{zz} + \mu z v_z + r(1-\gamma)v + v_t = 0 \quad \text{if} \quad 1 + z < \frac{(1-\gamma)v(t,z)}{v_z(t,z)} < \frac{1}{1-\varepsilon} + z. \tag{12}$$

Now, suppose that the no-trade region $\{(t,z) : 1 + z \leq \frac{(1-\gamma)v(t,z)}{v_z(t,z)} \leq \frac{1}{1-\varepsilon} + z\}$ coincides with some interval $l(t) \leq z \leq u(t)$ to be found. At $l(t)$ the left inequality in (12) holds as equality, while at $u(t)$ the right inequality holds as equality, leading to the boundary conditions:

$$(1 + l)v_z(t,l) - (1-\gamma)v(t,l) = 0, \tag{13}$$

$$(1/(1-\varepsilon) + u)v_z(t,u) - (1-\gamma)v(t,u) = 0. \tag{14}$$

These conditions are not sufficient to identify the solution to the optimization problem, since they can be matched for any trading boundary $l(t), u(t)$. The optimal boundaries are identified as the ones that satisfy the smooth-pasting conditions. These conditions can be seen as limits of the optimality conditions for an impulse control problem with a infinitesimally small cost [16]. In practice, they are derived by differentiating (13) and (14) with respect to z at the respective boundaries $z = l$ and $z = u$:

$$(1 + l)v_{zz}(t,l) + \gamma v_z(t,l) = 0, \tag{15}$$

$$(1/(1-\varepsilon) + u)v_{zz}(t,u) + \gamma v_z(t,u) = 0. \tag{16}$$

This system defines a *two-dimensional, linear* free-boundary problem in $(t,z) \in [0,T] \times \mathbb{R}$, which is not tractable in general. Liu and Loewenstein [32] obtain a semiexplicit solution with the randomization approach used by [5] to price American options.

With utility maximization from infinite-horizon consumption, the value function depends only on the safe and risky positions X_t, Y_t—the problem is stationary. Calculations are similar, with some minor differences: first, the self-financing condition (6) must include the term $-c_t dt$ in the cash balance, to account for consumption expenditures. Then, the martingale optimality principle takes the following slightly different form. Since utility is not only incurred at maturity but from consumption along the way, not the value function itself but the sum of past consumption

$\int_0^t e^{-\delta u} c_u^{1-\gamma}/(1-\gamma)du$ and the value function $e^{-\delta t}V(X_t, Y_t)$, representing future consumption, should be a supermartingale for any policy and a martingale for the optimizer. Here, δ is the time-preference parameter in (2). Then, the term V_t in (8) and in turn (11) has to be replaced by $\frac{c_t^{1-\gamma}}{1-\gamma} - c_t V_x - \delta V$. Pointwise maximization yields the optimal consumption rate $c_t = V_x^{-1/\gamma}$. Plugging this expression back into the HJB equation and accounting for homogeneity in wealth, the corresponding free-boundary problem for the reduced value function $v(z)$ then reads as:

$$\frac{\sigma^2}{2} z^2 v_{zz} + \mu z v_z + ((1-\gamma)r - \delta)v + \frac{\gamma}{1-\gamma}((1-\gamma)v - z v_z)^{1-\frac{1}{\gamma}}$$

$$= 0 \quad \text{if} \quad 1 + z < \frac{(1-\gamma)v(z)}{v_z(z)} < \frac{1}{1-\varepsilon} + z. \tag{17}$$

This is the *one-dimensional, nonlinear* free-boundary problem studied by [15], who prove a verification theorem, and find a numerical solution. Still, this problem is nontrivial, because the free boundaries points l, u are not easy to identify in terms of the model parameters, and the second order, nonlinear equation (17) does not admit a known explicit solution for given initial conditions.

The long-run problem (3) gives the best of both worlds, and more. But it requires more audacious heuristics, and nonstandard arguments to be made precise. Puzzlingly enough, this approach was proposed very early in the transaction costs literature by [17, 44], but its potential has not become clear until recently.

We start from (11), derived for a fixed horizon T, and note that the value function V should grow exponentially with the horizon. This observation, combined with homogeneity in wealth, leads to guess a solution of the form $V(t, X_t, Y_t) = (X_t)^{1-\gamma} v(Y_t/X_t) e^{-(1-\gamma)(r+\beta)t}$. It is clear that such a guess in general does not solve the finite-horizon problem, as it fails to satisfy its terminal condition. But it is reasonable to expect that it governs the long-run problem, for which the horizon never approaches. With the above guess, the HJB equation reduces to

$$\frac{\sigma^2}{2} z^2 v''(z) + \mu z v'(z) - (1-\gamma)\beta v(z) = 0 \quad \text{if} \quad 1 + z < \frac{(1-\gamma)v(z)}{v'(z)} < \frac{1}{1-\varepsilon} + z, \tag{18}$$

which is a *one-dimensional, linear* free-boundary problem—the best of both worlds. Note that in this system β is not exogenous, but an unknown parameter that determines the growth rate of the value function, and which has to be found along with the free boundaries l, u. Indeed, $r + \beta$ is the *equivalent safe rate* which makes a long-term investor indifferent between the original market and this alternative rate alone.

The two crucial advantages of the long-run problem are that the free boundaries l, u have explicit formulas in terms of β, and that it reduces to solving a Cauchy problem for a *first-order* ordinary differential equation. Depending on the problem at hand, this equation may even have an explicit solution, a fact that is useful although not essential for asymptotics. To derive the free boundary l, substitute first (15) and then (13) into (18) to obtain

$$-\frac{\sigma^2}{2}(1-\gamma)\gamma\frac{l^2}{(1+l)^2}v+\mu(1-\gamma)\frac{l}{1+l}v-(1-\gamma)\beta v=0.$$

Now, observe that $\pi_- = l/(1+l)$ is precisely the risky portfolio weight at the buy boundary, evaluated at the ask price. Factoring out $(1-\gamma)v$, it follows that (cf. [17])

$$-\frac{\gamma\sigma^2}{2}\pi_-^2+\mu\pi_- - \beta = 0. \tag{19}$$

Likewise, a similar calculation for u shows that the other root of (19) is $\pi_+ = u(1-\varepsilon)/(1+u(1-\varepsilon))$, which coincides with the risky portfolio weight at the sell boundary, evaluated at the bid price.

After these calculations, the boundaries l, u, or equivalently π_-, π_+, are uniquely identified as solutions of the above equation, once the parameter β is found:

$$\pi_\pm = \frac{\mu \pm \sqrt{\mu^2-2\gamma\sigma^2\beta}}{\gamma\sigma^2}.$$

The formulas become even clearer by replacing the parameter β with $\lambda = \sqrt{\mu^2-2\gamma\sigma^2\beta}$. With this notation, in which $\lambda = 0$ corresponds to the frictionless setting, $\beta = (\mu^2-\lambda^2)/2\gamma\sigma^2$ and the buy and sell boundaries have the intuitive representation

$$\pi_\pm = \frac{\mu\pm\lambda}{\gamma\sigma^2}, \tag{20}$$

from which $l = \frac{\pi_-}{1-\pi_-}$ and $u = \frac{1}{1-\varepsilon}\frac{\pi_+}{1-\pi_+}$ are obtained directly. Thus, it remains to find λ to identify both the free-boundaries and the equivalent safe rate $r+\beta$. To this end, it is convenient to apply the substitution

$$v(z) = e^{(1-\gamma)\int_0^{\log(z/l(\lambda))} w(y)dy}, \quad \text{i.e.,} \quad w(y) = \frac{l(\lambda)e^y v'(l(\lambda)e^y)}{(1-\gamma)v(l(\lambda)e^y)},$$

which reduces the free-boundary problem to a Cauchy problem with a terminal condition:

$$w'(y)+(1-\gamma)w(y)^2+\left(\frac{2\mu}{\sigma^2}-1\right)w(y)-\gamma\left(\frac{\mu-\lambda}{\gamma\sigma^2}\right)\left(\frac{\mu+\lambda}{\gamma\sigma^2}\right)=0,$$

$$y \in [0, \log u(\lambda)/l(\lambda)], \tag{21}$$

$$w(0) = \frac{\mu-\lambda}{\gamma\sigma^2}, \tag{22}$$

$$w(\log(u(\lambda)/l(\lambda))) = \frac{\mu+\lambda}{\gamma\sigma^2}. \tag{23}$$

In other words, the correct value of λ is identified as the one for which the above first-order Riccati equation satisfies both the initial and the terminal value conditions. For a fixed ε, such a value is the solution of a scalar equation obtained from the explicit solution of the Riccati equation (cf. Lemma 3.1). For $\varepsilon \sim 0$, the asymptotic expansion of $\lambda(\varepsilon)$ follows from the implicit function theorem—and some patient calculations (see Lemma 3.2 below).

Now, one could argue that the advantage of the nonlinear, first-order equation (21) over the linear, second-order equation (18) is only marginal. In fact, the variable w has the additional advantage, albeit still hidden at this point, that it coincides with the optimal *shadow* risky portfolio weight (cf. Lemma 4.4), a fact that is hinted at by its boundary conditions. Furthermore, and as a result, for $\gamma = 1$ (21) recovers the case of logarithmic utility [19], while (18) does not.

3.3 Explicit Formulas

Let us now show that the reduced value function w and the quantity λ are indeed well-defined. To this end, first determine, for a given small $\lambda > 0$, an explicit expression for the solution w of the ODE (21), complemented by the initial condition (22).

Lemma 3.1. *Let $0 < \mu/\gamma\sigma^2 \neq 1$. Then for sufficiently small $\lambda > 0$, the function*

$$w(\lambda, y) = \begin{cases} \frac{a(\lambda)\tanh[\tanh^{-1}(b(\lambda)/a(\lambda)) - a(\lambda)y] + (\frac{\mu}{\sigma^2} - \frac{1}{2})}{\gamma - 1}, & \text{if } \gamma \in (0,1) \text{ and } \frac{\mu}{\gamma\sigma^2} < 1 \text{ or } \gamma > 1 \text{ and } \frac{\mu}{\gamma\sigma^2} > 1, \\ \frac{a(\lambda)\tan[\tan^{-1}(b(\lambda)/a(\lambda)) + a(\lambda)y] + (\frac{\mu}{\sigma^2} - \frac{1}{2})}{\gamma - 1}, & \text{if } \gamma > 1 \text{ and } \frac{\mu}{\gamma\sigma^2} \in \left(\frac{1}{2} - \frac{1}{2}\sqrt{1 - \frac{1}{\gamma}}, \frac{1}{2} + \frac{1}{2}\sqrt{1 - \frac{1}{\gamma}}\right), \\ \frac{a(\lambda)\coth[\coth^{-1}(b(\lambda)/a(\lambda)) - a(\lambda)y] + (\frac{\mu}{\sigma^2} - \frac{1}{2})}{\gamma - 1}, & \text{otherwise}, \end{cases}$$

with

$$a(\lambda) = \sqrt{\left|(\gamma - 1)\frac{\mu^2 - \lambda^2}{\gamma\sigma^4} - \left(\frac{1}{2} - \frac{\mu}{\sigma^2}\right)^2\right|} \quad \text{and} \quad b(\lambda) = \frac{1}{2} - \frac{\mu}{\sigma^2} + (\gamma - 1)\frac{\mu - \lambda}{\gamma\sigma^2},$$

is a local solution of

$$w'(y) + (1-\gamma)w^2(y) + \left(\frac{2\mu}{\sigma^2} - 1\right)w(y) - \frac{\mu^2 - \lambda^2}{\gamma\sigma^4} = 0, \quad w(0) = \frac{\mu - \lambda}{\gamma\sigma^2}. \qquad (24)$$

Moreover, $y \mapsto w(\lambda, y)$ is increasing (resp. decreasing) for $\mu/\gamma\sigma^2 \in (0, 1)$ (resp. $\mu/\gamma\sigma^2 > 1$).

Proof. The first part of the assertion is easily verified by taking derivatives. The second follows by inspection of the explicit formulas. □

Next, establish that the crucial constant λ, which determines both the no-trade region and the equivalent safe rate, is well-defined. For small transaction costs $\varepsilon \sim 0$, its asymptotics are readily computed by means of the implicit function theorem.

Lemma 3.2. *Let $0 < \mu/\gamma\sigma^2 \neq 1$ and $w(\lambda,\cdot)$ be defined as in Lemma 3.1, and set*

$$l(\lambda) = \frac{\mu-\lambda}{\gamma\sigma^2-(\mu-\lambda)}, \quad u(\lambda) = \frac{1}{(1-\varepsilon)}\frac{\mu+\lambda}{\gamma\sigma^2-(\mu+\lambda)}.$$

Then, for sufficiently small $\varepsilon > 0$, there exists a unique solution λ of

$$w\left(\lambda, \log\left(\frac{u(\lambda)}{l(\lambda)}\right)\right) - \frac{\mu+\lambda}{\gamma\sigma^2} = 0. \tag{25}$$

As $\varepsilon \downarrow 0$, it has the asymptotics

$$\lambda = \gamma\sigma^2\left(\frac{3}{4\gamma}\left(\frac{\mu}{\gamma\sigma^2}\right)^2\left(1-\frac{\mu}{\gamma\sigma^2}\right)^2\right)^{1/3}\varepsilon^{1/3} + O(\varepsilon).$$

Proof. Write the boundary condition (25) as $f(\lambda,\varepsilon) = 0$, where:

$$f(\lambda,\varepsilon) = w(\lambda, \log(u(\lambda)/l(\lambda))) - \frac{\mu+\lambda}{\gamma\sigma^2}.$$

Of course, $f(0,0) = 0$ corresponds to the frictionless case. The implicit function theorem then suggests that for sufficiently small ε there exists a unique zero $\lambda(\varepsilon)$ with the asymptotics $\lambda(\varepsilon) \sim -\varepsilon f_\varepsilon/f_\lambda$, but the difficulty is that $f_\lambda = 0$, because λ is not of order ε. Heuristic arguments [39, 40] suggest that λ is of order $\varepsilon^{1/3}$. Thus, setting $\lambda = \delta^{1/3}$ and $\hat{f}(\delta,\varepsilon) = f(\delta^{1/3},\varepsilon)$, and computing the derivatives of the explicit formula for $w(\lambda,x)$ (cf. Lemma 3.1) shows that:

$$\hat{f}_\varepsilon(0,0) = -\frac{\mu\left(\mu-\gamma\sigma^2\right)}{\gamma^2\sigma^4}, \qquad \hat{f}_\delta(0,0) = \frac{4}{3\mu^2\sigma^2-3\gamma\mu\sigma^4}.$$

As a result:

$$\delta(\varepsilon) \sim -\frac{\hat{f}_\varepsilon(0,0)}{\hat{f}_\delta(0,0)}\varepsilon = \frac{3\mu^2\left(\mu-\gamma\sigma^2\right)^2}{4\gamma^2\sigma^2}\varepsilon \quad \text{whence}$$

$$\lambda(\varepsilon) \sim \left(\frac{3\mu^2\left(\mu-\gamma\sigma^2\right)^2}{4\gamma^2\sigma^2}\right)^{1/3}\varepsilon^{1/3}.$$

□

Henceforth, consider small transaction costs $\varepsilon > 0$, and let λ denote the constant in Lemma 3.2. Moreover, set $w(y) = w(\lambda, y)$, $a = a(\lambda)$, $b = b(\lambda)$, and $u = u(\lambda)$, $l = l(\lambda)$. In all cases, the function w can be extended smoothly to an open neighborhood of $[0, \log(u/l)]$ (resp. $[\log(u/l), 0]$ if $\mu/\gamma\sigma^2 > 1$). By continuity, the ODE (24) then also holds at 0 and $\log(u/l)$; inserting the boundary conditions for w yields the following counterparts for the derivative w':

Lemma 3.3. *Let $0 < \mu/\gamma\sigma^2 \neq 1$. Then, in all three cases,*

$$w'(0) = \frac{\mu - \lambda}{\gamma\sigma^2} - \left(\frac{\mu - \lambda}{\gamma\sigma^2}\right)^2, \quad w'\left(\log\left(\frac{u}{l}\right)\right) = \frac{\mu + \lambda}{\gamma\sigma^2} - \left(\frac{\mu + \lambda}{\gamma\sigma^2}\right)^2.$$

3.4 Discussion

The above heuristics offer a practical approach to portfolio choice problems with transaction costs, and can be adapted to accommodate additional model features. More importantly, they yield results that are robust to the model specification. In view of (20) and the asymptotics for λ in Lemma 3.2, the no-trade boundaries have the expansion:

$$\pi_\pm = \frac{\mu}{\gamma\sigma^2} \pm \left(\frac{3}{4\gamma}\left(\frac{\mu}{\gamma\sigma^2}\right)^2\left(1 - \frac{\mu}{\gamma\sigma^2}\right)^2\right)^{1/3} \varepsilon^{1/3} + O(\varepsilon). \tag{26}$$

This expansion coincides with the one obtained by [24] in the model with consumption. In other words, the long-run and the consumption models yield exactly the same solution at the leading order for small transaction costs. The expansions do differ at the second order, but such differences tend to have a modest effect for typical parameter values.

A major advantage of the long-run objective is the possibility to reduce the solution to a single algebraic equation for the parameter λ, in terms of which the free boundaries are found explicitly. In principle, one could attempt the same reduction in the consumption problem, by substituting equations (13) and (15) into (12). The result is a scalar equation for l in terms of $v(l)$, the value of the reduced value function at the trading boundary. Alas, the equation does not have an explicit solution. Also, the value of $v(l)$ is identified as the only one for which the solution to the differential equation (which also has no explicit solution) matches the analogous boundary condition at u. The situation is disappointingly more complicated than (19), which immediately identifies both boundaries in terms of a single parameter. In summary, the consumption problem yields a solution which is strikingly similar to the long-run problem, but in a much less tractable setting. Vice versa, the long-run solution provides a tractable first-order approximation to the consumption problem.

In the same vein, the long-run optimal portfolio is not far from optimal for utility maximization with terminal wealth. Indeed, Gerhold [18] show that the wealth corresponding to the long-run optimal portfolio matches the value function for any *finite* horizon T [3] at the leading order $\varepsilon^{2/3}$ for small transaction costs ε (compare Theorem 4.8 below). Hence, finite horizons—like consumption—only have a second-order effect on portfolio choice with transaction costs.

To apply the heuristic steps above to more complex problems with transaction costs, it is worth distinguishing the aspects that are special to the specific problem at hand from the ones that are flexible enough to be useful in other models. First, in general one cannot expect that a single, simple equation like (19) identifies both free boundaries. But the same argument that leads to this equation (the substitution of the boundary and smooth pasting conditions into the HJB equation) will generally lead in a long-run problem to some scalar equation for each boundary, in terms of the equivalent safe rate β of the problem. Such equations may be solved explicitly (as in the case of (19)) or not, but in the latter case an asymptotic solution will still be available, expanding the scalar equation around the frictionless values of (π, β).

Second, the reduced HJB equation may not be autonomous or have an explicit solution, which are two special features of (21). If the equation is not autonomous, the Cauchy problem cannot be started at some arbitrary point (zero in the previous example) without a further change of variable. If the free boundaries admit explicit solutions in terms of β, sometimes a careful choice of notation can lead to a simple expression for at least one boundary, which is a natural choice for the starting point of the Cauchy problem. The correct value of β is then identified as the one for which the remaining boundary condition is satisfied. Even if the differential equation has an explicit solution, this condition in general involves a scalar equation that cannot be solved explicitly. Regardless of an explicit formula, asymptotic expansions can be derived by substituting a series expansion for w in the differential equation.

4 Shadow Prices and Verification

We justify the heuristic arguments in the previous section by reducing the portfolio choice problem with transaction costs to another portfolio choice problem, without transaction costs. To do so, the bid and ask prices are replaced by a single "shadow price" $\tilde{S}_t$ evolving within the bid-ask spread, which yields the same optimal policy and utility. Evidently, *any* frictionless market extension with values in the bid-ask spread leads to more favorable terms of trade than the original market with transaction costs. To achieve equality, the particularly unfavorable shadow price must match the trading prices whenever its optimal policy transacts. The latter is then also feasible and in turn optimal in the original market with transaction costs, motivating the following notion.

Definition 4.1. A *shadow price* is a frictionless price process $\tilde{S}_t$ evolving within the bid-ask spread ($(1-\varepsilon)S_t \leq \tilde{S}_t \leq S_t$ a.s. for all t), such that there is an optimal strategy for $\tilde{S}_t$ which is of finite variation and entails buying only when the shadow price $\tilde{S}_t$ equals the ask price S_t, and selling only when $\tilde{S}_t$ equals the bid price $(1-\varepsilon)S_t$.

Once a candidate for such a shadow price is identified, long-run verification results for frictionless models (cf. Guasoni and Robertson [21]) deliver the optimality of the guessed policy.

4.1 Derivation of a Candidate Shadow Price

With a smooth candidate value function at hand, a candidate shadow price is identified as follows. By definition, trading the shadow price should not allow the investor to outperform the original market with transaction costs. In particular, if $\tilde{S}_t$ is the value of the shadow price at time t, then allowing the frictional investor to carry out at single trade at time t at this *frictionless* price should not allow her to increase her utility. A trade of ν risky shares at the frictionless price $\tilde{S}_t$ moves the investor's safe position X_t to $X_t - \nu\tilde{S}_t$ and her risky position (valued at the ask price S_t) from Y_t to $Y_t + \nu S_t$. Then—recalling that the second and third arguments of the candidate value functions V from the previous section were precisely the investor's safe and risky positions—the requirement that such a trade does not increase the investor's utility is tantamount to:

$$V(t, X_t - \nu\tilde{S}_t, Y_t + \nu S_t) \leq V(t, X_t, Y_t), \quad \forall \nu \in \mathbb{R}.$$

A Taylor expansion of the left-hand side for small ν then implies that $-\nu\tilde{S}_t V_x + \nu S_t V_y \leq 0$. Since this inequality has to hold both for positive and negative values of ν, it implies that

$$\tilde{S}_t = \frac{V_y}{V_x} S_t. \tag{27}$$

That is, the multiplicative deviation of the shadow price from the ask price should be the marginal rate of substitution of risky for safe assets for the optimal frictional investor. In particular, this formula immediately yields a candidate shadow price, once a smooth candidate value function has been identified. For the long-run problem, we derived the following candidate value function in the previous section:

$$V(t, X_t, Y_t) = e^{-(1-\gamma)(r+\beta)t}(X_t)^{1-\gamma} e^{(1-\gamma)\int_0^{\log(Y_t/lX_t)} w(y)dy}.$$

Using this equality to calculate the partial derivatives in (27), the candidate shadow price becomes:

$$\tilde{S}_t = \frac{w(\Upsilon_t)}{le^{\Upsilon_t}(1 - w(\Upsilon_t))} S_t, \tag{28}$$

where $\Upsilon_t = \log(X_t/lX_t^0)$ denotes the logarithm of the stock-cash ratio, centered in its value at the lower buying boundary l. If this candidate is indeed the right one, then its optimal strategy and value function should coincide with their frictional counterparts derived heuristically above. In particular, the optimal risky fraction $\tilde{\pi}_t$ should correspond to the same numbers φ_t^0 and φ_t of safe and risky shares, but now measured in terms of $\tilde{S}_t$ instead of the ask price S_t. As a consequence:

$$\tilde{\pi}_t = \frac{\varphi_t \tilde{S}_t}{\varphi_t^0 S_t^0 + \varphi_t \tilde{S}_t} = \frac{\varphi_t S_t \frac{w(\Upsilon_t)}{l e^{\Upsilon_t}(1-w(\Upsilon_t))}}{\varphi_t^0 S_t^0 + \varphi_t S_t \frac{w(\Upsilon_t)}{l e^{\Upsilon_t}(1-w(\Upsilon_t))}} = \frac{\frac{w(\Upsilon_t)}{1-w(\Upsilon_t)}}{1 + \frac{w(\Upsilon_t)}{1-w(\Upsilon_t)}} = w(\Upsilon_t), \quad (29)$$

where, for the third equality, we have used that the optimal frictional stock-cash ratio $\varphi_t S_t / \varphi_t^0 S_t^0$ equals $l e^{\Upsilon_t}$ by definition of Υ_t. We now turn to the corresponding value function $\tilde{V}$. By the definition of shadow price, it should coincide with its frictional counterpart V. In the frictionless case, it is more convenient to factor out the total wealth $\tilde{X}_t = \varphi_t^0 S_t^0 + \varphi_t \tilde{S}_t$ (in terms of the frictionless risky price $\tilde{S}_t$) instead of the safe position $X_t = \varphi_t^0 S_t^0$, giving

$$\tilde{V}(t, \tilde{X}_t, \Upsilon_t) = V(t, X_t, Y_t) = e^{-(1-\gamma)(r+\beta)t} \tilde{X}_t^{1-\gamma} \left(\frac{X_t}{\tilde{X}_t}\right)^{1-\gamma} e^{(1-\gamma)\int_0^{\Upsilon_t} w(y)dy}.$$

Since $X_t / \tilde{X}_t = 1 - w(\Upsilon_t)$ by definition of $\tilde{S}_t$, one can rewrite the last two factors as

$$\begin{aligned} &\left(\frac{X_t}{\tilde{X}_t}\right)^{1-\gamma} e^{(1-\gamma)\int_0^{\Upsilon_t} w(y)dy} \\ &= \exp\left((1-\gamma)\left[\log(1 - w(\Upsilon_t)) + \int_0^{\Upsilon_t} w(y)dy\right]\right) \\ &= (1 - w(0))^{\gamma-1} \exp\left((1-\gamma)\int_0^{\Upsilon_t} \left(w(y) - \frac{w'(y)}{1 - w(y)}\right) dy\right). \end{aligned}$$

Then, setting $\tilde{w} = w - \frac{w'}{1-w}$, the candidate long-run value function for $\tilde{S}$ becomes

$$\tilde{V}(t, \tilde{X}_t, \Upsilon_t) = e^{-(1-\gamma)(r+\beta)t} \tilde{X}_t^{1-\gamma} e^{(1-\gamma)\int_0^{\Upsilon_t} \tilde{w}(y)dy} (1 - w(0))^{\gamma-1}.$$

Starting from the candidate value function and optimal policy for $\tilde{S}$, we can now proceed to verify that they are indeed optimal for $\tilde{S}_t$, by adapting the argument from [21]. But before we do that, we have to construct the respective shadow processes.

4.2 Construction of the Shadow Price

The above heuristic arguments suggest that the optimal stock-cash ratio $Y_t / X_t = \varphi_t S_t / \varphi_t^0 S_t^0$ should take values in the interval $[l, u]$. Hence, $\Upsilon_t = \log(Y_t / l X_t)$ should be $[0, \log(u/l)]$-valued if the lower trading boundary l for the stock-cash ratio X_t/X_t^0 is positive. If the investor shorts the safe asset to leverage her risky position, the stock-cash ratio becomes negative. In the frictionless case, and also for small transaction costs, this happens if the Merton proportion $\mu/\gamma\sigma^2$ is bigger than 1. Then, the trading boundaries $l \leq u$ are both negative, so that the centered log-stock-cash ratio Υ_t should take values in $[\log(u/l), 0]$. In both cases, trading should only take place when the stock-cash ratio reaches the boundaries of this region. Hence, the numbers of safe and risky units φ_t^0 and φ_t should remain constant and $\Upsilon_t = \log(\varphi_t/l\varphi_t^0) + \log(S_t/S_t^0)$ should follow a Brownian motion with drift as long as Υ_t moves in $(0, \log(u/l))$ (resp. in $(\log(u/l), 0)$ if $\mu/\gamma\sigma^2 > 1$). This motivates to *define* the process Υ_t as reflected Brownian motion:

$$d\Upsilon_t = (\mu - \sigma^2/2)dt + \sigma dW_t + dL_t - dU_t, \quad \Upsilon_0 \in [0, \log(u/l)], \tag{30}$$

for continuous, adapted local time processes L and U which are nondecreasing (resp. nonincreasing if $\mu/\gamma\sigma^2 > 1$) and increase (resp. decrease if $\mu/\gamma\sigma^2 > 1$) only on the sets $\{\Upsilon_t = 0\}$ and $\{\Upsilon_t = \log(u/l)\}$, respectively. Starting from this process, whose existence is a classical result of [41], the process $\tilde{S}$ is defined in accordance with (28):

Lemma 4.2. *Let $(\xi^0, \xi) \in \mathbb{R}_+^2$ be the investor's initial endowment in units of the safe and risky asset. Define*

$$y = \begin{cases} 0, & \text{if } l\xi^0 S_0^0 \geq \xi S_0, \\ \log(u/l), & \text{if } u\xi^0 S_0^0 \leq \xi S_0, \\ \log\left[\xi S_0/(\xi^0 S_0^0 l)\right], & \text{otherwise}, \end{cases} \tag{31}$$

and let Υ be defined as in (30), starting at $\Upsilon_0 = y$. Then, $\tilde{S} = S\frac{w(\Upsilon)}{le^{\Upsilon}(1-w(\Upsilon))}$, with w as in Lemma 3.1, has the dynamics

$$d\tilde{S}(\Upsilon_t)/\tilde{S}(\Upsilon_t) = (\tilde{\mu}(\Upsilon_t) + r)\,dt + \tilde{\sigma}(\Upsilon_t)dW_t,$$

where $\tilde{\mu}(\cdot)$ and $\tilde{\sigma}(\cdot)$ are defined as

$$\tilde{\mu}(y) = \frac{\sigma^2 w'(y)}{w(y)(1-w(y))}\left(\frac{w'(y)}{1-w(y)} - (1-\gamma)w(y)\right), \quad \tilde{\sigma}(y) = \frac{\sigma w'(y)}{w(y)(1-w(y))}.$$

Moreover, the process $\tilde{S}$ takes values within the bid-ask spread $[(1-\varepsilon)S, S]$.

Note that the first two cases in (31) arise if the initial stock-cash ratio $\xi S_0/(\xi^0 S_0^0)$ lies outside of the interval $[l, u]$. Then, a jump from the initial position $(\varphi_{0-}^0, \varphi_{0-}) = (\xi^0, \xi)$ to the nearest boundary value of $[l, u]$ is required. This transfer requires the purchase resp. sale of the risky asset and hence the initial price $\tilde{S}_0$ is defined to match the buying resp. selling price of the risky asset.

Proof. The dynamics of $\tilde{S}_t$ result from Itô's formula, the dynamics of Υ_t, and the identity

$$w''(y) = 2(\gamma - 1)w'(y)w(y) - (2\mu/\sigma^2 - 1)w'(y), \tag{32}$$

obtained by differentiating the ODE (24) for w with respect to x. Therefore it remains to show that $\tilde{S}_t$ indeed takes values in the bid-ask spread $[(1-\varepsilon)S_t, S_t]$. To this end, notice that—in view of the ODE (24) for w—the derivative of the function $g(y) := w(y)/le^y(1 - w(y))$ is given by

$$g'(y) = \frac{w'(y) - w(y) + w^2(y)}{le^y(1 - w(y))^2} = \frac{\gamma(w^2 - 2\frac{\mu}{\gamma\sigma^2}w) + (\mu^2 - \lambda^2)/\gamma\sigma^4}{le^y(1 - w(y))^2}.$$

Due to the boundary conditions for w, the derivative g' vanishes at 0 and $\log(u/l)$. Differentiating its numerator gives $2\gamma w'(y)(w(y) - \frac{\mu}{\gamma\sigma^2})$. For $\frac{\mu}{\gamma\sigma^2} \in (0,1)$ (resp. $\frac{\mu}{\gamma\sigma^2} > 1$), w is increasing from $\frac{\mu-\lambda}{\gamma\sigma^2} < \frac{\mu}{\gamma\sigma^2}$ to $\frac{\mu+\lambda}{\gamma\sigma^2} > \frac{\mu}{\gamma\sigma^2}$ on $[0, \log(u/l)]$ (resp. decreasing from $\frac{\mu+\lambda}{\gamma\sigma^2}$ to $\frac{\mu-\lambda}{\gamma\sigma^2}$ on $[\log(u/l), 0]$); hence, w' is nonnegative (resp. nonpositive). Moreover, g' starts at zero for $y = 0$ (resp. $\log(u/l)$), then decreases (resp. increases), and eventually starts increasing (resp. decreasing) again, until it reaches level zero again for $y = \log(u/l)$ (resp. $y = 0$). In particular, g' is nonpositive (resp. nonnegative), so that g is decreasing on $[0, \log(u/l)]$ (resp. increasing on $[\log(u/l), 0]$ for $\frac{\mu}{\gamma\sigma^2} > 1$). Taking into account that $g(0) = 1$ and $g(\log(u/l)) = 1 - \varepsilon$, by the boundary conditions for w and the definition of u and l in Lemma 3.2, the proof is now complete. □

4.3 Verification

The long-run optimality of the candidate risky weight $\tilde{\pi}(\Upsilon_t) = w(\Upsilon_t)$ from (29) in the frictionless market with price process $\tilde{S}_t$ can now be verified by adapting the argument in [21]. The first step is to determine finite-horizon bounds, which provide lower and upper estimates for the maximal expected utility on any horizon T, by focusing on the values of the candidate long-run optimal policy and long-run optimal martingale measure.

These bounds are based on the concept of the (long-run) myopic probability, the hypothetical probability measure under which a logarithmic investor would adopt the same policy as the original power investor under the physical probability. The advantage of this probability is to decompose expected *power* utility (and its dual) into a long-run component *times* a transient component. This decomposition is similar in spirit to the separation of *logarithmic* utility into a long-run component *plus* a transitory component. To see the analogy, consider the logarithmic utility of a portfolio $\pi(\Theta_t)$ traded in a frictionless market with expected excess return $\tilde{\mu}(\Theta_t)$ and volatility $\tilde{\sigma}(\Theta_t)$ driven by some state variable Θ_t:

$$\log \tilde{X}_T^\pi = x + \int_0^T \left(\tilde{\mu}(\Theta_t)\pi(\Theta_t) - \frac{\tilde{\sigma}^2(\Theta_t)}{2}\pi^2(\Theta_t) \right) dt + \int_0^T \tilde{\sigma}(\Theta_t)\pi(\Theta_t) dW_t.$$

Now, if Θ_t follows an autonomous diffusion $d\Theta_t = b(\Theta_t)dt + dW_t$, the above stochastic integral can be replaced by applying Itô's formula to the function $\Pi(y) = \int_0^y \tilde{\sigma}(x)\pi(x)dx$:

$$\Pi(\Theta_T) - \Pi(\Theta_0) = \int_0^T \left(\tilde{\sigma}(\Theta_t)\pi(\Theta_t)b(\Theta_t) + \frac{1}{2}(\tilde{\sigma}\pi)'(\Theta_t) \right) dt + \int_0^T \tilde{\sigma}(\Theta_t)\pi(\Theta_t) dW_t.$$

Indeed, solving the second equation for the stochastic integral, and plugging it into the first equation yields:

$$\log \tilde{X}_T^\pi = x + \int_0^T \left((\tilde{\mu}(\Theta_t) - \tilde{\sigma}(\Theta_t)b(\Theta_t))\pi(\Theta_t) - \frac{\tilde{\sigma}^2(\Theta_t)}{2}\pi^2(\Theta_t) - \frac{(\tilde{\sigma}\pi)'(\Theta_t)}{2} \right) dt$$
$$+ (\Pi(\Theta_T) - \Pi(\Theta_0)).$$

This decomposes the logarithmic utility into an integral, which represents the long-run component, and a residual transitory term, which depends only on the initial and terminal values of the state variable. If the function Π is integrable with respect to the invariant measure of Θ, the contribution of the transitory component to the equivalent safe rate $\frac{1}{T}E[\log X_T^\pi]$ is negligible for long horizons.

The myopic probability is key to perform a similar decomposition with power utility. Again, denote by $\tilde{\mu}$ the risky asset's drift under the original measure, and by $\hat{\mu}$ its counterpart under the myopic probability; the corresponding volatility $\tilde{\sigma}$ of course has to be the same under both equivalent measures. With logarithmic utility, the optimal portfolio is $\hat{\pi}_t = \hat{\mu}_t/\tilde{\sigma}_t^2$ even if $\hat{\mu}_t$ and $\tilde{\sigma}_t$ are stochastic [37]. As the definition of the myopic probability requires that the corresponding log-optimal portfolio $\hat{\pi}_t$ coincides with the optimal portfolio $\tilde{\pi}_t$ for power utility under the original probability, Girsanov's theorem dictates that the measure change from the original to the myopic probability is governed by the stochastic exponential of $\int_0^T (-\frac{\tilde{\mu}}{\tilde{\sigma}} + \tilde{\sigma}\tilde{\pi})dW_t$. This measure change shifts the asset's drift by the same amount, times $\tilde{\sigma}$, thereby yielding a myopic drift of $\tilde{\sigma}^2\tilde{\pi}$, which yields the same optimal policy. Given this guess for the myopic probability, the finite-horizon bounds follow by routine calculations carried out in the proof of the following lemma:

Lemma 4.3. *For a fixed time horizon $T > 0$, let $\beta = \frac{\mu^2-\lambda^2}{2\gamma\sigma^2}$ and let the function w be defined as in Lemma 3.1. Then, for the shadow payoff $\tilde{X}_T$ corresponding to the policy $\tilde{\pi}(\Upsilon_t) = w(\Upsilon_t)$ and the shadow discount factor $\tilde{M}_T = e^{-rT}\mathcal{E}(-\int_0^\cdot \frac{\tilde{\mu}}{\tilde{\sigma}}dW_t)_T$, the following bounds hold true:*

$$E[\tilde{X}_T^{1-\gamma}] = \tilde{X}_0^{1-\gamma} e^{(1-\gamma)(r+\beta)T} \hat{E}[e^{(1-\gamma)(\tilde{q}(\Upsilon_0)-\tilde{q}(\Upsilon_T))}], \tag{33}$$

$$E\left[\tilde{M}_T^{1-\frac{1}{\gamma}}\right]^\gamma = e^{(1-\gamma)(r+\beta)T} \hat{E}\left[e^{(\frac{1}{\gamma}-1)(\tilde{q}(\Upsilon_0)-\tilde{q}(\Upsilon_T))}\right]^\gamma, \tag{34}$$

where $\tilde{q}(y) := \int_0^y (w(z) - \frac{w'(z)}{1-w(z)})dz$ *and* $\hat{E}\,[\cdot]$ *denotes the expectation with respect to the* myopic probability $\hat{P}$, *defined by*

$$\frac{d\hat{P}}{dP} = \exp\left(\int_0^T \left(-\frac{\tilde{\mu}(\Upsilon_t)}{\tilde{\sigma}(\Upsilon_t)} + \tilde{\sigma}(\Upsilon_t)\tilde{\pi}(\Upsilon_t)\right) dW_t \right.$$
$$\left. -\frac{1}{2}\int_0^T \left(-\frac{\tilde{\mu}(\Upsilon_t)}{\tilde{\sigma}(\Upsilon_t)} + \tilde{\sigma}(\Upsilon_t)\tilde{\pi}(\Upsilon_t)\right)^2 dt\right).$$

Proof. First note that $\tilde{\mu}, \tilde{\sigma}$, and w are functions of Υ_t, but the argument is omitted throughout to ease notation. Now, to prove (33), notice that the frictionless shadow wealth process $\tilde{X}_t$ with dynamics $\frac{d\tilde{X}_t}{\tilde{X}_t} = w\frac{d\tilde{S}_t}{\tilde{S}_t} + (1-w)\frac{dS_t^0}{S_t^0}$ satisfies:

$$\tilde{X}_T^{1-\gamma} = \tilde{X}_0^{1-\gamma} e^{(1-\gamma)\int_0^T (r+\tilde{\mu}w-\frac{\tilde{\sigma}^2}{2}w^2)dt+(1-\gamma)\int_0^T \tilde{\sigma}w dW_t}.$$

Hence:

$$\tilde{X}_T^{1-\gamma} = \tilde{X}_0^{1-\gamma}\frac{d\hat{P}}{dP} e^{\int_0^T ((1-\gamma)(r+\tilde{\mu}w-\frac{\tilde{\sigma}^2}{2}w^2)+\frac{1}{2}(-\frac{\tilde{\mu}}{\tilde{\sigma}}+\tilde{\sigma}w)^2)dt+\int_0^T ((1-\gamma)\tilde{\sigma}w-(-\frac{\tilde{\mu}}{\tilde{\sigma}}+\tilde{\sigma}w))dW_t}.$$

Inserting the definitions of $\tilde{\mu}$ and $\tilde{\sigma}$, the second integrand simplifies to $(1-\gamma)\sigma(\frac{w'}{1-w} - w)$. Similarly, the first integrand reduces to $(1-\gamma)(r + \frac{\sigma^2}{2}(\frac{w'}{1-w})^2 - (1-\gamma)\sigma^2\frac{w'w}{1-w} + (1-\gamma)\frac{\sigma^2}{2}w^2)$. In summary:

$$\tilde{X}_T^{1-\gamma} = \tilde{X}_0^{1-\gamma}\frac{d\hat{P}}{dP} e^{(1-\gamma)\int_0^T (r+\frac{\sigma^2}{2}(\frac{w'}{1-w})^2-(1-\gamma)\sigma^2\frac{w'w}{1-w}+(1-\gamma)\frac{\sigma^2}{2}w^2)dt+(1-\gamma)\int_0^T \sigma(\frac{w'}{1-w}-w)dW_t}. \tag{35}$$

The boundary conditions for w and w' imply $w(0) - \frac{w'(0)}{1-w(0)} = w(\log(u/l)) - \frac{w'(\log(u/l))}{1-w(\log(u/l))} = 0$; hence, Itô's formula yields that the local time terms vanish in the dynamics of $\tilde{q}(\Upsilon_t)$:

$$\tilde{q}(\Upsilon_T) - \tilde{q}(\Upsilon_0) = \int_0^T \left(\mu - \frac{\sigma^2}{2}\right)\left(w - \frac{w'}{1-w}\right)$$
$$+\frac{\sigma^2}{2}\left(w' - \frac{w''(1-w)+w'^2}{(1-w)^2}\right)dt + \int_0^T \sigma\left(w - \frac{w'}{1-w}\right)dW_t. \tag{36}$$

Substituting the second derivative w'' according to the ODE (32) and using the resulting identity to replace the stochastic integral in (35) yields

$$\tilde{X}_T^{1-\gamma} = \tilde{X}_0^{1-\gamma}\frac{d\hat{P}}{dP}e^{(1-\gamma)\int_0^T(r+\frac{\sigma^2}{2}w'+(1-\gamma)\frac{\sigma^2}{2}w^2+(\mu-\frac{\sigma^2}{2})w)dt}e^{(1-\gamma)(\tilde{q}(\Upsilon_0)-\tilde{q}(\Upsilon_T))}.$$

After inserting the ODE (24) for w, the first bound thus follows by talking the expectation.

The argument for the second bound is similar. Plugging in the definitions of $\tilde{\mu}$ and $\tilde{\sigma}$, the shadow discount factor $\tilde{M}_T = e^{-rT}\mathcal{E}(-\int_0^\cdot \frac{\tilde{\mu}}{\tilde{\sigma}}dW)_T$ and the myopic probability $\hat{P}$ satisfy:

$$\begin{aligned}\tilde{M}_T^{1-\frac{1}{\gamma}} &= e^{\frac{1-\gamma}{\gamma}\int_0^T\frac{\tilde{\mu}}{\tilde{\sigma}}dW_t+\frac{1-\gamma}{\gamma}\int_0^T(r+\frac{\tilde{\mu}^2}{2\tilde{\sigma}^2})dt}\\ &= \frac{d\hat{P}}{dP}e^{\frac{1-\gamma}{\gamma}\int_0^T(\frac{\tilde{\mu}}{\tilde{\sigma}}-\frac{\gamma}{1-\gamma}(-\frac{\tilde{\mu}}{\tilde{\sigma}}+\tilde{\sigma}w))dW_t+\frac{1-\gamma}{\gamma}\int_0^T(r+\frac{\tilde{\mu}^2}{2\tilde{\sigma}^2}+\frac{\gamma}{2(1-\gamma)}(-\frac{\tilde{\mu}}{\tilde{\sigma}}+\tilde{\sigma}w)^2)dt}\\ &= \frac{d\hat{P}}{dP}e^{\frac{1-\gamma}{\gamma}\int_0^T\sigma(\frac{w'}{1-w}-w)dW_t+\frac{1-\gamma}{\gamma}\int_0^T(r+\frac{\sigma^2}{2}(\frac{w'}{1-w})^2-(1-\gamma)\sigma^2\frac{w'w}{1-w}+(1-\gamma)\frac{\sigma^2}{2}w^2)dt}.\end{aligned}$$

Again replace the stochastic integral using (36) and the ODE (32), obtaining

$$\tilde{M}_T^{1-\frac{1}{\gamma}} = \frac{d\hat{P}}{dP}e^{\frac{1-\gamma}{\gamma}\int_0^T(r+\frac{\sigma^2}{2}w'+(1-\gamma)\frac{\sigma^2}{2}w^2+(\mu-\frac{\sigma^2}{2})w)dt}e^{\frac{1-\gamma}{\gamma}(\tilde{q}(\Upsilon_0)-\tilde{q}(\Upsilon_T))}.$$

Inserting the ODE (24) for w, taking the expectation, and raising it to power γ, the second bound follows. □

With the finite horizon bounds at hand, it is now straightforward to establish that the policy $\tilde{\pi}(\Upsilon_t)$ is indeed long-run optimal in the frictionless market with price $\tilde{S}_t$.

Lemma 4.4. *Let $0 < \mu/\gamma\sigma^2 \neq 1$ and let w be defined as in Lemma 3.1. Then, the risky weight $\tilde{\pi}(\Upsilon_t) = w(\Upsilon_t)$ is long-run optimal with equivalent safe rate $r+\beta$ in the frictionless market with price process $\tilde{S}_t$. The corresponding wealth process (in terms of $\tilde{S}_t$), and the numbers of safe and risky units are given by*

$$\tilde{X}_t = (\xi^0 S_0^0 + \xi\tilde{S}_0)\mathcal{E}\left(\int_0^\cdot (r + w(\Upsilon_s)\tilde{\mu}(\Upsilon_s))ds + \int_0^\cdot w(\Upsilon_s)\tilde{\sigma}(\Upsilon_s)dW_s\right)_t,$$

$$\varphi_{0-} = \xi, \quad \varphi_t = w(\Upsilon_t)\tilde{X}_t/\tilde{S}_t \quad \textit{for } t \geq 0,$$

$$\varphi_{0-}^0 = \xi^0, \quad \varphi_t^0 = (1-w(\Upsilon_t))\tilde{X}_t/S_t^0 \quad \textit{for } t \geq 0.$$

Proof. The formulas for the wealth process and the corresponding numbers of safe and risky units follow directly from the standard frictionless definitions. Now let $\tilde{M}_t$ be the shadow discount factor from Lemma 4.3. Then, standard duality arguments for power utility (cf. Lemma 5 in [21]) imply that the shadow payoff $\tilde{X}_t^\phi$ corresponding to *any* admissible strategy ϕ_t satisfies the inequality

$$E\left[(\tilde{X}_T^{\phi})^{1-\gamma}\right]^{\frac{1}{1-\gamma}} \leq E\left[\tilde{M}_T^{\frac{\gamma-1}{\gamma}}\right]^{\frac{\gamma}{1-\gamma}}. \tag{37}$$

This inequality in turn yields the following upper bound, valid for any admissible strategy ϕ_t in the frictionless market with shadow price $\tilde{S}_t$:

$$\liminf_{T\to\infty} \frac{1}{(1-\gamma)T} \log E\left[(\tilde{X}_T^{\phi})^{1-\gamma}\right] \leq \liminf_{T\to\infty} \frac{\gamma}{(1-\gamma)T} \log E\left[\tilde{M}_T^{\frac{\gamma-1}{\gamma}}\right]. \tag{38}$$

Since the function $\tilde{q}$ is bounded on the compact support of Υ_t, the second bound in Lemma 4.3 implies that the right-hand side equals $r+\beta$. Likewise, the first bound in the same lemma implies that the shadow payoff $\tilde{X}_t$ (corresponding to the policy φ_t) attains this upper bound, concluding the proof. □

The next Lemma establishes that the candidate $\tilde{S}_t$ is indeed a shadow price.

Lemma 4.5. *Let $0 < \mu/\gamma\sigma^2 \neq 1$. Then, the number of shares $\varphi_t = w(\Upsilon_t)\tilde{X}_t/\tilde{S}_t$ in the portfolio $\tilde{\pi}(\Upsilon_t)$ in Lemma 4.4 has the dynamics*

$$\frac{d\varphi_t}{\varphi_t} = \left(1 - \frac{\mu-\lambda}{\gamma\sigma^2}\right) dL_t - \left(1 - \frac{\mu+\lambda}{\gamma\sigma^2}\right) dU_t. \tag{39}$$

Thus, φ_t increases only when $\Upsilon_t = 0$, that is, when $\tilde{S}_t$ equals the ask price, and decreases only when $\Upsilon_t = \log(u/l)$, that is, when $\tilde{S}_t$ equals the bid price.

Proof. Itô's formula and the ODE (32) yield

$$dW(\Upsilon_t) = -(1-\gamma)\sigma^2 w'(\Upsilon_t)w(\Upsilon_t)dt + \sigma w'(\Upsilon_t)dW_t + w'(\Upsilon_t)(dL_t - dU_t).$$

Integrating $\varphi_t = w(\Upsilon_t)\tilde{X}_t/\tilde{S}_t$ by parts twice, inserting the dynamics of $w(\Upsilon_t)$, $\tilde{X}_t$, $\tilde{S}_t$, and simplifying yields:

$$\frac{d\varphi_t}{\varphi_t} = \frac{w'(\Upsilon_t)}{w(\Upsilon_t)} d(L_t - U_t).$$

Since L_t and U_t only increase (resp. decrease when $\mu/\gamma\sigma^2 > 1$) on $\{\Upsilon_t = 0\}$ and $\{\Upsilon_t = \log(u/l)\}$, respectively, the assertion now follows from the boundary conditions for w and w'. □

The optimal growth rate for any frictionless price within the bid-ask spread must be greater or equal than in the original market with bid-ask process $((1-\varepsilon)S_t, S_t)$, because the investor trades at more favorable prices. For a *shadow price*, there is an optimal strategy that only entails buying (resp. selling) stocks when $\tilde{S}_t$ coincides with the ask- resp. bid price. Hence, this strategy yields the same payoff when executed at bid-ask prices, and thus is also optimal in the original model with transaction costs. The corresponding equivalent safe rate must also be the same, since the difference due to the liquidation costs vanishes as the horizon grows in (3):

Proposition 4.6. *For a sufficiently small spread ε, the strategy (φ_t^0, φ_t) from Lemma 4.4 is also long-run optimal in the original market with transaction costs, with the same equivalent safe rate.*

Proof. As φ_t only increases (resp. decreases) when $\tilde{S}_t = S_t$ (resp. $\tilde{S}_t = (1-\varepsilon)S_t$), the strategy (φ_t^0, φ_t) is also self-financing for the bid-ask process $((1-\varepsilon)S_t, S_t)$. Since $S_t \geq \tilde{S}_t \geq (1-\varepsilon)S_t$ and the number φ_t of risky shares is always positive, it follows that

$$\varphi_t^0 S_t^0 + \varphi_t \tilde{S}_t \geq \varphi_t^0 S_t^0 + \varphi_t^+ (1-\varepsilon) S_t - \varphi_t^- S_t \geq (1 - \tfrac{\varepsilon}{1-\varepsilon} \tilde{\pi}(Y_t))(\varphi_t^0 S_t^0 + \varphi_t \tilde{S}_t). \quad (40)$$

The shadow risky fraction $\tilde{\pi}(\Upsilon_t) = w(\Upsilon_t)$ is bounded from above by $(\mu + \lambda)/\gamma\sigma^2 = \mu/\gamma\sigma^2 + O(\varepsilon^{1/3})$. For a sufficiently small spread ε, the strategy (φ_t^0, φ_t) is therefore also admissible for $((1-\varepsilon)S_t, S_t)$. Moreover, (40) then also yields

$$\begin{aligned} &\liminf_{T\to\infty} \frac{1}{(1-\gamma)T} \log E\left[(\varphi_T^0 S_T^0 + \varphi_T^+(1-\varepsilon)S_T - \varphi_T^- S_T)^{1-\gamma}\right] \\ &= \liminf_{T\to\infty} \frac{1}{(1-\gamma)T} \log E\left[(\varphi_T^0 S_T^0 + \varphi_T \tilde{S}_T)^{1-\gamma}\right], \end{aligned} \quad (41)$$

that is, (φ_t^0, φ_t) has the same growth rate, either with $\tilde{S}_t$ or with $[(1-\varepsilon)S_t, S_t]$.

For any admissible strategy (ψ_t^0, ψ_t) for the bid-ask spread $[(1-\varepsilon)S_t, S_t]$, set $\tilde{\psi}_t^0 = \psi_{0-}^0 - \int_0^t \tilde{S}_s/S_s^0 d\psi_s$. Then, $(\tilde{\psi}_t^0, \psi_t)$ is a self-financing trading strategy for $\tilde{S}_t$ with $\tilde{\psi}_t^0 \geq \psi_t^0$. Together with $\tilde{S}_t \in [(1-\varepsilon)S_t, S_t]$, the long-run optimality of (φ_t^0, φ_t) for $\tilde{S}_t$, and (41), it follows that:

$$\begin{aligned} &\liminf_{T\to\infty} \frac{1}{T}\frac{1}{(1-\gamma)} \log E\left[(\psi_T^0 S_T^0 + \psi_T^+(1-\varepsilon)S_T - \psi_T^- S_T)^{1-\gamma}\right] \\ &\leq \liminf_{T\to\infty} \frac{1}{T}\frac{1}{(1-\gamma)} \log E\left[(\tilde{\psi}_T^0 S_T^0 + \psi_T \tilde{S}_T)^{1-\gamma}\right] \\ &\leq \liminf_{T\to\infty} \frac{1}{T}\frac{1}{(1-\gamma)} \log E\left[(\varphi_T^0 S_T^0 + \varphi_T \tilde{S}_T)^{1-\gamma}\right] \\ &= \liminf_{T\to\infty} \frac{1}{T}\frac{1}{(1-\gamma)} \log E\left[(\varphi_T^0 S_T^0 + \varphi_T^+(1-\varepsilon)S_T - \varphi_T^- S_T)^{1-\gamma}\right]. \end{aligned}$$

Hence (φ_t^0, φ_t) is also long-run optimal for $((1-\varepsilon)S_t, S_t)$. □

By putting together the above statements we obtain the following main result:

Theorem 4.7. *For a small spread $\varepsilon > 0$, and $0 < \mu/\gamma\sigma^2 \neq 1$, the process $\tilde{S}_t$ in Lemma 4.2 is a shadow price. A long-run optimal policy—both for the frictionless market with price $\tilde{S}_t$ and in the market with bid-ask prices $(1-\varepsilon)S_t, S_t$—is to keep the risky weight $\tilde{\pi}_t$ (in terms of $\tilde{S}_t$) in the no-trade region*

$$[\pi_-, \pi_+] = \left[\frac{\mu - \lambda}{\gamma\sigma^2}, \frac{\mu + \lambda}{\gamma\sigma^2}\right].$$

As $\varepsilon \downarrow 0$, its boundaries have the asymptotics

$$\pi_\pm = \frac{\mu}{\gamma\sigma^2} \pm \left(\frac{3}{4\gamma}\left(\frac{\mu}{\gamma\sigma^2}\right)^2\left(1 - \frac{\mu}{\gamma\sigma^2}\right)^2\right)^{1/3} \varepsilon^{1/3} + O(\varepsilon).$$

The corresponding equivalent safe rate is:

$$r + \beta = r + \frac{\mu^2 - \lambda^2}{\gamma\sigma^2} = r + \frac{\mu^2}{2\gamma\sigma^2}$$
$$- \frac{\gamma\sigma^2}{2}\left(\frac{3}{4\gamma}\left(\frac{\mu}{\gamma\sigma^2}\right)^2\left(1 - \frac{\mu}{\gamma\sigma^2}\right)^2\right)^{2/3} \varepsilon^{2/3} + O(\varepsilon^{4/3}).$$

If $\mu/\gamma\sigma^2 = 1$, then $\tilde{S}_t = S_t$ is a shadow price, and it is optimal to invest all wealth in the risky asset at time $t = 0$, never to trade afterwards. In this case, the equivalent safe rate is the frictionless value $r + \beta = r + \mu^2/2\gamma\sigma^2$.

Proof. First let $0 < \mu/\gamma\sigma^2 \neq 1$. Optimality of the strategy (φ_t^0, φ_t) associated to $\tilde{\pi}(\Upsilon_t)$ for $\tilde{S}_t$ has been shown in Lemma 4.4. The asymptotic expansions are an immediate consequence of their counterpart for λ (cf. Lemma 3.2) and Taylor expansion. Next, Lemma 4.5 shows that $\tilde{S}_t$ is a shadow price process in the sense of Definition 4.1. Proposition 4.6 shows that, for small transaction costs ε, the same policy is also optimal, with the same equivalent safe rate, in the original market with bid-ask prices $(1 - \varepsilon)S_t, S_t$.

Consider now the degenerate case $\mu/\gamma\sigma^2 = 1$. Then the optimal strategy in the frictionless model $\tilde{S}_t = S_t$ transfers all wealth to the risky asset at time $t = 0$, never to trade afterwards ($\varphi_t^0 = 0$ and $\varphi_t = \xi + \xi^0 S_0^0/S_0$ for all $t \geq 0$). Hence it is of finite variation and the number of shares never decreases from the unlevered initial position, and increases only at time $t = 0$, where the shadow price coincides with the ask price. Thus, $\tilde{S}_t = S_t$ is a shadow price. The remaining assertions then follow as in Proposition 4.6 above. □

The trading boundaries in this paper are optimal for a long investment horizon, but are also approximately optimal for finite horizons. The following theorem, which complements the main result, makes this point precise:

Theorem 4.8. *Fix a time horizon $T > 0$. Then, the finite-horizon equivalent safe rate of the liquidation value $\Xi_T^\phi = \phi_T^0 S_T^0 + \phi_T^+(1 - \lambda)S_T - \phi_T^- S_T$ associated to* any *strategy (ϕ^0, ϕ) satisfies the upper bound*

$$\frac{1}{T}\log E\left[(\Xi_T^\phi)^{1-\gamma}\right]^{\frac{1}{1-\gamma}} \leq r + \frac{\mu^2 - \lambda^2}{2\gamma\sigma^2} + \frac{1}{T}\log(\phi_{0-}^0 + \phi_{0-} S_0) + \frac{\mu}{\gamma\sigma^2}\frac{\epsilon}{T} + O(\varepsilon^{4/3}), \quad (42)$$

and the finite-horizon equivalent safe rate of our long-run optimal strategy (φ^0, φ) *satisfies the lower bound*

$$\frac{1}{T}\log E\left[(\Xi_T^{\varphi})^{1-\gamma}\right]^{\frac{1}{1-\gamma}} \geq r + \frac{\mu^2-\lambda^2}{2\gamma\sigma^2} + \frac{1}{T}\log(\varphi_{0-}^0 + \varphi_{0-}S_0)$$
$$-\left(\frac{2\mu}{\gamma\sigma^2} + \frac{\varphi_{0-}S_0}{\varphi_{0-}^0 + \varphi_{0-}S_0}\right)\frac{\varepsilon}{T} + O(\varepsilon^{4/3}). \qquad (43)$$

In particular, for the same unlevered initial position ($\phi_{0-} = \varphi_{0-} \geq 0, \phi_{0-}^0 = \varphi_{0-}^0 \geq 0$)*, the equivalent safe rates of* (ϕ^0, ϕ) *and of the optimal policy* (φ^0, φ) *for horizon* T *differ by at most*

$$\frac{1}{T}\left(\log E\left[(\Xi_T^{\phi})^{1-\gamma}\right]^{\frac{1}{1-\gamma}} - \log E\left[(\Xi_T^{\varphi})^{1-\gamma}\right]^{\frac{1}{1-\gamma}}\right) \leq \left(\frac{3\mu}{\gamma\sigma^2}+1\right)\frac{\varepsilon}{T} + O(\varepsilon^{4/3}). \qquad (44)$$

This result implies that the horizon, like consumption, only has a second order effect on portfolio choice with transaction costs, because the finite-horizon equivalent safe rate matches, at the leading order $\epsilon^{2/3}$, the equivalent safe rate of the stationary long-run optimal policy, and recovers, in particular, the first-order asymptotics for the finite-horizon value function obtained by Bichuch [3, Theorem 4.1].

Proof (Proof of Theorem 4.8). Let (ϕ^0, ϕ) be any admissible strategy starting from the initial position $(\varphi_{0-}^0, \varphi_{0-})$. Then as in the proof of Proposition 4.6, we have $\Xi_T^{\phi} \leq \tilde{X}_T^{\phi}$ for the corresponding shadow payoff, that is, the terminal value of the wealth process $\tilde{X}_t^{\phi} = \phi_0^0 + \phi_0\tilde{S}_0 + \int_0^t \phi_s d\tilde{S}_s$ corresponding to trading ϕ in the frictionless market with price process $\tilde{S}_t$. Hence, Lemma 5 in [21] and the second bound in Lemma 4.3 imply that

$$\frac{1}{(1-\gamma)T}\log E\left[(\Xi_T^{\phi})^{1-\gamma}\right] \leq r + \beta + \frac{1}{T}\log(\varphi_{0-}^0 + \varphi_{0-}S_0)$$
$$+\frac{\gamma}{(1-\gamma)T}\log \hat{E}\left[e^{(\frac{1}{\gamma}-1)(\tilde{q}(\Upsilon_0)-\tilde{q}(\Upsilon_T))}\right]. \qquad (45)$$

For the strategy (φ^0, φ) from Lemma 4.5, we have $\Xi_T^{\varphi} \geq (1-\frac{\varepsilon}{1-\varepsilon}\frac{\mu+\lambda}{\gamma\sigma^2})\tilde{X}_T^{\varphi}$ by the proof of Proposition 4.6. Hence the first bound in Lemma 4.3 yields

$$\frac{1}{(1-\gamma)T}\log E\left[(\Xi_T^{\varphi})^{1-\gamma}\right] \geq r + \beta + \frac{1}{T}\log(\varphi_{0-}^0 + \varphi_{0-}\tilde{S}_0)$$
$$+\frac{1}{(1-\gamma)T}\log \hat{E}\left[e^{(1-\gamma)(\tilde{q}(\Upsilon_0)-\tilde{q}(\Upsilon_T))}\right]$$
$$+\frac{1}{T}\log\left(1-\frac{\varepsilon}{1-\varepsilon}\frac{\mu+\lambda}{\gamma\sigma^2}\right). \qquad (46)$$

To determine explicit estimates for these bounds, we first analyze the sign of $\tilde{w}(y) = w - \frac{w'}{1-w}$ and hence the monotonicity of $\tilde{q}(y) = \int_0^y \tilde{w}(z)dz$. Whenever $\tilde{w} = 0$, i.e., $w' = w(1-w)$, the derivative of $\tilde{w}$ is

$$\tilde{w}' = w' - \frac{w''(1-w) + w'^2}{(1-w)^2} = \frac{(1-2\gamma)w'w + \frac{2\mu}{\sigma^2}w'}{1-w} - \left(\frac{w'}{1-w}\right)^2 = 2\gamma w\left(\frac{\mu}{\gamma\sigma^2} - w\right),$$

where we have used the ODE (32) for the second equality. Since $\tilde{w}$ vanishes at 0 and $\log(u/l)$ by the boundary conditions for w and w', this shows that the behaviour of $\tilde{w}$ depends on whether the investor's position is leveraged or not. In the absence of leverage, $\mu/\gamma\sigma^2 \in (0,1)$, $\tilde{w}$ is defined on $[0, \log(u/l)]$. It vanishes at the left boundary 0 and then increases since its derivative is initially positive by the initial condition for w. Once the function w has increased to level $\mu/\gamma\sigma^2$, the derivative of $\tilde{w}$ starts to become negative; as a result, $\tilde{w}$ begins to decrease until it reaches level zero again at $\log(u/l)$. In particular, $\tilde{w}$ is nonnegative for $\mu/\gamma\sigma^2 \in (0,1)$.

In the leverage case $\mu/\gamma\sigma^2 > 1$, the situation is reversed. Then, $\tilde{w}$ is defined on $[\log(u/l), 0]$ and, by the boundary condition for w at $\log(u/l)$, therefore starts to decrease after starting from zero at $\log(u/l)$. Once w has decreased to level $\mu/\gamma\sigma^2$, $\tilde{w}$ starts increasing until it reaches level zero again at 0. Hence, $\tilde{w}$ is nonpositive for $\mu/\gamma\sigma^2 > 1$.

Now, consider Case 2 of Lemma 3.1; the calculations for the other cases follow along the same lines with minor modifications. Then $\mu/\gamma\sigma^2 \in (0,1)$ and $\tilde{q}$ is positive and increasing. Hence,

$$\frac{\gamma}{(1-\gamma)T}\log \hat{E}\left[e^{(\frac{1}{\gamma}-1)(\tilde{q}(\Upsilon_0)-\tilde{q}(\Upsilon_T))}\right] \leq \frac{1}{T}\int_0^{\log(u/l)} \tilde{w}(y)dy \tag{47}$$

and likewise

$$\frac{1}{(1-\gamma)T}\log \hat{E}\left[e^{(1-\gamma)(\tilde{q}(\Upsilon_0)-\tilde{q}(\Upsilon_T))}\right] \geq -\frac{1}{T}\int_0^{\log(u/l)} \tilde{w}(y)dy. \tag{48}$$

Since $\tilde{w}(y) = w(y) - w'/(1-w)$, the boundary conditions for w imply

$$\int_0^{\log(u/l)} \tilde{w}(y)dy = \int_0^{\log(u/l)} w(y)dy - \log\left(\frac{\mu-\lambda-\gamma\sigma^2}{\mu+\lambda-\gamma\sigma^2}\right). \tag{49}$$

By elementary integration of the explicit formula in Lemma 3.1 and using the boundary conditions from Lemma 3.3 for the evaluation of the result at 0 resp. $\log(u/l)$, the integral of w can also be computed in closed form:

$$\int_0^{\log(u/l)} w(y)dy = \frac{\frac{\mu}{\sigma^2}-\frac{1}{2}}{\gamma-1}\log\left(\frac{1}{1-\varepsilon}\frac{(\mu+\lambda)(\mu-\lambda-\gamma\sigma^2)}{(\mu-\lambda)(\mu+\lambda-\gamma\sigma^2)}\right) + \frac{1}{2(\gamma-1)}\log\left(\frac{(\mu+\lambda)(\mu+\lambda-\gamma\sigma^2)}{(\mu-\lambda)(\mu-\lambda-\gamma\sigma^2)}\right). \tag{50}$$

As $\epsilon \downarrow 0$, a Taylor expansion and the power series for λ then yield

$$\int_0^{\log(u/l)} \tilde{w}(y)dy = \frac{\mu}{\gamma\sigma^2}\varepsilon + O(\varepsilon^{4/3}).$$

Likewise,

$$\log\left(1 - \frac{\varepsilon}{1-\varepsilon}\frac{\mu-\lambda}{\gamma\sigma^2}\right) = -\frac{\mu}{\gamma\sigma^2}\varepsilon + O(\varepsilon^{4/3}),$$

as well as

$$\log(\varphi^0_{0-} + \varphi_{0-}\tilde{S}_0) \geq \log(\varphi^0_{0-} + \varphi_{0-}S_0) - \frac{\varphi_{0-}S_0}{\varphi^0_{0-} + \varphi_{0-}S_0}\varepsilon + O(\varepsilon^2),$$

and the claimed bounds follow from (45) and (47) resp. (46) and (48). □

5 Open Problems

In this section we mention three problems for which, in our view, the above approach holds promise, and the effect of transaction costs is likely to be substantial. Of course, only future research can shed light on this point.

5.1 Multiple Assets

In sharp contrast to frictionless models, passing from one to several risky assets is far from trivial with transaction costs. The reason is that, since in the free boundary problem the unknown boundary has one dimension less than the number of risky assets, with one asset it reduces to two points only, but with two assets it already becomes an unknown curve. More importantly, multiple assets introduce novel effects, which defy the one-dimensional intuition, as we now argue. For example, consider a market with two risky assets with prices S^1_t and S^2_t:

$$\frac{dS^1_t}{S^1_t} = \mu_1 dt + \sigma_1 dW^1_t \tag{51}$$

$$\frac{dS^2_t}{S^2_t} = \mu_2 dt + \varrho\sigma_2 dW^1_t + \sigma_2\sqrt{1-\varrho^2}dW^2_t \tag{52}$$

where $\mu_1, \sigma_1, \mu_2, \sigma_2 > 0$, $\varrho \in [-1, 1]$, and W^1, W^2 are two independent Brownian motions. Even for this simple model with power utility, the solution to the portfolio choice problem is unknown. Some recent papers, e.g., [1, 2, 4, 29, 38], offer some insights—and raise a number of questions.

Recall that the frictionless portfolio in the above model is $\pi = \frac{1}{\gamma}\Sigma^{-1}\mu$, where $\mu = (\mu_1, \mu_2)$ is the vector of excess returns, and Σ is the covariance matrix defined as $\Sigma_{11} = \sigma_1^2$, $\Sigma_{12} = \Sigma_{21} = \varrho\sigma_1\sigma_2$, and $\Sigma_{22} = \sigma_2^2$. In other words:

$$\pi_1 = \frac{\mu_1 - \beta_1\mu_2}{\gamma(1-\varrho^2)\sigma_1^2} \qquad \pi_2 = \frac{\mu_2 - \beta_2\mu_1}{\gamma(1-\varrho^2)\sigma_2^2}$$

where $\beta_i = (\varrho\sigma_1\sigma_2)/\sigma_i^2$ are the betas of each asset with respect to the other. In particular, for two uncorrelated assets the portfolio separates, in that the optimal weight for each risky asset in the market with all assets equals the optimal weight for the risky asset in a market with that risky asset only. This separation property is intuitive and appealing, and reduces the analysis of frictionless portfolio choice problems with multiple uncorrelated assets to the single asset case. Liu [30] and Guasoni and Muhle-Karbe [20] show that such a separation carries over to transaction cost models with exponential utility.

Surprisingly enough, separation seems to fail with constant relative risk aversion, in that the width of the no-trade region for each asset is affected by the presence of the other, even with zero correlation and logarithmic utility. For example, the heuristics in [29] yield the following width for the no-trade region of the first asset, compare their equation (50):

$$H_1 = \left(\frac{3\varepsilon}{2\sigma_1^2}\left[\left(\frac{1}{2}\mu'\Sigma^{-1}\mu + \sigma_1^2\right)\pi_1^2 - \mu_1\pi_1^2\right]\right)^{1/3}. \tag{53}$$

This quantity clearly depends also on μ_2 and σ_2 through the total squared Sharpe ratio $\mu'\Sigma^{-1}\mu$, even with zero correlation, and hence differs from the width of the no-trade region with a single risky asset:

$$h_1 = \left(\frac{3\varepsilon}{4}\pi_1^2(1-\pi_1)^2\right)^{1/3}. \tag{54}$$

Further, a simple calculation shows that, if $\varrho = 0$, then:

$$H_1^3 - h_1^3 = \frac{1}{2\sigma_1^4}\left(\frac{\mu_1\mu_2}{\sigma_1\sigma_2}\right)^2. \tag{55}$$

In other words, the no-trade region in the larger market is always wider than the no-trade region with one asset, and they coincide only if either asset is useless ($\mu_1 = 0$ or $\mu_2 = 0$). In all other cases, the presence of an independent asset increases the no-trade region of the others, presumably because the variation of the position in each asset becomes less important for the overall welfare of the investor than with a single asset. This observation clearly runs against the common wisdom of fund-separation results for frictionless markets, and has potential implications for intermediation and welfare.

Note that in a frictionless market an investor with power utility is indifferent between trading two uncorrelated assets with Sharpe ratios μ_1/σ_1, μ_2/σ_2, and a single asset with Sharpe ratio $\sqrt{(\mu_1/\sigma_1)^2 + (\mu_2/\sigma_2)^2}$, that is, squared Sharpe ratios and in turn equivalent safe rates add across independent shocks. The above observation suggests that this property no longer holds with transaction costs, and an important open question is to understand the welfare difference between the two markets. If the two-asset market is more attractive, then investors benefit from access to individual securities rather than only to a limited number of funds, in contrast to classic fund-separation results. Of course, the question is whether this effect is indeed present and large enough to be relevant.

5.2 *Predictability*

Can future stock returns be predicted with public information? And what increase in welfare can one expect from this information? Predictably enough, these questions have generated a voluminous literature, which evaluates the statistical significance as well as the in-sample and out-of-sample performance of several predictors that focus either on stock characteristics, such as the dividend-yield and earnings-price ratio, or interest rates, such as the term-spread and the corporate-spread.

Perhaps less predictably, this voluminous literature remains divided between the weak statistical significance of several models, and the strong economic significance of parameter estimates. On the one hand, the standard errors of the predictability parameters are of the same order of magnitude as the parameter estimates themselves; on the other hand, these estimates—if valid—imply a substantial welfare increase. These opposing viewpoints are discussed in [45], who offer a critical view of the empirical literature, and find that most models have poor out-of-sample performance, and [7], who argues that the absence of predictability in dividend growth implies the presence of return predictability.

Kim and Omberg [27] introduce a basic model with predictable returns, based on one asset with price S_t, and one state variable θ_t:

$$dS_t/S_t = (\mu + \alpha\theta_t)dt + \sigma dW_t, \tag{56}$$

$$d\theta_t = -\kappa\theta_t dt + dB_t. \tag{57}$$

Here θ_t represents a state variable, like the dividend yield, that helps predict future returns, in that the conditional distribution of S_T/S_t at time t depends on θ_t. The two Brownian motions W and B typically have a substantial negative correlation ϱ. The parameter α controls the predictability of returns, with $\alpha = 0$ corresponding to the classical case of IID returns.

In such a market, expected returns change over time, mean-reverting to the average μ. Such variation is detrimental for an investor who adopts the constant policy $\pi = \frac{\mu}{\gamma\sigma^2}$, which is optimal for $\alpha = 0$, because time-varying returns increase the dispersion of the final payoff. However, the investor can benefit from *market timing*, that is the ability to adopt an investment policy that depends on the current value of the state variable θ_t. This point is easily seen for logarithmic utility, for which the optimal portfolio is $\pi_t = (\mu + \alpha\theta_t)/\sigma^2$, and the corresponding equivalent safe rate has the simple formula:

$$\lim_{T\to\infty} \frac{1}{T} E\left[\log X_T^\pi\right] = \frac{\mu^2}{2\sigma^2} + \frac{\alpha^2}{4\kappa\sigma^2}. \tag{58}$$

This expression shows that the investor benefits from stronger signals (larger α) and from slower mean reversion of the return rate (smaller κ), and the same conclusion broadly applies to power utility, even though the formulas become clumsier, as the optimal portfolio includes an intertemporal hedging component that is absent in the logarithmic case.

The above calculation underlies most estimates of the economic significance of predictability, but obviously ignores transaction costs. This omission may be especially important, as market timing requires active trading, which in turn entails higher costs. In short, while the potential benefit of predicability is clear from the frictionless theory, its potential costs are blissfully ignored, but may be substantial, and a priori may or may not offset benefits.

Remarkably enough, the above model with transaction costs has never been solved, even for logarithmic utility. Intuitively, the solution of this model should lead to a buy curve $\pi_-(\theta)$ and to a sell curve $\pi_+(\theta)$, which describe the no-trade region for each value of the state variable θ_t. Still at an intuitive level, the width of the no-trade region should be wider for values of θ that are farther from zero, since the portfolio is increasingly likely to return towards the frictionless optimum without trading.

At the technical level, the model includes two state variables: the predictor θ_t, and the current risky weight π_t. The presence of two state variables in turn implies that the value function satisfies an elliptic linear partial differential equation within the no-trade region, along with the boundary and smooth-pasting conditions at the boundary. The difficulty is to characterize the shape of the no-trade interval $[\pi_-(\theta), \pi_+(\theta)]$, as a function of the state θ_t, along with its implied equivalent safe rate.

Solving such a model can contribute to the predictability debate by clarifying the extent to which the ability to *forecast* future returns can translate into the ability to *deliver* higher returns by trading. When transaction costs are included, it may turn out that potential benefits of market timing are minimal, even if return predictability is statistically significant.

5.3 Options Spreads

Options listed on stock exchanges display much wider bid-ask spreads than their underlying assets. While the spread on a large capitalization stock is typically less than ten basis points, even the most liquid at-the-money options have spreads of several *percentage* points. To the best of our knowledge, there seems to be no theoretical work that links the bid-ask spread of an asset to the spread of its options. The closest research appears to be that on stochastic dominance bounds [9–11], which should be satisfied in equilibrium among utility maximizers. Interestingly, Constantinides et al. [12, 13] report frequent violations of these bounds, even among commonly traded options.

Of course, in a frictionless, complete market, both spreads are zero, and the option is replicated by a trading strategy in the underlying asset. The problem is that introducing a bid-ask spread for the underlying asset immediately makes the notion of option price ambiguous. Even if the asset follows a geometric Brownian motion, with transaction costs the superreplication price of any call option equals the stock price itself [43]. Similarly, the subreplication price is zero. Thus, one cannot interpret the bid and ask prices of the option as replication bounds, if the intention is to obtain a realistic spread.

In contrast to the previous two problems, in which the model is clear and the challenges are mathematical, this question poses some conceptual issues at the outset. One possibility is to interpret the bid and ask prices of the options as marginal prices in a partial equilibrium setting. For example, suppose that the bid and ask prices of the asset are exogenous, and follow geometric Brownian motion, with a constant relative bid-ask spread. Suppose also that a representative investor freely trades this asset, and a European option with maturity T, as to maximize utility from terminal wealth, either at the same maturity, or at some long horizon.

Since options, unlike stocks, exist in zero net supply, assume that the representative investor's optimal policy is to keep a zero position in the option at all times. In a complete frictionless market, this condition uniquely identifies the option price as the unique arbitrage-free price. With transaction costs, it leaves more flexibility in option price dynamics. Indeed, consider the shadow price corresponding to the utility maximization problem. Since the shadow market is complete, the shadow asset price uniquely identifies a shadow price for the option as the conditional expectation under the risk-neutral probability. For an option of European type with payoff $G(S_T)$, the latter will then be a function $g(t, S_t, Y_t)$ of time, the current stock price, and the current value of the state variable measuring the ratio of risky and safe positions.

Now, suppose that to the original (not shadow) market one adds the option, with a price dynamics equal to the shadow option price, and zero spread. This market is equivalent to the one with the asset only: by contradiction, if some trading strategy delivered a higher utility than the optimum in the asset-only market, the same strategy would also deliver the same or higher utility in the shadow market (by domination), thereby contradicting the definition of a shadow market. Now, the shadow option price depends on the state variable, which is unobservable since market makers cannot see the private positions of market participants. However, taking the pointwise maxima $\overline{g}(t, S_t) = \max_{y\in[0,\log(u/l)]} g(t, S_t, y)$ and minima $\underline{g}(t, S_t) = \min_{y\in[0,\log(u/l)]} g(t, S_t, y)$ over *all* values of $Y_t \in [0, \log(u/l)]$, one can obtain observable upper and lower bounds on the option price, which depend on the asset price alone. Such bounds are natural candidates for bid and ask prices of the option, because they are the minimal observable bounds that an option price needs to satisfy if its net demand has to be zero.

The question is whether this construction can predict bid-ask spreads that are consistent with the ones observed in reality, hence much wider than those of the underlying asset.

References

1. C. Atkinson, P. Ingpochai, The influence of correlation on multi-asset portfolio optimization with transaction costs. J. Comput. Financ. **10**(2), 53–96 (2006)
2. C. Atkinson, S. Mokkhavesa, Multi-asset portfolio optimization with transaction cost. Appl. Math. Financ. **11**(2), 95–123 (2004)
3. M. Bichuch, Asymptotic analysis for optimal investment in finite time with transaction costs. SIAM J. Financ. Math. **3**(1), 433–458 (2012)
4. M. Bichuch, S.E. Shreve, Utility maximization trading two futures with transaction costs. Preprint (2011)
5. P. Carr, Randomization and the American put. Rev. Financ. Stud. **11**(3), 597–626 (1998)
6. J. Choi, M. Sirbu, G. Žitković, Shadow prices and well-posedness in the problem of optimal investment and consumption with transaction costs. Preprint (2012)
7. J.H. Cochrane, The dog that did not bark: A defense of return predictability. Rev. Financ. Stud. **21**(4), 1533–1575 (2008)
8. G.M. Constantinides, Capital market equilibrium with transaction costs. J. Polit. Econ. **94**(4), 842–862 (1986)
9. G.M. Constantinides, S. Perrakis, Stochastic dominance bounds on derivatives prices in a multiperiod economy with proportional transaction costs. J. Econ. Dyn. Control **26**(7–8), 1323–1352 (2002)
10. G.M. Constantinides, S. Perrakis, Stochastic dominance bounds on american option prices in markets with frictions. Rev. Financ. **11**(1), 71–115 (2007)
11. G.M. Constantinides, T. Zariphopoulou, Bounds on prices of contingent claims in an intertemporal economy with proportional transaction costs and general preferences. Financ. Stoch. **3**(3), 345–369 (1999)
12. G.M. Constantinides, J.C. Jackwerth, S. Perrakis, Mispricing of S&P 500 index options. Rev. Financ. Stud. **22**(3), 1247–1277 (2009)
13. G.M. Constantinides, M. Czerwonko, J. Jackwerth, S. Perrakis, Are options on index futures profitable for risk-averse investors? empirical evidence. J. Financ. **66**(4), 1407–1437 (2011)
14. J. Cvitanić, I. Karatzas, Hedging and portfolio optimization under transaction costs: a martingale approach. Math. Financ. **6**(2), 133–165 (1996)
15. M.H.A. Davis, A.R. Norman, Portfolio selection with transaction costs. Math. Oper. Res. **15**(4), 676–713 (1990)
16. B. Dumas, Super contact and related optimality conditions. J. Econ. Dyn. Control **15**(4), 675–685 (1991)

17. B. Dumas, E. Luciano, An exact solution to a dynamic portfolio choice problem under transactions costs. J. Financ. **46**(2), 577–595 (1991)
18. S. Gerhold, J. Muhle-Karbe, W. Schachermayer, The dual optimizer for the growth-optimal portfolio under transaction costs. Financ. Stoch. **17**(2), 325–354 (2013). doi:10.1007/s00780-011-0165-9
19. S. Gerhold, P. Guasoni, J. Muhle-Karbe, W. Schachermayer, Transaction costs, trading volume, and the liquidity premium. Financ. Stoch. (May 2013). doi:10.1007/s00780-013-0210-y
20. P. Guasoni, J. Muhle-Karbe, Long horizons, high risk-aversion, and endogenous spreads. Preprint (2011)
21. P. Guasoni, S. Robertson, Portfolios and risk premia for the long run. Ann. Appl. Probab. **22**(1), 239–284 (2012)
22. P. Guasoni, M. Rásonyi, W. Schachermayer, Consistent price systems and face-lifting pricing under transaction costs. Ann. Appl. Probab. **18**(2), 491–520 (2008)
23. A. Herzegh, V. Prokaj, Shadow price in the power utility case. Preprint (2011)
24. K. Janeček, S.E. Shreve. Asymptotic analysis for optimal investment and consumption with transaction costs. Financ. Stoch. **8**(2), 181–206 (2004)
25. E. Jouini, H. Kallal, Martingales and arbitrage in securities markets with transaction costs. J. Econ. Theory **66**(1), 178–197 (1995)
26. J. Kallsen, J. Muhle-Karbe, On using shadow prices in portfolio optimization with transaction costs. Ann. Appl. Probab. **20**(4), 1341–1358 (2010)
27. T.S. Kim, E. Omberg, Dynamic nonmyopic portfolio behavior. Rev. Financ. Stud. **9**(1), 141–161 (1996)
28. D. Kramkov, W. Schachermayer, The asymptotic elasticity of utility functions and optimal investment in incomplete markets. Ann. Appl. Probab. **9**(3), 904–950 (1999)
29. S.L. Law, C.F. Lee, S. Howison, J.N. Dewynne, Correlated multi-asset portfolio optimisation with transaction cost. Preprint (2007)
30. H. Liu, Optimal consumption and investment with transaction costs and multiple risky assets. J. Financ. **59**(1), 289–338 (2004)
31. J. Liu, Portfolio selection in stochastic enviroments. Rev. Financ. Stud. **20**(1), 1–39 (2007)
32. H. Liu, M. Loewenstein, Optimal portfolio selection with transaction costs and finite horizons. Rev. Financ. Stud. **15**(3), 805–835 (2002)
33. R. Liu, J. Muhle-Karbe, Portfolio selection with small transaction costs and binding portfolio constraints. Preprint (2012)
34. M. Loewenstein, On optimal portfolio trading strategies for an investor facing transactions costs in a continuous trading market. J. Math. Econ. **33**(2), 209–228 (2000)
35. M.J.P. Magill, G.M. Constantinides, Portfolio selection with transactions costs. J. Econ. Theory **13**(2), 245–263 (1976)
36. R.C. Merton, Lifetime portfolio selection under uncertainty: The continuous-time case. Rev. Econ. Stat. **51**(3), 247–257 (1969)
37. R.C. Merton, Optimum consumption and portfolio rules in a continuous-time model. J. Econ. Theory **3**(4), 373–413 (1971)
38. K. Muthuraman, S. Kumar, Multidimensional portfolio optimization with proportional transaction costs. Math. Financ. **16**(2), 301–335 (2006)
39. L.C.G. Rogers, Why is the effect of proportional transaction costs $O(\delta^{2/3})$? in *Mathematics of Finance*. Contemporary Mathematics, vol. 351 (American Mathematical Society, Providence, 2004), pp. 303–308
40. S.E. Shreve, H.M. Soner, Optimal investment and consumption with transaction costs. Ann. Appl. Probab. **4**(3), 609–692 (1994)
41. A.V. Skorohod, Stochastic equations for diffusion processes with boundaries, II. Teor. Verojatnost. i Primenen. **7**, 5–25 (1962)
42. H.M. Soner, N. Touzi, Homogenization and asymptotics for small transaction costs. Preprint (2012)
43. H.M. Soner, S.E. Shreve, J. Cvitanić, There is no nontrivial hedging portfolio for option pricing with transaction costs. Ann. Appl. Probab. **5**(2), 327–355 (1995)

44. M. Taksar, M.J. Klass, D. Assaf, A diffusion model for optimal portfolio selection in the presence of brokerage fees. Math. Oper. Res. **13**(2), 277–294 (1988)
45. I. Welch, A. Goyal, A comprehensive look at the empirical performance of equity premium prediction. Rev. Financ. Stud. **21**(4), 1455–1508 (2008)
46. A.E. Whalley, P. Wilmott, An asymptotic analysis of an optimal hedging model for option pricing with transaction costs. Math. Financ. **7**(3), 307–324 (1997)

Cubature Methods and Applications

D. Crisan, K. Manolarakis, and C. Nee

Abstract We present an introduction to a new class of numerical methods for approximating distributions of solutions of stochastic differential equations. The convergence results for these methods are based on certain sharp gradient bounds established by Kusuoka and Stroock under non-Hörmader constraints on diffusion semigroups. These bounds and some other subsequent refinements are covered in these lectures. In addition to the description of the new class of methods and the corresponding convergence results, we include an application of these methods to the numerical solution of backward stochastic differential equations. As it is well-known, backward stochastic differential equations play a central role in pricing financial derivatives.

1 Introduction

Stochastic differential equations (SDEs) are ideal models for the evolution of randomly perturbed dynamical systems. Such systems pervade a variety of areas of human activity, including biology, communications, engineering, finance and physics.

The solution of an SDE is amenable to numerical approximations even in high dimensions. Classical methods such as the Euler method work well provided the distribution of the SDE and the function that we wish to integrate are sufficiently smooth. In particular, when the SDE is driven by non-singular noise, the convergence properties of classical numerical methods are well understood. However, in the 1980s, Kusuoka and Stroock [34] relaxedthe conditions under which some of

D. Crisan (✉) · K. Manolarakis · C. Nee
Department of Mathematics, Imperial College London, 180 Queen's Gate, London SW7 2AZ, UK
e-mail: d.crisan@imperial.ac.uk

F.E. Benth et al., *Paris-Princeton Lectures on Mathematical Finance 2013*, Lecture Notes in Mathematics 2081, DOI 10.1007/978-3-319-00413-6_4,

the smoothness properties of the semigroup associated to the solution of the SDE remain valid. They replaced the classical Hörmander condition requirement by a weaker condition: the so-called UFG condition. Essentially, this condition states that the Lie algebra generated by the vector fields appearing in the noise term of the equation is finite dimensional when viewed as a module over the space of bounded infinitely differentiable functions. Kusuoka and Stroock showed that the semigroup remains smooth in any direction belonging to the above algebra. This fundamental result forms the theoretical basis of a recently developed class of high accuracy numerical methods. In the last 10 years, Kusuoka, Lyons, Ninomiya and Victoir [29,36,49] developed several numerical algorithms based on Chen's iterated integrals expansion. These new algorithms generate approximations to the solution of the SDE in the form of the empirical distribution of a cloud of particles with deterministic trajectories. They work under a weaker condition (termed the UFG condition, see Sect. 2.3 for details) rather than the ellipticity/Hörmander condition and are faster than the corresponding classical methods. Let us describe briefly the framework and structure of these methods:

In the following, let $(\Omega, \mathcal{W}, \mathbb{P})$ be the standard (d-dimensional) Wiener space:

$$\Omega = \{\omega \in \mathcal{C}([0,\infty);\mathbb{R}^d), \omega(0) = 0\}, \qquad \mathcal{W} = \mathcal{B}(C([0,\infty);\mathbb{R}^d)),$$

where $\mathcal{C}([0,\infty);\mathbb{R}^d)$ is the set of $\mathbb{R}^d$-valued continuous paths endowed with the corresponding Borel σ-algebra $\mathcal{B}(C([0,\infty);\mathbb{R}^d))$ and $\mathbb{P}$ is the probability measure such that the coordinate mapping process:

$$B = \{B_t = (B_t^i)_{i=1}^d, t \in [0,\infty)\}, \qquad B_t(\omega) := \omega(t) := (\omega_i(t) : i = 1,\dots,d)$$

is a d-dimensional Brownian motion under $\mathbb{P}$. We define $B_t^0 := t$ for notational simplicity.

Let $V_0, V_1, \dots, V_d \in \mathcal{C}(\mathbb{R}^N;\mathbb{R}^N)$ be $d+1$ Lipschitz vector fields and

$$X = \{X_t^x, t \in [0,\infty), x \in \mathbb{R}^N\}$$

be the solution of the following stochastic differential equation

$$X_t^x = x + \sum_{i=0}^{d} \int_0^t V_i\left(X_s^x\right) dB_s^i. \tag{1}$$

Equation (1) has a unique solution (see, for example, Theorem 2.9 page 289 in[26]). To be more precise, there exists a unique stochastic process adapted with respect to the augmented filtration generated by the Brownian motion B for which identity (1) holds true. The measurability property of X_t is crucial. However, this condition is sometimes overlooked and treated as a rather meaningless theoretical requirement. In effect, the condition means that there is a $\mathcal{B}(C([0,t];\mathbb{R}^d))/\mathcal{B}(\mathbb{R}^N)$-measurable mapping $\alpha_{t,x} : \mathcal{C}([0,t];\mathbb{R}^d) \to \mathbb{R}^N$ such that

$$X_t^x = \alpha_{t,x}(B_{[0,t]}), \quad \mathbb{P} - \text{a.s.} \tag{2}$$

Hence X_t^x is determined by the driving noise $\{B_s, s \in [0,t]\}$. Put differently, if we know B then (theoretically) we will also know the value of X_t^x.[1]

Example 1. For the following equations, one can explicitly write the solution of the SDE as a function of the Brownian motion B:

$$X_t^x = x + \int_0^t aX_s^x dB_s^0 + \int_0^t bX_s^x dB_s^1, \; X_t^x = x\exp\left(bB_t^1 + (a - b^2/2)B_t^0\right) \tag{3}$$

$$X_t^x = x + \int_0^t aX_s^x dB_s^0 + \int_0^t b dB_s^1, \; X_t = xe^{aB_t^0} + b\int_0^t e^{a(B_t^0 - B_s^0)} dB_s^1 \tag{4}$$

$$\begin{pmatrix} X_t^{x,1} \\ X_t^{x,2} \end{pmatrix} = \begin{pmatrix} x^1 \\ x^2 \end{pmatrix} + \int_0^t \begin{pmatrix} a \\ 0 \end{pmatrix} dB_s^1 + \int_0^t \begin{pmatrix} 0 \\ bX_s^1 \end{pmatrix} dB_s^2 \tag{5}$$

$$\begin{pmatrix} X_t^{x,1} \\ X_t^{x,2} \end{pmatrix} = \begin{pmatrix} x^1 + aB_t^1 \\ x^2 + \int_0^t b(x^1 + aB_s^1) dB_s^2 \end{pmatrix} \tag{6}$$

In general it is not possible to have explicit formulae for the solution of the stochastic differential equation, in other words the mapping $\alpha_{t,x}$ appearing in the representation (2) is not known. Hence accurate numerical approximations of X are highly desirable. In particular, we are interested in computing quantities of the form

$$\mathbb{E}[\varphi(X_{t,x})] = \mathbb{E}[\varphi \circ \alpha_{t,x}(B_{[0,t]})] = \int_\Omega \varphi \circ \alpha_{t,x}(\omega)\, \mathbb{P}(d\omega), \tag{7}$$

where φ is a given test function and $\mathbb{P}$ is the probability distribution of the Brownian motion (the Wiener measure). The computation of expectations of the form (7) has particular relevance in mathematical finance through the pricing of financial contracts. Indeed, calculating the expected value of functionals of the solution of a stochastic differential equation (which would be assumed as the model of the underlying price process) in a very short time is a standard problem in finance and is one which has ruled more exotic models out of practical implementation in industry.

Computing quantities of the form (7) is also relevant for the estimation of infinite dimensional random dynamical systems. The theory of infinite dimensional

[1]The process X is uniquely identified by (1) only up to a set of measure 0. Two processes X^1 and X^2 satisfying (1) are indistinguishable: the set $\{\omega \in \Omega | \exists t \in [0,\infty) \text{ such that } X_t^1(\omega) = X_t^2(\omega)\}$ is a $\mathbb{P}$-null set (has probability zero). Similarly, the identity (2) holds $\mathbb{P}$-almost surely, i.e. there can be a subset of Ω of probability zero where (2) does not hold.

random dynamical systems shares many of the concepts and results with their finite dimensional counter-parts. Many examples are determined by stochastic and deterministic partial differential equations. These partial differential equations have solutions $u(t,x)$ that admit certain representations, called Feynman–Kac representations, in terms of certain functionals integrated with respect to the law of a stochastic process:

$$u(t,x) = \mathbb{E}[\Lambda_{t,x}\left(X^x_{[0,t]}\right)]. \tag{8}$$

A large class of such PDEs exhibit the common feature that the process X appearing in (8) has a representation of the form (2) hence their solution $u(t,x)$ can be represented as

$$u(t,x) = \mathbb{E}[(\Lambda_{t,x} \circ \alpha_{t,x})\left(B_{[0,t]}\right)], \tag{9}$$

where the functional $\Lambda'_{t,x} \equiv \Lambda_{t,x} \circ \alpha_{t,x}$ is nonlinear and, possibly, implicitly defined. Examples, include linear PDEs, semilinear PDEs such as those appearing in the pricing of financial derivatives under trading constraints, McKean–Vlasov equations, Navier–Stokes equation, Burgers equation, Zakai equation, etc.

It follows that the computation of $u(t,x)$ requires the approximation of the law of the process X if the Feynman–Kac formula (8) is used, or the law of the Brownian motion B if one uses (9) instead. However this is not enough. The functionals $\Lambda_{t,x}$ respectively $\Lambda'_{t,x}$ do not have a closed form, in other words they cannot be explicitly described and, more importantly, integrated with respect to the approximating law. One needs to approximate them with versions whose integral with respect to the corresponding approximating law can be easily computed. Obviously, the error of the approximation of the solution of the PDE obtained in this manner will depend on both the error introduced when approximating the functional and that introduced when approximating the law of the process. Care must be taken so as not to compound the corresponding errors. In practice both approximations are performed simultaneously. Nevertheless, when it comes to estimating the approximation error it helps to separate them. The numerical methods discussed in the following entail the following three steps:

- Replacing the law of B with the law of a simpler process $\tilde{B}$. The process $\tilde{B}$ will have bounded variation paths and its so-called "signature" will approximate that of the original B. The support of the law of the process $\tilde{B}_{[0,t]}$ is chosen to have finite support. In other words, there are only a finite number of paths, $\omega_i : [0,t] \to \mathbb{R}^d$, $i = 1,\ldots,n_t$ such that

$$\lambda_{i,t} := \mathbb{P}(\tilde{B}_{[0,t]} = \omega_i) > 0.^2$$

[2] Of course the sum of the weights $\lambda_{i,t}$ is 1, i.e., $\sum_{i=1}^{n_t} \lambda_{i,t} = 1$.

- Approximating $\Lambda'_{t,x}$ with an explicit/simple version $\tilde{\Lambda}'_{t,x}$. Here we will exploit the smoothness properties of the functional $\Lambda'_{t,x}$. Such properties will be analyzed in the next chapter by using Malliavin Calculus techniques.
- Integrate $\tilde{\Lambda}'_{t,x}$ with respect to the law of $\tilde{B}$. This step consists in computing the average of $\tilde{\Lambda}'_{t,x}$ estimated over the n_t realizations of $\tilde{B}_{[0,t]}$. In essence, we will have

$$u(t,x) \simeq \mathbb{E}\left[\tilde{\Lambda}'_{t,x}\left(\tilde{B}_{[0,t]}\right)\right] = \sum_{i=1}^{n_t} a_{i,t}\tilde{\Lambda}'_{t,x}(\omega_i).$$

If the number of paths n_t contained in the support of $\tilde{B}_{[0,t]}$ is above a threshold that depends on the capabilities of the hardware on which the algorithm is run, then an additional procedure is required to reduce n_t to a manageable size. One can employ a Monte Carlo procedure similar to that used in the classical schemes (e.g. Euler–Maruyama) or the so-called "tree based branching algorithm" [14], a minimal variance selection procedure analysed in Sect. 3.

To understand the choice of the simple process $\tilde{B}$, let us introduce briefly the classical Euler–Maruyama method.[3] For this we choose a partition Π of a generic interval, say, $[0,T]$

$$\Pi\ :\ \ 0 = \tau_0 < \tau_1 < \ldots < \tau_n < \ldots \tau_N = T.$$

and we denote by δ the mesh of the partition $\delta = \max_{i=1,\ldots,N}(\tau_i - \tau_{i-1})$. We do not specify the choice of the partition. However if the partition is equidistant, then $\delta = T/N$ and it is also called the time step. Let $Y^x = \{Y_t^x,\ t \in [0,T]\}$ be the continuous time process satisfying the evolution equation

$$Y_t^x = Y_{\tau_n}^x + \int_{\tau_n}^t V_0\left(Y_s^x\right) ds + \sum_{i=1}^d \frac{1}{\sqrt{\tau_{n+1}-\tau_n}} \int_{\tau_n}^t V_i\left(Y_s^x\right) \xi_n^i ds, \quad t \in \left[\tau_n, \tau_{n+1}\right], \tag{10}$$

where $\{\xi_n^i,\ i = 1,\ldots,d,\ n = 0,\ldots,N-1\}$ are mutually independent random variables whose moments match the moments of a standard Gaussian random variable up to order 3 and with initial value $Y_0^x = x$. More precisely we require that the random variables ξ_n^i must be independent, with moments satisfying,

$$\mathbb{E}\left[\xi_n^i\right] = \mathbb{E}\left[\left(\xi_n^i\right)^3\right] = 0, \quad \mathbb{E}\left[\left(\xi_n^i\right)^2\right] = 1. \tag{11}$$

In particular, ξ_n^i can be chosen to have the Bernoulli distribution

---

[3] To be more precise, following the phraseology of [27], we describe here the *simplified weak* Euler scheme for a scalar SDE driven by a multi-dimensional noise.

$$P\left(\xi_n^i = \pm 1\right) = \frac{1}{2}. \tag{12}$$

We can recast the evolution equation of the process Y^x in a similar manner to that of X^x. Let $\tilde{B} = \{\tilde{B}_t, t \in [0, \infty)\}$ be the d-dimensional stochastic process[4]

$$\tilde{B}_t = \xi_{[Nt]}(t - \tau_{[NT]}) + \sum_{n=1}^{[Nt]} \xi_{n-1}\sqrt{\tau_n - \tau_{n-1}}, \tag{13}$$

where the last term is chosen to be 0 if $[NT] = 0$ and $\{\xi_n,\ n = 0, \ldots, N\}$ are d-dimensional random vectors with corresponding entries $\xi_n = \left(\xi_n^1, \ldots, \xi_n^d\right)$. Then $\tilde{B}$ has piecewise-linear trajectories and, if we use an equidistant partition, the support of $\tilde{B}_t$ has $n_t = 2^n$ paths for $t \in (\tau_{n-1}, \tau_n], n = 1, \ldots, [NT]$. Then Y is the solution of the following *ordinary* differential equation

$$Y_t^x = x + \sum_{i=0}^{d} \int_0^t V_i\left(Y_s^x\right) d\tilde{B}_s^i, \tag{14}$$

where, as in (1), we defined $\tilde{B}_t^0 := t$. Under suitable conditions, the process Y^x, is a first order approximation of the equation (1) associated with the partition Π (see, for example Theorem 14.1.5, page 460 in [27]). More precisely, we have

$$|\mathbb{E}[\varphi(X_t^x)] - \mathbb{E}[\varphi(Y_t^x)]| \leq C_\varphi \delta,\ \ t \in [0, T].$$

The paths $\{\omega_1, \ldots, \omega_{n_t}\}$ in the support of $\tilde{B}$ are the realizations of a random walk (linearly interpolated between jumps). Then $\lambda_{i,t} := \mathbb{P}(\tilde{B}_{[0,t]} = \omega_i) = \frac{1}{n_t}$ and

$$\mathbb{E}[\varphi(Y_t^x)] = \sum_{i=1}^{n_t} \lambda_{i,t}\varphi(Y_t^{x,i}),$$

where $Y^{x,i}$ is the solution of the ordinary differential equation (14) corresponding to the path ω_i. That is

$$Y_t^{x,i} = x + \sum_{j=0}^{d} \int_0^t V_j\left(Y_s^{x,i}\right) d\omega_i^j(s).$$

If the ordinary differential equation (14) has no explicit solution, one can choose without loss of accuracy, a process Z^x which satisfies an explicit/implicit

[4] In (13) and subsequently, $[z]$ denotes the integer part of $z \in \mathbb{R}$.

discretization of (14). For example

$$Z^x_{\tau_{n+1}} = Z^x_{\tau_n} + V_0\left(Z^x_{\tau_n}\right)(\tau_{n+1} - \tau_n) + \sum_{i=1}^{d} V_i\left(Z^x_{\tau_n}\right)\xi^i_n\sqrt{\tau_{n+1} - \tau_n}. \tag{15}$$

The solution of (15) is customarily called the Euler–Maruyama approximation of X and has the same order of approximation as Y (order 1). The ODE (14) has solutions that evolve in the support of the original diffusion so it manifests good numerical stability conditions. Classical higher order approximations of (1) such as those described in Chaps. 14 and 15 in [27] no longer have this property. The question that arises is whether it would be possible to produce a high order approximation that still has this property. The answer is yes and this is exactly what a cubature method does. One can replace the process $\tilde{B}$ by a "better" approximation of B which, in turn, will lead to a high order approximation of the solution of (1). To understand in what sense $\tilde{B}$ is an approximation of B and how can it be improved we need to explain in brief the concept of a signature of a path. Let

$$T\left(\mathbb{R}^d\right) = \bigoplus_{i=0}^{\infty}\left(\mathbb{R}^d\right)^{\otimes i}, \quad T^{(m)}\left(\mathbb{R}^d\right) = \bigoplus_{i=0}^{m}(\mathbb{R}^d)^{\otimes i}$$

be the tensor algebra of all non-commutative polynomials over $\mathbb{R}^d$ and, respectively, the tensor algebra of all non-commutative polynomials of degree less than $m+1$. For a path $\omega : [0,\infty) \to \mathbb{R}^d$ with finite variation we define its signature $S_{s,t}(\omega) \in T\left(\mathbb{R}^d\right)$ to be the corresponding Chen's iterated integrals expansion:

$$S_{s,t}(\omega) = \sum_{k=0}^{\infty}\int_{s<t_1\ldots t_k<t} d\omega_{t_1}\otimes\ldots\otimes d\omega_{t_k},$$

where

$$\int_{0<t_1\ldots t_k<t} d\omega_{t_1}\otimes\ldots\otimes d\omega_{t_k} := \sum_{i_1,\ldots,i_k}\left(\int_{0<t_1\ldots t_k<t} d\omega^{i_1}_{t_1}\ldots d\omega^{i_k}_{t_k}\right)e_{i_1}\otimes\ldots\otimes e_{i_k},$$

and $(e_{i_1}\otimes\ldots\otimes e_{i_k}), i_1,\ldots,i_k \in \{1,\ldots,d\}$, is the canonical basis of $(\mathbb{R}^d)^{\otimes k}$. Similarly we define its truncated signature $S_{s,t}(\omega) \in T^{(m)}\left(\mathbb{R}^d\right)$ to be

$$S^m_{s,t}(\omega) = \sum_{k=0}^{m}\int_{s<t_1\ldots t_k<t} d\omega_{t_1}\otimes\ldots\otimes d\omega_{t_k}.$$

Similarly the (random) signature and, respectively, the truncated signature of the Brownian motion are

$$S_{s,t}(B) = \sum_{k=0}^{\infty} \int_{s<t_1\ldots t_k<t} dB_{t_1} \otimes \ldots \otimes dB_{t_k}, \; S_{s,t}^m(B) = \sum_{k=0}^{m} \int_{s<t_1\ldots t_k<t} dB_{t_1} \otimes \ldots \otimes dB_{t_k}. \tag{16}$$

In (16), the stochastic (iterated) integrals are of Stratonovitch type.

The expected value of $S_{s,t}(B)$ uniquely identifies the law of B, i.e., the Wiener measure.[5] Moreover, if $\hat{B}$ is a process such that

$$\mathbb{E}\left[S_{k\delta,(k+1)\delta}^m(B)\right] = \mathbb{E}[S_{k\delta,(k+1)\delta}^m(\hat{B})], k = 0, 1, \ldots, N-1, \tag{17}$$

then for certain classes of functionals Λ', $\mathbb{E}[\Lambda'(B')]$ is a high order approximation of $\mathbb{E}[\Lambda'(B)]$. In particular, if $\Lambda'_{t,x}$ is the functional that gives the solution of the SDE (1) for $t = N\delta$, i.e., $\Lambda'_{t,x}(B) = \varphi(X_t^x)$, then

$$|\mathbb{E}[\varphi(X_t^x)] - \mathbb{E}[\varphi(Y_t^x)]| \le C_\varphi \delta^{\frac{m-1}{2}}, \tag{18}$$

where Y_t^x is the solution of the ordinary differential equation (14) driven by $\tilde{B}$. We prove this result in Sect. 3 of the current lecture notes. In particular, the process $\tilde{B}$ as defined (13) satisfies (17) with $m = 3$.

The proof of (18), requires the smoothness of the (diffusion) semigroup $\{P_t, \; t \in [0, \infty)\}$ defined as

$$(P_t\varphi)(x) = \mathbb{E}[\varphi(X_t^x)], \quad x \in \mathbb{R}^d, \quad t \ge 0,$$

where φ is an appropriately chosen test function. If the vector fields satisfy the *ellipticity* or, more generally, the *uniform Hörmander* condition, $P_t\varphi$ is smooth for any bounded measurable function φ and $t > 0$. Many of the classical numerical schemes rely on this property and so Hörmander's paper [24] is a major contribution to this field. A probabilistic version of this result led Malliavin [40] to develop his celebrated stochastic calculus of variations through which one can prove, probabilistically, the sufficiency of Hörmander's condition.

The work of Kusuoka and Stroock [32–34] in the 1980s provided an extension of Malliavin's results. In it, they proved precise gradient bounds that are valid under a general condition termed the UFG condition, see Sect. 2.3 for details. The UFG condition imposed on the vector fields $\{V_i, i = 0, \ldots, d\}$ essentially states that the $C_b^\infty(\mathbb{R}^d)$-module $\mathcal{M}$ generated by the vector fields $\{V_i, i = 1, \ldots, d\}$ within the Lie algebra generated by $\{V_i, i = 0, \ldots, d\}$ is finite dimensional. The UFG condition implies Hörmander's hypoellipticity condition, but not viceversa. There are explicit examples for which Hörmander's condition fails to hold, but for which the UFG condition is satisfied (see Example 15). In particular, the condition does not require that the vector space $\{W(x)|W \in \mathcal{M}\}$ is homeomorphic to $\mathbb{R}^d$ for

[5] See Proposition 118 in [18].

any $x \in \mathbb{R}^d$. Moreover, under the UFG condition, the dimension of the space $\{W(x)|W \in \mathcal{M}\}$ is not required to be constant over $\mathbb{R}^d$. Such generality makes any Frobenius type approach to prove smoothness of the solution very difficult. Indeed the authors are not aware of any alternative proof of the smoothness results of the solution of $P_t\varphi$ (under the UFG condition) other than that given by Kusuoka and Stroock. Kusuoka and Stroock prove that, under the UFG condition, $P_t\varphi$ is differentiable in the direction of any vector field W belonging to $\mathcal{M}$ and deduce precise gradient bounds of the form:

$$\|W_1 \ldots W_k P_t\varphi\|_\infty \leq \frac{C^k}{t^l}\|\varphi\|_p, \tag{19}$$

where l is a constant that depends explicitly on the vector fields $W_i \in \mathcal{M}$, $i = 1, \ldots, k$ and $\|\varphi\|_p$ is the standard L_p norm of the function φ.

Whilst the Kusuoka–Stroock result does not suffice to justify the convergence of classical numerical schemes, it is tailor-made for the cubature methods. The global error of numerical schemes depends intrinsically on the smoothness of $P_t\varphi$, but only in the direction of the vector fields W belonging to $\mathcal{M}$. As a result, the cubature methods are proved to work under the more general UFG condition, unlike the classical numerical methods.

The lecture notes are structured as follows: In the following section, we provide a "clean" treatment of the (sharp) gradient bounds of the type (19) deduced under the minimal smoothness requirements on imposed on the vector fields $\{V_i, i = 0, \ldots, d\}$. Such results are intrinsically related to the solution of the linear parabolic partial differential equation

$$\partial_t u(t,x) = \frac{1}{2}\sum_{i=1}^{d} V_i^2 u(t,x) + V_0 u(t,x), \quad (t,x) \in (0,\infty) \times \mathbb{R}^d. \tag{20}$$

We show how the Kusuoka–Stroock approach can be used to recover the smoothness of the solution of (20) under the Hörmander condition. In the Hörmander case, it is straightforward to show that $P_t\varphi$ is indeed the (unique) classical solution of (20) with φ being the initial condition of the PDE. In particular we show that u is differentiable in any direction including direction V_0. The situation is more delicate in the absence of the Hörmander condition. Under the UFG condition, (20) may not have a solution in the classical sense. As explained in [44], it turns out that $P_t\varphi$ remains differentiable in the direction $\mathcal{V}_0 := \partial_t - V_0$ when viewed as a function $(t,x) \to P_t\varphi(x)$ over the product space $(0,\infty) \times \mathbb{R}^d$. This together with the continuity at $t = 0$ implies that $P_t\varphi$ is the unique (classical) solution of the equation

$$\mathcal{V}_0 u(t,x) = \frac{1}{2}\sum_{i=1}^{d} V_i^2 u(t,x), \quad (t,x) \in (0,\infty) \times \mathbb{R}^d. \tag{21}$$

In Sect. 3, we incorporate cubature methods into a larger class of methods, and deduce their convergence rates under the UFG condition and an additional constraint called *the* V_0 *condition*. We also deduce the convergence rates of the cubature methods combined with an algorithm for controlling the computational effort—the tree based branching algorithm (or TBBA for short). The section is concluded with an application of the cubature and TBBA method to the approximation of a call option on a Heston model price process.

Section 4 is dedicated to the application of cubature methods to the numerical solution of backward stochastic differential equations.

The lecture notes are concluded with an appendix comprising a number of technical lemmas and a proof of the convergence of the cubature method in the absence of the V_0 condition.

2 Sharp Gradient Bounds

In this chapter we give a full and self-contained proof of Kusuoka's gradient bounds (cf. [30]). The main difference between what is done there and what is presented here, is that we relax the restrictive assumptions on the SDE coefficients (in [30] they are assumed to be smooth and uniformly bounded). In later chapters, we shall apply these results to prove convergence of the cubature method.

2.1 Framework

Recall that $(\Omega, \mathcal{W}, \mathbb{P})$ is the standard (d-dimensional) Wiener space:

$$\Omega = \{\omega \in \mathcal{C}([0,\infty);\mathbb{R}^d), \omega(0) = 0\}, \qquad \mathcal{W} = \mathcal{B}(\mathcal{C}([0,\infty);\mathbb{R}^d)),$$

where $\mathcal{C}([0,\infty);\mathbb{R}^d)$ is the set of $\mathbb{R}^d$-valued continuous paths endowed with the uniform norm topology, $\mathcal{W}$ is the corresponding Borel σ-algebra $\mathcal{B}(\mathcal{C}([0,\infty);\mathbb{R}^d))$ and $\mathbb{P}$ is the probability measure such that the coordinate mapping process:

$$B = \{B_t, t \in [0,\infty)\}, \qquad B_t(\omega) := \omega(t) := (\omega_i(t) : i = 1,\ldots,d)$$

is a d-dimensional Brownian motion under $\mathbb{P}$. We define $B_t^0 := t$ for notational simplicity.

Let k be a positive integer to be determined at a later stage. Assume that $V_1,\ldots,V_d \in \mathcal{C}_b^{k+1}(\mathbb{R}^N;\mathbb{R}^N)$[6] and $V_0 \in \mathcal{C}_b^k(\mathbb{R}^N;\mathbb{R}^N)$ are $d+1$ vector fields and let

[6]For any positive integer m, the set $\mathcal{C}_b^m(\mathbb{R}^a;\mathbb{R}^b)$ is the set of all bounded continuous functions $\varphi : \mathbb{R}^a \to \mathbb{R}^b$, m-times continuously differentiable with all derivatives bounded.

$X = \{X_t^x, t \in [0,\infty), x \in \mathbb{R}^N\}$ be the following stochastic flow

$$X_t^x = x + \sum_{i=0}^{d} \int_0^t V_i(X_s^x) \circ dB_s^i. \tag{22}$$

In (22) the stochastic integrals $\int_0^t V_i(X_s^x) \circ dB_s^i$, $i = 1, \ldots, d$ are Stratonovitch integrals whereas $\int_0^t V_0(X_s^x) \circ dB_s^0$ is a standard Riemann integral.

Remark 2. In the following, we will view the vector fields $V_0, V_1, \ldots, V_d$ as both vector-valued functions and first order differential operators defined as follows: for $V_i(x) = (V_i^1(x), \ldots, V_i^N(x))^\top$ the corresponding first order differential operator will be

$$V_i = \sum_{j=1}^{N} V_i^j \partial_j, \qquad V_i f(x) = \nabla f(x) V_i(x), \quad \textit{where} \quad \nabla f(x) = (\partial_1 f(x), \ldots, \partial_N f(x)).$$

Using this notation, from (22) we have the standard chain rule

$$f(X_t^x) = f(x) + \sum_{i=0}^{d} \int_0^t V_i f(X_s^x) \circ dB_s^i$$

for any $f \in C_b^3(\mathbb{R}^N, \mathbb{R})$. We remark that the different levels of differentiability chosen for V_0 and $V_1, \ldots, V_d$ ensure that the corresponding Itô equation has $C_b^k(\mathbb{R}^N; \mathbb{R}^N)$ coefficients.

It is a classical result that the stochastic flow $X = \{X_t^x, t \in [0,\infty), x \in \mathbb{R}^N\}$ is differentiable in the space variable x. See for example Kunita [28] or Nualart [51, Theorem 2.2.1, p. 119]. We state the required result in the following:

Theorem 3. *Let $X = \{X_t^x, t \in [0,\infty), x \in \mathbb{R}^N\}$ be the solution of (22). Then X has a modification (again denoted by X) such that the mapping*

$$x \in \mathbb{R}^N \longrightarrow X_t^x \in \mathbb{R}^N$$

is k-times continuously differentiable, for each t, P-almost surely. Moreover the Jacobian of $X_t^{(\cdot)}$ at x, $J_t^{(\cdot)} := (\partial_j X_t^{i,(\cdot)})_{1 \le i,j \le N}$ satisfies the matrix stochastic differential equation[7]*:*

[7]In (23) and subsequently, ∂V_i is the matrix valued map $\partial V_i := (\partial_n V_i^m)_{1 \le n,m \le N}$.

$$\begin{cases} dJ_t^x = \sum_{i=0}^d \partial V_i(X_t^x) J_t^x \circ dB_t^i, \\ J_0^x = I. \end{cases} \tag{23}$$

The Jacobian is almost surely invertible (as a matrix) and its inverse, $(J_t^x)^{-1}$, satisfies the SDE

$$\begin{cases} d(J_t^x)^{-1} = -\sum_{i=0}^d (J_t^x)^{-1} \partial V_i(X_t^x) \circ dB_t^i, \\ (J_0^x)^{-1} = I. \end{cases} \tag{24}$$

In addition, the following integrability result holds

$$\sup_{t\in[0,T]} \mathbb{E}\left[\left|\frac{\partial^{|\gamma|} X_t^x}{\partial x^\gamma}\right|^p\right] < C_{T,p}, \quad \forall\, p \geq 1, T > 0, 0 < |\gamma| \leq k, \forall x \in \mathbb{R}^N. \tag{25}$$

2.2 Malliavin Differentiation

For an absolutely continuous path $h \in \mathcal{C}([0,\infty);\mathbb{R}^d)$, we denote by h' its derivative. Let H be the space

$$H = \{h \in \Omega,\ h \text{ absolutely continuous},\ h' \in L^2([0,\infty);\mathbb{R}^d)\} \subset \Omega.$$

H is endowed with a Hilbert structure under the inner product

$$\langle h, g\rangle_H := \langle h', g'\rangle_{L^2([0,\infty);\mathbb{R}^d)} := \int_0^\infty h'(u)\cdot g'(u)du$$

and is called the *Cameron–Martin* space. We use this space to define the Malliavin derivative.

Definition 4 (Malliavin Derivative). Let $f \in C_b^\infty(\mathbb{R}^n,\mathbb{R})$, $h_1,\ldots,h_n \in H$ and $F : \Omega \to \mathbb{R}$ be the functional given by:

$$F(\omega) = f\left(\int_0^\infty h_1'(t)dB_t(\omega),\ldots,\int_0^\infty h_n'(t)dB_t(\omega)\right), \tag{26}$$

where, for any $h_i' = (h_{i,1}',\ldots,h_{i,d}')$,

$$\int_0^\infty h_i'(t)dB_t := \sum_{j=0}^d \int_0^\infty h_{i,j}'(t)dB_t^j.$$

Any functional of the form (26) is called *smooth* and we denote the class of all such functionals by $\mathcal{S}$. Then the Malliavin derivative of F, denoted by $DF \in L^2(\Omega; H)$ is given by:

$$DF = \sum_{i=1}^{n} \partial_i f\left(\int_0^\infty h_1'(u)dB_u, \dots, \int_0^\infty h_n'(u)dB_u\right) h_i \tag{27}$$

We will often make use of the notation: $D_h F := \langle DF, h\rangle_H$ for $h \in H$. Observe that $D_h F$ is the directional derivative of F in the direction h as

$$\begin{aligned} D_h F(\omega) &= \sum_{i=1}^{d} \partial_i f\left(\int_0^\infty h_1'(u)dB_u(\omega), \dots, \int_0^\infty h_n'(u)dB_u(\omega)\right)\langle h_i, h\rangle_H \\ &= \frac{d}{d\epsilon} f\left(\int_0^\infty h_1'(u)dB_u(\omega) + \epsilon\langle h_1', h'\rangle_{L^2([0,\infty))},\right. \\ &\qquad \left.\dots, \int_0^\infty h_n'(u)dB_u(\omega) + \epsilon\langle h_n', h'\rangle_{L^2([0,\infty))}\right)\bigg|_{\epsilon=0}. \end{aligned}$$

and, since $B_t(\omega + \varepsilon h) = B_t(\omega) + \varepsilon h(t)$, this yields

$$dB_t(\omega + \varepsilon h) = dB_t(\omega) + \varepsilon h'(t)\, dt.$$

Hence

$$\begin{aligned} D_h F(\omega) &= \frac{d}{d\varepsilon} f\left(\int_0^\infty h_1'(u)\, dB_u(\omega + \varepsilon h), \dots, \int_0^\infty h_n'(u)\, dB_u(\omega + \varepsilon h)\right)\bigg|_{\varepsilon=0}. \\ &= \frac{d}{d\epsilon} F(\omega + \epsilon h)\bigg|_{\epsilon=0}. \end{aligned} \tag{28}$$

If $F \in \mathcal{S}$ and $h \in H$, then the following basic integration by parts formula holds

$$\mathbb{E}\left[F \int_0^\infty h'(t)dB_t\right] = \mathbb{E}[\langle DF, h\rangle_H]. \tag{29}$$

The proof of this formula is very simple: It uses an integration by parts formula for the finite dimensional Gaussian density (see, e.g., Lemma 1.2.1 in Nualart [51]).

The set of smooth functionals (random variables) $\mathcal{S}$ is dense in $L^p(\Omega)$, for any $p \geq 1$. That is, for any $F \in L^p(\Omega)$ there exists $\{F_n\} \subset \mathcal{S}$ such that

$$\|F_n - F\|_{L^p(\Omega)} \to 0.$$

This result is available in, for example, Nualart [51]. Its proof relies on showing that a subset of $\mathcal{S}$ (the Wiener polynomials) is dense in $L^p(\Omega)$. This is done by using

Hermite polynomials and the Wiener–Itô chaos expansion. The density property of $\mathcal{S}$ is used to extend the definition of the Malliavin derivative to the set of all square integrable random variable for which there exist an approximating sequence of smooth random variables such that the corresponding Malliavin derivatives converge too. This approach works as the Malliavin derivatives of two convergent sequences of smooth random variables converging to the same $L^2(\Omega)$-limit have the same $L^2([0,\infty)\times\Omega)$-limit. To be more precise we have the following (see, e.g., Nualart [51]) :

Proposition 5 (Closability of the Malliavin Derivative operator). *The Malliavin derivative, a linear unbounded operator* $D : \mathcal{S} \to L^2([0,\infty)\times\Omega;\mathbb{R}^d)$ *is closable as an operator from* $L^2(\Omega;\mathbb{R}^d)$ *into* $L^2([0,\infty)\times\Omega;\mathbb{R}^d)$. *In other words if* $\{F_n\}\subset\mathcal{S}$ *is a sequence of smooth random variables such that:* $\|F_n\|_{L^2(\Omega)}\to 0$ *and* $\|DF_n\|_{L^2([0,\infty)\times\Omega)}$ *is convergent then it follows that*

$$\|DF_n\|_{L^2([0,\infty)\times\Omega)}\to 0.$$

More generally, the Malliavin derivative operator is closable as an operator from $L^p(\Omega;\mathbb{R}^d)$ *into* $L^p(\Omega;H)$ *for any* $p\geq 1$. *For* $p\neq 2$ *we use with the norm:*

$$\|DF\|^p_{L^p(\Omega;H)} := \mathbb{E}\left[\|DF\|^p_H\right]. \tag{30}$$

The proof of the closability of the Malliavin operator relies on the basic integration by parts formula (29).

We denote by $\mathbb{D}^{1,p}$ the domain of the Malliavin derivative operator as an operator from $L^p(\Omega;\mathbb{R}^d)$ into $L^p(\Omega;H)$ for any $p\geq 1$. More precisely, $\mathbb{D}^{1,p}$ is the closure of the set $\mathcal{S}$ within $L^p(\Omega;\mathbb{R}^d)$ with respect to the norm:

$$\|F\|_{\mathbb{D}^{1,p}} = \left(\mathbb{E}[|F|^p]+\mathbb{E}[\|DF\|^p_H]\right)^{\frac{1}{p}}.$$

The higher order Malliavin derivatives are defined in a similar manner. For smooth random variables, the iterated derivative D^kF, $k\geq 2$, is a random variable with values in $H^{\otimes k}$ defined as

$$D^kF := \sum_{i_1,\ldots,i_k=1}^{n} \partial_{i_1,\ldots,i_k} f\left(\int_0^\infty h_1'(u)dB_u,\ldots,\int_0^\infty h_n'(u)dB_u\right) h_{i_1}\otimes\ldots\otimes h_{i_k},$$

where $h_i(.) := \int_0^\cdot h_i'(s)ds$. The above expression for D^kF coincides with that obtained by iteratively applying the Malliavin differential operation. Indeed, for $h\in H$, $F\in\mathcal{S}$, it is easily seen that $D_hF\in\mathcal{S}$. As per (28), it can be shown that,

$$D_{h_k}D_{h_{k-1}}\ldots D_{h_1}F = \langle D^kF, h_1\otimes\ldots\otimes h_k\rangle_{H^{\otimes k}}.$$

In an analogous way, one can close the operator D^k from $L^p(\Omega)$ to $L^p(\Omega; H^{\otimes k})$. So, for any $p \geq 1$ and natural $k \geq 1$, we define $\mathbb{D}^{k,p}$ to be the closure of $\mathcal{S}$ with respect to the norm:

$$\|F\|^p_{\mathbb{D}^{k,p}} := \mathbb{E}[|F|^p] + \sum_{j=1}^{k} \mathbb{E}[\|D^j F\|^p_{H^{\otimes j}}].$$

Note that for $p = 2$ the following isometry holds $L^p(\Omega \times [0,\infty)^k; \mathbb{R}^d) \simeq L^2(\Omega; H^{\otimes k})$. Hence one may identify $D^k F$ as a process: $D^k_{t_1,\ldots,t_k} F$.

A random variable F is said to be *smooth in the Malliavin sense* if $F \in \mathbb{D}^{k,p}$ for all $p \geq 1$ and all $k \geq 1$. We denote by $\mathbb{D}^\infty$ the set of all smooth random variables in the Malliavin sense. For example, the solution X^x_t to (22) satisfies $X^i_t \in \mathbb{D}^{k,p}$ for all $t \in [0,\infty)$ and $p \geq 1$ provided $V_0, \ldots, V_d \in \mathcal{C}^\infty_b(\mathbb{R}^N; \mathbb{R}^N)$ (see Theorem 8 below).

Moreover, there is nothing which restricts consideration to $\mathbb{R}^d$-valued random variables. Indeed, one can consider more general Hilbert space-valued random variables, and the theory would extend in an appropriate way. To this end, denote $\mathbb{D}^{k,p}(E)$ to be the appropriate space of E-valued random variables, where E is some separable Hilbert space. For more details, see [51], where also the proof of the following chain rule formula can be found:

Proposition 6 (Chain Rule for the Malliavin Derivative). *If $\varphi : \mathbb{R}^m \to \mathbb{R}$ is a continuously differentiable function with bounded partial derivatives, and $F = (F_1, \ldots, F_m)$ is a random vector with components belonging to $\mathbb{D}^{1,p}$ for some $p \geq 1$. Then $\varphi(F) \in \mathbb{D}^{1,p}$, with*

$$D\varphi(F) = (\nabla\varphi)(F)DF = \sum_{i=1}^{m} \partial_i \varphi(F) DF_i,$$

where $\nabla\varphi$ is the row vector $(\partial_1\varphi, \ldots, \partial_n\varphi)$ and DF is the (column) vector $(DF_1, \ldots, DF_n)^\top$.

Lemma 7 (The Malliavin derivative and integration). *Assume that E is a separable real Hilbert space. Consider $f : [0,\infty) \times \Omega \to E$, and suppose that for each $t \in [0,T]$ we have $f(t) \in \mathbb{D}^{1,2}(E)$ and $t \to f(t)$ is adapted with respect to the natural filtration of B.*[8] *Moreover, suppose that:*

$$\mathbb{E}\int_0^T \|f(t)\|^2_E dt < \infty \qquad \mathbb{E}\int_0^T \|Df(t)\|^2_{E\otimes H} dt < \infty \tag{31}$$

[8] Although not used in the sequel, the result holds for general $f : [0,\infty) \times \Omega \to E$ such that $f(t) \in \mathbb{D}^{1,2}(E)$ for any $t \in [0,T]$, i.e., not necessarily adapted with respect to the natural filtration of B. In this case, the $F_i(T)$ is the Skorohod integral and not the Itô integral of f. See, for example, Proposition 1.38 page 43 in [51].

Then $F_i(T) := \int_0^T f(t)dB_t^i \in \mathbb{D}^{1,2}(E)$ *for all* $i = 0, 1, \ldots, d$, *with*

$$DF_0(T) = \int_0^T Df(t)dB_t^0$$

$$DF_i(T) = \int_0^T Df(t)dB_t^i + \int_0^{T\wedge\cdot} f(s)ds, \quad i = 1, \ldots, d.$$

Also

$$D_h F_0(T) = \int_0^T D_h f(t)dB_t^0$$

$$D_h F_i(T) = \int_0^T D_h f(t)dB_t^i + \int_0^T f(t)h_i'(t)dt, \quad i = 1, \ldots, d.$$

Moreover, assuming that

$$\mathbb{E}\int_0^T \|D^{k-1} f(t)\|^2_{E\otimes H^{\otimes(k-1)}} dt < \infty, \qquad \mathbb{E}\int_0^T \|D^k f(t)\|^2_{E\otimes H^{\otimes k}} dt < \infty$$

one has for the iterated Malliavin derivative operator D^k:

$$D^k F_0 = \int_0^T D^k f(t)dB_t^0$$

$$D^k F_i(T) = \int_0^T D^k f(t)dB_t^i + \int_0^{T\wedge\cdot} D^{k-1} f(s)ds, \quad i = 1, \ldots, d.$$

Proof. The proof is done using an induction argument. See Kusuoka and Stroock [32] for details. □

Theorem 8. *Assume* X *is the stochastic flow which solves* (22), *where the coefficients* $V_1, \ldots, V_d \in C_b^{k+1}(\mathbb{R}^N; \mathbb{R}^N)$ *and* $V_0 \in C_b^k(\mathbb{R}^N; \mathbb{R}^N)$. *Then* $X_t^{x,i} \in \mathbb{D}^{k,p}$ *for all* $t \in [0, \infty)$, $i = 1, \ldots, N$ *and* $p \geq 1$. *Furthermore, the matrix valued process* $DX_t^x := (D^j X_t^{x,i})_{i=1,\ldots,N; j=1,\ldots,d}$ *satisfies the stochastic differential equation:*

$$DX_t^x = \sum_{i=0}^d \int_0^t \partial V_i(X_u^x) DX_u^x \circ dB_u^i + \left(\int_0^{t\wedge\cdot} V_j(X_u^x)du\right)_{j=1,\ldots,d}. \tag{32}$$

Hence,

$$D_h X_t^x = \sum_{i=0}^d \int_0^t \partial V_i(X_u^x) D_h X_u^x \circ dB_u^i + \sum_{k=1}^d \int_0^t V_k(X_u^x)h_i'(u)du. \tag{33}$$

If the vector fields $V_0, \ldots, V_d$ are uniformly bounded then the following bound on the norms of the derivatives can be shown to hold:

$$\sup_{\substack{t\in[0,T]\\ x\in\mathbb{R}^N}} \mathbb{E}\left[\| D^k X_t^x \|_{H^{\otimes k}}^p\right] < C_{k,p}, \quad \forall\, p \in [1,\infty),\ T > 0. \tag{34}$$

If, however, the vector fields $V_0, \ldots, V_d$ are globally Lipschitz continuous but not necessarily bounded, then it may only be deduced that the following holds:

$$\sup_{t\in[0,T]} \mathbb{E}\left[\| D^k X_t^x \|_{H^{\otimes k}}^p\right] < C_{k,p}(1+|x|)^p, \quad \forall\, p \in [1,\infty),\ T > 0. \tag{35}$$

Proof. See Nualart [51, pp. 119–124]. □

Corollary 9. *For any $(t,x) \in [0,\infty) \times \mathbb{R}^N$, we have that*

$$(J_t^x)^{-1} D X_t^x = \left(\int_0^{t\wedge\cdot} (J_s^x)^{-1} V_j(X_s^x) ds\right)_{j=1,\ldots,d}. \tag{36}$$

Proof. This is a simple result of applying integration by parts to the product $(J_t^x)^{-1} D X_t^x$, using the SDEs from the respective processes. For a complete proof see, for example, Nualart [51, Sect. 2.3.1]. □

Definition 10 (Lie Bracket of Vector Fields). Let $V, W \in \mathcal{C}^1(\mathbb{R}^N;\mathbb{R}^N)$ be two vector fields. The Lie bracket of V and W is a third vector field, $[V,W]$, defined by:

$$[V,W] := \partial W.V - \partial V.W,$$

where $\partial V := (\partial_j V^i)_{1\le i,j\le N}$ and the multiplication is that of a matrix by a vector.

The Lie bracket is a bilinear differential form $[.,.] : \mathcal{C}^{m_1} \times \mathcal{C}^{m_2} \to \mathcal{C}^{m_1\wedge m_2 - 1}$, where $1 \le m_1, m_2 \le \infty$, which satisfies the identities:

$$[V,W] = -[W,V] \quad \text{and} \quad [U,[V,W]] + [W,[U,V]] + [V,[W,U]] = 0.$$

The latter is known as the *Jacobi Identity*.

Corollary 11. *Let $W \in \mathcal{C}^3(\mathbb{R}^N;\mathbb{R}^N)$ then there holds:*

$$d\left[(J_t^x)^{-1} W(X_t^x)\right] = -\sum_{k=0}^{d} (J_t^x)^{-1}[W, V_k](X_t^x) \circ dB_t^k. \tag{37}$$

Proof. Note that

$$W(X_t^x) = W(x) + \sum_{k=0}^{d} \int_0^t \partial W(X_s^x) V_k(X_s^x) \circ dB_s^k.$$

Thus, by an analogous formula for matrix–vector SDEs we have:

$$\begin{aligned}
(J_t^x)^{-1} W(X_t^x) &= \int_0^t (J_s^x)^{-1} \circ dW(X_s^x) + \int_0^t d(J_s^x)^{-1} \circ W(X_s^x) \\
&= \sum_{k=0}^{d} \int_0^t (J_s^x)^{-1} \partial W(X_s^x) V_k(X_s^x) \circ dB_s^k \\
&\quad - \sum_{k=0}^{d} \int_0^t (J_s^x)^{-1} \partial V_k(X_s^x) W(X_s^x) \circ dB_s^k \\
&= - \sum_{k=0}^{d} \int_0^t (J_s^x)^{-1} [W, V_k](X_s^x) \circ dB_s^k.
\end{aligned}$$

□

The alternative representation (36) for $(J_t^x)^{-1} DX_t^x$ will be used in deriving the integration by parts formula and Lie brackets are a natural occurrence in this analysis. We may apply Corollary (11) iteratively to expand an expression for $(J_t^x)^{-1} V_i(X_t^x)$ for $i = 1, \ldots, d$, as far as the differentiability constraints on the vector fields permit. The divergence operator—which is the adjoint of the Malliavin derivative—plays a vital role in the construction of our integration by parts formula. This operator is also called the Skorohod integral. It coincides with a generalisation of the Itô integral to anticipating integrands. A detailed discussion of the divergence operator can be found in Nualart [51].

Definition 12 (Divergence operator). Denote by δ the adjoint of the operator D. That is, δ is an unbounded operator on $L^2(\Omega \times [0,\infty); \mathbb{R}^d)$ with values in $L^2(\Omega)$ such that:

1. $\mathrm{Dom}\, \delta = \{u \in L^2(\Omega \times [0,\infty); \mathbb{R}^d); |\mathbb{E}(\langle DF, u\rangle_H)| \leq c \|F\|_{L^2(\Omega)}, \ \forall F \in \mathbb{D}^{1,2}\}$.
2. For every $u \in \mathrm{Dom}\, \delta$, then $\delta(u) \in L^2(\Omega)$ satisfies:

$$\mathbb{E}(F\delta(u)) = \mathbb{E}(\langle DF, u\rangle_H).$$

The following important results are shown in Sect. 1.5 of Nualart [51]:

1. D is continuous from $\mathbb{D}^{k,p}(E)$ into $\mathbb{D}^{k-1,p}(H \otimes E)$
2. $\langle DF, DF\rangle_H \in \mathbb{D}^\infty$ if $F, G \in \mathbb{D}^\infty$
3. δ is continuous from $\mathbb{D}^\infty(H)$ into $\mathbb{D}^\infty$.

Remark 13. If $u = (u^1, \dots, u^d) \in \operatorname{Dom} \delta$ has the property that $t \to u(\cdot, t)$ is $\mathcal{F}_t$-adapted, then the adjoint $\delta(u)$, is nothing more than the Itô integral of u with respect to the d-dimensional Brownian motion $B_t = (B_t^1, \dots, B_t^d)$. i.e.

$$\delta(u) = \sum_{i=1}^{d} \int_0^\infty u^i(\cdot, s) dB_s^i.$$

2.3 The UFG Condition

Define $\mathcal{A}$ to be the set of all n-tuples of natural numbers of any size n with the following form: $\mathcal{A} := \{1, \dots, d\} \cup \bigcup_{k \in \mathbb{N}_0} \{0, 1, \dots, d\}^k$. We endow $\mathcal{A}$ with the product:

$$\alpha * \beta := (\alpha_1, \dots, \alpha_k, \beta_1, \dots, \beta_l), \text{ where } \alpha = (\alpha_1, \dots, \alpha_k), \ \beta = (\beta_1, \dots, \beta_l) \in \mathcal{A}.$$

Define $\mathcal{A}_{\emptyset,0} := \mathcal{A} \cup \{\emptyset, 0\}$. We consider the following n-tuples lengths:

$$|\alpha| := \begin{cases} k, & \text{if } \alpha = (\alpha_1, \dots, \alpha_k), \\ 0, & \text{if } \alpha = \emptyset. \end{cases}$$

$$\|\alpha\| := |\alpha| + \operatorname{card}\{i : \alpha_i = 0, i = 1, \dots, d\}.$$

We also introduce the sets

$$\mathcal{A}(m) = \{\alpha \in \mathcal{A} : \|\alpha\| \le m\} \quad \mathcal{A}_{\emptyset,0}(m) := \{\alpha \in \mathcal{A}_{\emptyset,0} : \|\alpha\| \le m\}.$$

We now define the vector field concatenation $V_{[\alpha]}$, $\alpha \in \mathcal{A}_{\emptyset,0}$ inductively, as follows:

$$\begin{aligned} V_{[\emptyset]} &:= 0, \\ V_{[i]} &:= V_i, & i = 0, 1, \dots, d, \\ V_{[\alpha * i]} &:= [V_{[\alpha]}, V_i], & i = 0, 1, \dots, d. \end{aligned}$$

In a similar vein to the above, we also define the Stratonovich integral concatenation,

$$\hat{B}_t^{\circ\alpha}, \ t \in [0, \infty), \ \alpha \in \mathcal{A}_{\emptyset,0}$$

inductively:

$$\hat{B}_t^{\circ\emptyset} \equiv 1, \quad \hat{B}_t^{\circ i} := B_t^i, \quad \hat{B}_t^{\circ\alpha * i} := \int_0^t \hat{B}_s^{\circ\alpha} \circ dB_s^i, \quad i = 0, 1, \dots, d.$$

We now introduce the main assumption for the gradient bounds analysis: the UFG condition. The purpose of the UFG condition, in its purest form, is to truncate the expansion obtained when considering the expression $(J_t^x)^{-1}V_i(X_t^x)$, for $i = 1, \ldots, d$. Recalling the work of the previous section, this appears when considering the product $(J_t^x)^{-1}DX_t^x$ between the Malliavin derivative and the inverse of the Jacobian of the stochastic flow. The UFG condition is a *"finite generation"* assumption, which helps to provide integration by parts formula.

Definition 14 (UFG Condition). Let $\{V_i : i = 0, \ldots, d\}$, be a system of vector fields such that $V_1, \ldots, V_d \in C_b^{k+1}(\mathbb{R}^N; \mathbb{R}^N)$ and $V_0 \in C_b^k(\mathbb{R}^N; \mathbb{R}^N)$. We say that $\{V_i : i = 0, \ldots, d\}$ satisfy the UFG condition if, there exists $m \in \mathbb{N}$, $m \leq k - 1$, such for any $\alpha \in \mathcal{A}$, $\alpha = \alpha' * i$, $\alpha' \in \mathcal{A}(m)$ and $i = 0, \ldots, d$, there exist uniformly bounded functions $\varphi_{\alpha,\beta} \in C_b^{k+1-|\alpha|}(\mathbb{R}^N, \mathbb{R})$, with $\beta \in \mathcal{A}(m)$ such that

$$V_{[\alpha]}(x) = \sum_{\beta \in \mathcal{A}(m)} \varphi_{\alpha,\beta}(x) V_{[\beta]}(x).$$

Heuristically, the UFG conditions states that higher order Lie brackets can be expressed as a linear combination of lower order Lie brackets, for some fixed order m. The uniform Hörmander condition implies the UFG condition, but not vice versa as we can see from the following example, taken from Kusuoka [30]:

Example 15. Assume $d = 1$ and $N = 2$. Let $V_0, V_1 \in C_b^\infty(\mathbb{R}^2; \mathbb{R}^2)$ be given by

$$V_0(x_1, x_2) = \sin x_1 \frac{\partial}{\partial x_1} \qquad V_1(x_1, x_2) = \sin x_1 \frac{\partial}{\partial x_2}$$

Then $\{V_0, V_1\}$ do not satisfy the Hörmander condition. However the UFG condition is satisfied with $m = 4$.

Remark 16.

1. The UFG condition is defined in such a way (i.e. with $m \leq k - 1$) that the elements $V_{[\alpha]}$ are well-defined and such that we may apply Corollary 11 to $V_{[\alpha]}$ for all $\alpha \in \mathcal{A}(m)$.
2. The regularity of the coefficients $\varphi_{\alpha,\beta}$ is chosen in accordance with what one would expect, given the regularity of $V_{[\alpha]}$.
3. We draw attention to the fact that we have assumed the coefficients are uniformly bounded. Although this assumption does not materially restrict the strength of the results, it does make them more presentable and reduces the complexity in the proof. Essentially the boundedness of the coefficients means there is a natural and elegant description for how the gradient bounds may increase as a function of $|x|$. We endeavor to draw attention to the effects of this assumption where appropriate. In many examples of interest, this assumption imposes no unnecessary restrictions.

2.4 The Central Representation Formula

From Corollary (11) and the UFG condition, we have, for each $\alpha \in \mathcal{A}(m)$,

$$\begin{aligned} d\left[(J_t^x)^{-1}V_{[\alpha]}(X_t^x)\right] &= -\sum_{i=0}^{d}(J_t^x)^{-1}[V_{[\alpha]},V_i](X_t^x)\circ dB_t^i \\ &= -\sum_{i=0}^{d}(J_t^x)^{-1}V_{[\alpha*i]}(X_t^x)\circ dB_t^i \\ &= \sum_{i=0}^{d}\sum_{\beta\in\mathcal{A}(m)} c_{\alpha,\beta}^i(X_t^x)(J_t^x)^{-1}V_{[\beta]}(X_t^x)\circ dB_t^i\ , \end{aligned} \tag{38}$$

where the coefficients $c^i_{\alpha,\beta}$, $\alpha,\beta \in \mathcal{A}(m)$, $i = 0,\dots,d$ are given by

$$c^i_{\alpha,\beta}(x) = \begin{cases} -1 & \text{if } \alpha * i \in \mathcal{A}(m) \text{ and } \beta = \alpha * i \\ 0 & \text{if } \alpha * i \in \mathcal{A}(m) \text{ and } \beta \neq \alpha * i \\ -\varphi_{\alpha * i,\beta} & \text{if } \alpha * i \notin \mathcal{A}(m) \end{cases}. \tag{39}$$

We note, in particular, that $c^i_{\alpha,\beta} \in C_b^{k+1-|\alpha|}(\mathbb{R}^N,\mathbb{R})$ are uniformly bounded. We obtained a representation of the vector fields $V_{[\alpha]}$, $\alpha \in \mathcal{A}(m)$ (estimated at $(X_\cdot^x)$) in terms of the Lie brackets $V_{[\alpha*i]} := [V_{[\alpha]}, V_i]$, $\alpha \in \mathcal{A}(m)$, $i = 0,\dots,d$, which where then reverted back to the original set of vector fields $V_{[\alpha]}$, $\alpha \in \mathcal{A}(m)$ via the UFG condition. Without the UFG condition, the resulting representation would potentially be infinite. Indeed, the Hörmander approach relies on showing that, after a certain number of iterations (taking Lie brackets of the resulting vector fields), the remainder term arising from the expansion becomes very small. The UFG condition is more general than Hörmander's (see [24]) famous criterion for hypoellipticity of linear differential operators and it allows us to take a different approach. We can view (38) as a linear system of SDEs whose coefficients are of suitably chosen differentiability whose solutions are the processes $t \to (J_t^x)^{-1}V_{[\alpha]}(X_t^x)$, $\alpha \in \mathcal{A}(m)$. This enables us to represent these processes in terms of their initial values $V_{[\alpha]}(x)$, $\alpha \in \mathcal{A}(m)$ and the corresponding representation facilitates the integration by parts formula. Moreover, we shall see how the same representation leads to the classical non-degeneracy result: The gradient bounds obtained under the UFG condition shall implicitly recover Hörmander's result, see Theorem 70.

By considering the above as a closed linear system of equations, we are able to equivalently view it as the matrix SDE:

$$Y(t,x) = Y(0,x) + \sum_{i=0}^{d}\int_0^t C^i(X_s^x)Y(s,x)\circ dB_s^i, \tag{40}$$

where $Y(0,x) = V(x) := \left(V_{[\alpha]}(x)\right)_{\alpha\in\mathcal{A}(m)} \in \mathbb{R}^{N_m}\times\mathbb{R}^N$, ($N_m = card(\mathcal{A}_m)$) and $C^i : \mathbb{R}^N \to \mathbb{R}^{N_m}\otimes\mathbb{R}^{N_m}$ are given by

$$C^i(x) := \left(c^i_{\alpha,\beta}(x)\right)_{\alpha,\beta\in\mathcal{A}(m)}.$$

We are able to take advantage of the linear nature of this system of SDEs by considering in more generality the matrix which produces such vectors. Namely,

Lemma 17. *Assume that $A(t,x)$, $(t,x)\in[0,\infty)\times\mathbb{R}^N$ is the $N_m\times N_m$-matrix which is the unique solution to the matrix stochastic differential equation*

$$dA(t,x) = \sum_{i=0}^{d} C^i(X^x_t)A(t,x)\circ dB^i_t, \tag{41}$$

where $A(0,x) = I$. Then $Y(t,x) = A(t,x)Y(0,x)$.

Proof. We need only show that $A(t,x)Y(0,x)$ solves equation (40), then by the uniqueness of SDE solutions (see, for example Karatzas and Shreve [26]), the result follows. But,

$$\begin{aligned} d(A(t,x)Y(0,x)) &= A(t,x)\circ dY(0,x) + \circ dA(t,x)Y(0,x) = \circ dA(t,x)Y(0,x) \\ &= \sum_{i=0}^{d} C^i(X^x_t)A(t,x)Y(0,x)\circ dB^i_t \end{aligned}$$

and, clearly, $A(0,x)Y(0,x) = Y(0,x)$. □

The above results show that all the relevant information about the solution (40) is captured by the solution (41). We can apply classical results about solutions of SDEs to obtain the following proposition.

Proposition 18. *The matrix stochastic differential equation (41) has a unique solution, $A = (a_{\alpha,\beta})_{\alpha,\beta\in\mathcal{A}(m)}$ with components $a_{\alpha,\beta} : [0,\infty)\times\mathbb{R}^N \to \mathbb{R}$, $\alpha,\beta\in\mathcal{A}(m)$ that satisfy the mutually dependent SDEs:*

$$a_{\alpha,\beta}(t,x) = \delta_{\alpha,\beta} + \sum_{i=0}^{d}\sum_{\gamma\in\mathcal{A}(m)}\int_0^t c^i_{\alpha,\gamma}(X^x_u)a_{\gamma,\beta}(u,x)\circ dB^i_u.$$

Moreover $a_{\alpha,\beta}(t,.) : \mathbb{R}^N \to \mathbb{R}$ are a.s. $k-m$ times differentiable in x for fixed $t\in[0,\infty)$ and $a_{\alpha,\beta}(.,.)$ is jointly continuous in $[0,\infty)\times\mathbb{R}^N$ with probability one, for each $\alpha,\beta\in\mathcal{A}(m)$ and

$$\sup_{\substack{x\in\mathbb{R}^N\\ t\in[0,T]}} \mathbb{E}\left[\left|\frac{\partial^{|\gamma|}}{\partial x^\gamma} a_{\alpha,\beta}(t,x)\right|^p\right] < \infty, \quad \forall\, p \in [1,\infty),\ T>0, \tag{42}$$

for any multi-index γ with $|\gamma| \leq k - m$. Finally, for any $l \leq k - m$

$$\sup_{t\in[0,T]} \mathbb{E}\left[\| D^l a_{\alpha,\beta}(t,x)\|^p_{H\otimes l}\right] < C_{l,p}(1+|\,x\,|)^p \quad \forall\, p \in [1,\infty),\ T>0, \tag{43}$$

Furthermore, the matrix $A = (a_{\alpha,\beta})_{\alpha,\beta\in\mathcal{A}(m)}$ is invertible, and its inverse $B = (b_{\alpha,\beta})_{\alpha,\beta\in\mathcal{A}(m)}$ satisfies the matrix SDE:

$$B(t,x) = I - \sum_{i=0}^{d} \int_0^t B(u,x) C^i(X_u^x) \circ dB_u^i.$$

Moreover, the components $b_{\alpha,\beta}$, $\alpha,\beta \in \mathcal{A}(m)$, are a.s. $k-m$ times differentiable in x for fixed $t \in [0,\infty)$, jointly continuous in (t,x) and

$$\sup_{\substack{t\in[0,T]\\ x\in\mathbb{R}^N}} \mathbb{E}\left[\left|\frac{\partial^{|\gamma|}}{\partial x^\gamma} b_{\alpha,\beta}(t,x)\right|^p\right] < C_{T,p}, \tag{44}$$

for each $p \in [1,\infty)$, $T > 0$, $|\,\gamma\,| \leq k - m$ and some constant $C_{T,p}$. Finally, for any $l \leq k - m$

$$\sup_{t\in[0,T]} \mathbb{E}\left[\| D^l b_{\alpha,\beta}(t,x)\|^p_{H\otimes l}\right] < C_{l,p}(1+|\,x\,|)^p \quad \forall\, p \in [1,\infty),\ T>0, \tag{45}$$

Proof. This is very similar to Theorem 8. The only difference here is that the bounds on the norms of the iterated Malliavin derivatives are now bounded only linearly in $|\,x\,|$. This is obvious once one considers Theorem 8 and, in particular, inequality (34). It is clear from this equation that the norm of the Malliavin derivatives inherits the linear growth of the vector fields. All higher order Malliavin derivatives inherit this linearity from the first order Malliavin derivative, but given the boundedness of the derivatives of the vector fields, have no worse than linear growth. □

Remark 19. (**a**) The above proposition highlights an idiosyncratic difference between the Malliavin derivative and the normal derivative for the solutions of such SDEs. It stems from the fact that the Malliavin derivative of X_t^x has an unbounded norm over $x \in \mathbb{R}^N$, as it has Lipschitz continuous coefficients. However, the same result for the norm of the classical derivative of X_t^x **is** bounded over $x \in \mathbb{R}^N$. Note this difference would not appear if we assumed the vector fields were uniformly bounded.

(**b**) Although not used in the sequel, identities (42)–(45) hold true with the supremum taken inside the expectation.

We now seek to study the solution to (41), whose elements will be absolutely fundamental to our analysis. We note initially, that although this matrix is potentially very large, with potentially significant mutual dependence, many of the terms which make up this mutual dependence are zero. This allows us to get a good handle on the matrix. Note that for fixed $\alpha, \beta \in \mathcal{A}(m)$ we have

$$a_{\alpha,\beta}(t,x) = \delta_{\alpha\beta} + \sum_{i=0}^{d} \sum_{\gamma \in \mathcal{A}(m)} \int_0^t c^i_{\alpha,\gamma}(X^x_s) a_{\gamma,\beta}(s,x) \circ dB^i_s. \tag{46}$$

The coefficients $c^i_{\alpha,\gamma}$ identified in (39) lead to the following:
For $\|\alpha\| \leq m-2$ there holds: $\|\alpha * i\| \leq m$ for all $i = 0, \dots, d$, so $c^i_{\alpha,\gamma} \neq 0$ only when $\gamma = \alpha * i$. In which case $c^i_{\alpha,\gamma} = -1$. i.e.

$$a_{\alpha,\beta}(t,x) = \delta_{\alpha\beta} - \sum_{i=0}^{d} \int_0^t a_{\alpha * i,\beta}(s,x) \circ dB^i_s.$$

For $\|\alpha\| = m-1$ there holds: $\|\alpha * i\| = m$ for $i = 1, \dots, d$, with $\|\alpha * 0\| = m+1$. Hence $\alpha * i \in \mathcal{A}(m)$ for $i = 1, \dots, d$, and $\alpha * 0 \notin \mathcal{A}(m)$. i.e.

$$a_{\alpha,\beta}(t,x) = \delta_{\alpha\beta} - \sum_{i=1}^{d} \int_0^t a_{\alpha * i,\beta}(s,x) \circ dB^i_s - \sum_{\gamma \in \mathcal{A}(m)} \int_0^t \varphi_{\alpha * 0,\gamma}(X^x_s) a_{\gamma,\beta}(s,x) ds.$$

For $\|\alpha\| = m$ there holds: $\|\alpha * i\| > m$ for $i = 0, \dots, d$. Hence $\alpha * i \notin \mathcal{A}(m)$ for $i = 0, \dots, d$. i.e.

$$a_{\alpha,\beta}(t,x) = \delta_{\alpha\beta} - \sum_{i=0}^{d} \sum_{\gamma \in \mathcal{A}(m)} \int_0^t \varphi_{\alpha * i,\gamma}(X^x_s) a_{\gamma,\beta}(s,x) \circ dB^i_s.$$

An explicit form for $a_{\alpha,\beta}$ is sought and is easy to identify from (46). In fact, each element of the matrix A can be split up into a sum of two terms: the term which arises from $\delta_{\alpha\beta}$—a iterated Stratonovich integral of a constant—and a remainder term. That is, for any $\alpha, \beta \in \mathcal{A}(m)$,

$$a_{\alpha,\beta}(t,x) = a^0_{\alpha,\beta}(t,x) + r_{\alpha,\beta}(t,x), \tag{47}$$

where

$$a^0_{\alpha,\beta}(t,x) = \begin{cases} (-1)^{|\gamma|} \hat{B}^{\circ\gamma}_t & \text{if } \beta = \alpha * \gamma \text{ for some } \gamma \in \mathcal{A}(m) \\ 0 & \text{otherwise} \end{cases}$$

and

$$r_{\alpha,\beta}(t,x) = \sum_{\substack{\gamma\in\mathcal{A},j=0,\dots,d \\ \text{s.t. } \|\alpha*\gamma\|\le m \\ \|\gamma*j\|\ge m+1-\|\alpha\|}} \sum_{\delta\in\mathcal{A}(m)} \int_0^t \int_0^{s_k} \cdots \int_0^{s_1} (-1)^{|\gamma|} c^j_{\alpha*\gamma,\delta}(X^x_s)$$

$$\times a_{\delta,\beta}(s,x) \circ dB^j_s \circ dB^{\gamma_1}_{s_1} \dots dB^{\gamma_k}_{s_k}$$

The following proposition is a good indicator of how we are able to identify the explicit short-time asymptotic rates in terms of time, t.

Proposition 20. *For any $T > 0$, $p \in [1,\infty)$, $\alpha, \beta \in \mathcal{A}(m)$ and $\gamma \in \mathcal{A}_0$, the following hold*

$$\sup_{t\in(0,T]} \mathbb{E}\left[\left(t^{-\|\gamma\|/2} \left| \hat{B}^{\circ\gamma}_t \right|\right)^p\right] < \infty, \tag{48}$$

$$\sup_{\substack{x\in\mathbb{R}^N \\ t\in(0,T]}} \mathbb{E}\left[\left(t^{-(m+1-\|\alpha\|)/2} \left| r_{\alpha,\beta}(t,x) \right|\right)^p\right] < \infty. \tag{49}$$

Proof. The proof of these result is left for the appendix. □

We are now ready to derive the integration by parts formula. Let $f \in C^\infty_b(\mathbb{R}^N,\mathbb{R})$, then, using (36), we get

$$\begin{aligned} Df(X^x_t) &= \nabla f(X^x_t) DX^x_t \\ &= \nabla(f\circ X_t)(x)(J^x_t)^{-1} DX^x_t \\ &= \nabla(f\circ X_t)(x) \left(\int_0^{t\wedge\cdot} (J^x_s)^{-1} V_i(X^x_s)ds\right)_{i=1,\dots,d}. \end{aligned}$$

The idea is to develop the preceding equality to isolate terms involving $\nabla(f \circ X_t)(x)$. Once isolated, the operators of the Malliavin calculus will be used to derive an integration by parts formula. Now we note that, from Lemma 17:

$$(J^x_s)^{-1} V_i(X^x_s) = (A(s,x)V(x))_i = \sum_{\beta\in\mathcal{A}(m)} a_{i,\beta}(t,x) V_{[\beta]}(x).$$

Hence,

$$\begin{aligned} Df(X^x_t) &= \nabla(f\circ X_t)(x) \left(\int_0^{t\wedge\cdot} \sum_{\beta\in\mathcal{A}(m)} a_{i,\beta}(s,x) V_{[\beta]}(x)ds\right)_{i=1,\dots,d} \\ &= \nabla(f\circ X_t)(x) \sum_{\beta\in\mathcal{A}(m)} V_{[\beta]}(x) \left(\int_0^{t\wedge\cdot} a_{i,\beta}(s,x)ds\right)_{i=1,\dots,d} \\ &= \sum_{\beta\in\mathcal{A}(m)} V_{[\beta]}(f\circ X_t)(x) k_\beta(t,x), \end{aligned}$$

where

$$k_\beta(t,x) := \left(\int_0^{t\wedge\cdot} a_{i,\beta}(s,x)ds\right)_{i=1,\ldots,d}.$$

We re-write the previous equation into a linear system of equations by taking the H inner product with $k_\alpha(t,x)$ for all $\alpha \in \mathcal{A}(m)$. i.e.

$$\left\langle Df(X_t^x), k_{(1)}(t,x)\right\rangle_H = \sum_{\beta\in\mathcal{A}(m)} V_{[\beta]}(f\circ X_t)(x)\left\langle k_\beta(t,x), k_{(1)}(t,x)\right\rangle_H$$

$$\vdots$$

$$\langle Df(X_t^x), k_\alpha(t,x)\rangle_H = \sum_{\beta\in\mathcal{A}(m)} V_{[\beta]}(f\circ X_t)(x)\left\langle k_\beta(t,x), k_\alpha(t,x)\right\rangle_H$$

$$\vdots$$

$$\left\langle Df(X_t^x), k_{\alpha(N_m)}(t,x)\right\rangle_H = \sum_{\beta\in\mathcal{A}(m)} V_{[\beta]}(f\circ X_t)(x)\left\langle k_\beta(t,x), k_{\alpha(N_m)}(t,x)\right\rangle_H$$

Define, for $\alpha \in \mathcal{A}(m)$:

$$\begin{aligned}
D^{(\alpha)}f(X_t^x) &:= \langle Df(X_t^x), k_\alpha(t,x)\rangle_H \\
M_{\alpha,\beta}(t,x) &:= t^{-(\|\alpha\|+\|\beta\|)/2}\left\langle k_\alpha(t,x), k_\beta(t,x)\right\rangle_H \\
&= t^{-(\|\alpha\|+\|\beta\|)/2}\sum_{i=1}^{d}\int_0^t a_{i,\alpha}(s,x)a_{i,\beta}(s,x)ds.
\end{aligned}$$

This leaves us with

$$D^{(\alpha)}f(X_t^x) = \sum_{\beta\in\mathcal{A}(m)} t^{(\|\alpha\|+\|\beta\|)/2}M_{\alpha,\beta}(t,x)V_{[\beta]}(f\circ X_t)(x).$$

The above can be seen as a linear system of equations driven by a random matrix

$$M(t,x) = (M_{\alpha,\beta}(t,x))_{\alpha,\beta}.$$

The invertibility of this matrix is a major step forward towards an integration by parts. For then there would hold, $\mathbb{P}$-a.s:

$$V_{[\alpha]}(f \circ X_t)(x) = t^{-\|\alpha\|/2} \sum_{\beta \in \mathcal{A}(m)} t^{-\|\beta\|/2} M^{-1}_{\alpha,\beta}(t,x) D^{(\beta)} f(X^x_t).$$

Proposition 21. *$M(t,x)$ is $\mathbb{P}$-a.s. invertible. Moreover, for $p \in [1,\infty)$ and $\alpha, \beta \in \mathcal{A}(m)$, there holds*

$$\sup_{t \in (0,1], x \in \mathbb{R}^N} \mathbb{E}\left[(M^{-1}_{\alpha,\beta}(t,x))^p \right] < \infty \tag{50}$$

Proof. The proof of invertibility is lengthy and is left to the appendix. □

2.5 Kusuoka–Stroock Functions

We introduce now a class of functions which we shall call Kusuoka–Stroock functions. Such functions play the central role in the deduction of the integration by parts formulae (IBPF) and the control of the derivatives of the semigroup P_t. In particular we will show that if $(t,x) \to \Phi(t,x)$ is a Kusuoka–Stroock function, then there exists another Kusuoka–Stroock function $(t,x) \to \Phi_\alpha(t,x)$, $\alpha \in \mathcal{A}(m)$ such that:

$$\mathbb{E}[\Phi(t,x) V_{[\alpha]}(f \circ X_t)(x)] = t^{-\|\alpha\|/2} \mathbb{E}[\Phi_\alpha(t,x) f(X^x_t)].$$

This class of functions is closed under the operations which are taken during the formation of the IBPF. As a result this space supports iterative applications of the above formula.

Definition 22 (Local Kusuoka–Stroock functions). Let E be a separable Hilbert space and let $r \in \mathbb{R}$, $n \in \mathbb{N}$. We denote by $\mathcal{K}^{\mathrm{loc}}_r(E,n)$ the set of functions: $f : (0,T] \times \mathbb{R}^N \to \mathbb{D}^{n,\infty}(E)$ satisfying the following:

1. $f(t,.)$ is n-times continuously differentiable and $\frac{\partial^\alpha f}{\partial x^\alpha}(.,.)$ is continuous in $(t,x) \in (0,T] \times \mathbb{R}^N$ a.s. for any $\alpha \in \mathcal{A}$ satisfying $|\alpha| \leq n$
2. For any K compact subset of $\mathbb{R}^N$ and $k \in \mathbb{N}$, $p \in [1,\infty)$ with $k \leq n - |\alpha|$, we have

$$\sup_{t \in (0,T], x \in K} t^{-r/2} \left\| \frac{\partial^\alpha f}{\partial x^\alpha} \right\|_{\mathbb{D}^{k,p}(E)} < \infty. \tag{51}$$

If (51) holds globally over $\mathbb{R}^N$, we write $f \in \mathcal{K}_r(E,n)$ and denote $\mathcal{K}^{loc}_r(n) := \mathcal{K}^{loc}_r(\mathbb{R},n)$ and, respectively, $\mathcal{K}_r(n) := \mathcal{K}_r(\mathbb{R},n)$.

The functions belonging to the set $\mathcal{K}^{loc}_r(E,n)$ satisfy the following properties which form the basis of the integration by parts formula.

Lemma 23 (Properties of local Kusuoka–Stroock functions). *The following hold*

1. *Suppose* $f \in \mathcal{K}_r^{loc}(E,n)$, *where* $r \geq 0$. *Then, for* $i = 1,\ldots,d$,

$$\int_0^{\cdot} f(s,x)dB_s^i \in \mathcal{K}_{r+1}^{loc}(E,n) \quad and \quad \int_0^{\cdot} f(s,x)ds \in \mathcal{K}_{r+2}^{loc}(E,n).$$

2. $a_{\alpha,\beta}, b_{\alpha,\beta} \in \mathcal{K}^{loc}_{(\|\beta\|-\|\alpha\|)\vee 0}(k-m)$ *for any* $\alpha,\beta \in \mathcal{A}(m)$.
3. $k_\alpha \in \mathcal{K}^{loc}_{\|\alpha\|}(H,k-m)$ *for any* $\alpha \in \mathcal{A}(m)$.
4. $D^{(\alpha)}u := \langle Du(t,x), k_\alpha\rangle_H \in \mathcal{K}^{loc}_{r+\|\alpha\|}(n \wedge [k-m])$ *where* $u \in \mathcal{K}_r^{loc}(n)$ *and* $\alpha \in \mathcal{A}(m)$.
5. *If* $M^{-1}(t,x)$ *is the inverse matrix of* $M(t,x)$, *then* $M^{-1}_{\alpha,\beta} \in \mathcal{K}_0^{loc}(k-m)$, $\alpha,\beta \in \mathcal{A}(m)$.
6. *If* $f_i \in \mathcal{K}_{r_i}^{loc}(n_i)$ *for* $i = 1,\ldots,N$, *then*

$$\prod_{i=1}^N f_i \in \mathcal{K}^{loc}_{r_1+\ldots+r_N}(\min_i n_i) \quad and \quad \sum_{i=1}^N f_i \in \mathcal{K}^{loc}_{\min_i r_i}(\min_i n_i).$$

Moreover, if we assume the vector fields $V_0,\ldots,V_d$ *are also uniformly bounded, then (2)–(5) hold with* $\mathcal{K}^{loc}$ *replaced by* $\mathcal{K}$.

Proof. This is proved in the appendix. □

2.6 Integration by Parts Formulae

In this section we synthesise the developed results to obtain various integration by parts formulae, in a way which should now be familiar. We note that some of the stated results are for iterated derivatives of the semigroup P_t (cf. Corollary 28) along vector fields of the Lie algebra. Seeing as the purpose of this section is to look at derivatives of the semigroup, we shall always assume that $V_{[\alpha_1]},\ldots,V_{[\alpha_{N+M}]}$ have sufficient smoothness for this operation to be well-defined.

Theorem 24 (Integration by Parts formula I). *Under the UFG condition, for any* $\Phi \in \mathcal{K}_r^{loc}(n)$ *and for any* $\alpha \in \mathcal{A}(m)$, *there exists* $\Phi_\alpha \in \mathcal{K}_r^{loc}((n-1)\wedge(k-m-1))$ *such that:*

$$\mathbb{E}\left[\Phi(t,x)V_{[\alpha]}(f\circ X_t)(x)\right] = t^{-\|\alpha\|/2}\mathbb{E}\left[\Phi_\alpha(t,x)f(X_t^x)\right], \quad t>0, x\in\mathbb{R}^N \tag{52}$$

for any $f \in C_b^\infty(\mathbb{R}^N;\mathbb{R})$. *In addition, for any* $q > p$

$$\sup_{t\in(0,T]} \mathbb{E}\left[|\,\Phi_\alpha(t,x)\,|^p\right] \leq C_{p,q}(1+|\,x\,|)^p \sup_{t\in(0,T]} \mathbb{E}[\|\Phi(t,x)\|^p_{\mathbb{D}^{2,q}}]. \tag{53}$$

Moreover, if $\Phi \in \mathcal{K}_r(n)$ and the vector fields V_i, $i = 0, 1, \ldots, d$ are uniformly bounded, then $\Phi_\alpha \in \mathcal{K}_r((n-1) \wedge (k-m-1))$. In particular,

$$\sup_{t\in(0,T]} \sup_{x\in\mathbb{R}^N} \mathbb{E}\left[| \Phi_\alpha(t,x) |^p\right] < \infty. \tag{54}$$

Proof. We showed in the previous section that

$$V_{[\alpha]}(f \circ X_t)(x) = t^{-\|\alpha\|/2} \sum_{\beta\in\mathcal{A}(m)} t^{-\|\beta\|/2} M^{-1}_{\alpha,\beta}(t,x) D^{(\beta)}(f(X_t^x))$$

holds $\mathbb{P}$-a.s. By the product rule for the Malliavin derivative:

$$\begin{aligned} D^{(\beta)}(\Phi(t,x)\, M^{-1}_{\alpha,\beta}(t,x)\, f(X_t^x)) &= D^{(\beta)}\Phi(t,x)\, M^{-1}_{\alpha,\beta}(t,x)\, f(X_t^x) \\ &\quad + \Phi(t,x)\, D^{(\beta)} M^{-1}_{\alpha,\beta}(t,x)\, f(X_t^x) \\ &\quad + \Phi(t,x)\, M^{-1}_{\alpha,\beta}(t,x)\, D^{(\beta)} f(X_t^x). \end{aligned}$$

Then

$$\begin{aligned} &\mathbb{E}\left[\Phi(t,x) V_{[\alpha]}(f \circ X_t)(x)\right] \\ &= t^{-\frac{\|\alpha\|}{2}} \sum_{\beta\in\mathcal{A}(m)} t^{-\frac{\|\beta\|}{2}} \mathbb{E}\left[\Phi(t,x) M^{-1}_{\alpha,\beta}(t,x) D^{(\beta)}(f(X_t^x))\right] \\ &= t^{-\frac{\|\beta\|}{2}} \mathbb{E}\left[\Phi_\alpha(t,x) f(X_t^x)\right], \end{aligned}$$

where

$$\begin{aligned} \Phi_\alpha(t,x) = \sum_{\beta\in\mathcal{A}(m)} t^{-\|\beta\|/2} \Big\{ &\Phi(t,x)\, M^{-1}_{\alpha,\beta}(t,x) \delta(k_\beta(t,x)) \\ &- \Phi(t,x) D^{(\beta)} M^{-1}_{\alpha,\beta}(t,x) - D^{(\beta)}\Phi(t,x)\, M^{-1}_{\alpha,\beta}(t,x) \Big\}. \end{aligned}$$

The claim $\Phi_\alpha \in \mathcal{K}_r^{\mathrm{loc}}((n-1) \wedge (k-m-1))$ follows from a diligent application of Lemma 23, namely, parts 3–6. Note that the only term unbounded in x in the expression for Φ_α is $D^{(\beta)} M^{-1}_{\alpha,\beta}(t,x)$ which has linear growth in x. Finally, the bound (53) can be proved by observing that, due to (43)

$$\sup_{t\in(0,T]} \mathbb{E} \left| D^{(\gamma)} M^{-1}_{\beta,\gamma}(t,x) \right|^p \le C(1 + | x |)^p. \tag{55}$$

Hence, the bound

$$\sup_{t\in(0,T]} \mathbb{E}\,|\,\Phi_\alpha(t,x)\,|^p \le C_p(1+|\,x\,|)^p \sup_{t\in(0,T]} \|\Phi(t,x)\|_{2,q}^p$$

follows by applying the following to the expression for $\Phi_\alpha(t,x)$: (55), Hölder's inequality, and the uniform boundedness of the L^r norm of M^{-1} and k_γ over $(t,x)\in(0,T]\times\mathbb{R}^N$ for each $r\ge 1$. □

Remark 25. Following from Remark 13, the adjoint $\delta(k_\gamma(t,x))$ is the Itô integral of $k_\gamma(t,x)$ with respect to the d-dimensional Brownian motion $B_t=(B_t^1,\ldots,B_t^d)$ as the process $s\to k_\gamma(t,x)(s)$ is $\mathcal{F}_s$-adapted for almost all $(t,x)\in(0,T]\times\mathbb{R}^N$. That is, we have that

$$\delta(k_\gamma(t,x))=\sum_{i=1}^d\int_0^1 k_\gamma(t,x)^i(s)dB_s^i.$$

It follows that for processes with values in $K^r(E)$ which are a.e. adapted as stochastic processes in H, that $\delta(f):=\delta(f(.,.))\in\mathcal{K}_{r+1}(E)$.

Corollary 26 (Integration by Parts formula II). *Under the UFG condition, for any $\Phi\in\mathcal{K}_r^{loc}(n)$ and for any $\alpha\in\mathcal{A}(m)$, there exists $\Phi'_\alpha\in\mathcal{K}_r^{loc}((n-1)\wedge(k-m-1))$ such that:*

$$\mathbb{E}[\Phi(t,x)(V_{[\alpha]}f)(X_t^x)]=t^{-\|\alpha\|/2}\mathbb{E}[\Phi'_\alpha(t,x)f(X_t^x)],\ \ t>0, x\in\mathbb{R}^N \tag{56}$$

for any $f\in C_b^\infty(\mathbb{R}^N;\mathbb{R})$. In addition, for any $q>p$

$$\sup_{t\in(0,T]}\mathbb{E}\left[|\,\Phi'_\alpha(t,x)\,|^p\right]\le C_{p,q}(1+|\,x\,|)^p\sup_{t\in(0,T]}\mathbb{E}\left[\|\Phi(t,x)\|_{\mathbb{D}^{2,q}}^p\right]. \tag{57}$$

Moreover, if $\Phi\in\mathcal{K}_r(n)$ and the vector fields V_i, $i=0,1,\ldots,d$ are uniformly bounded, then $\Phi'_\alpha\in\mathcal{K}_r((n-1)\wedge(k-m-1))$. In particular,

$$\sup_{t\in(0,T]}\sup_{x\in\mathbb{R}^N}\mathbb{E}\left[|\,\Phi'_\alpha(t,x)\,|^p\right]<\infty. \tag{58}$$

Proof. The first observation is the following relationship:

$$\begin{aligned}(V_{[\alpha]}f)(X_t^x)&=\nabla f(X_t^x)V_{[\alpha]}(X_t^x)\\&=(J_t^x)^{-T}\nabla(f\circ X_t)(x)V_{[\alpha]}(X_t^x)\\&=\nabla(f\circ X_t)(x)(J_t^x)^{-1}V_{[\alpha]}(X_t^x),\end{aligned}$$

where $(J_t^x)^{-T} := ((J_t^x)^{-1})^T$. At this point refer back to the closed linear system of equations, which induced the expression:

$$(J_t^x)^{-1} V_{[\alpha]}(X_t^x) = \sum_{\beta \in \mathcal{A}(m)} a_{\alpha,\beta}(t,x) V_{[\beta]}(x).$$

Again, the central position of the UFG condition is emphasised, as

$$\begin{aligned}\nabla(f \circ X_t)(x)(J_t^x)^{-1} V_{[\alpha]}(X_t^x) &= \sum_{\beta \in \mathcal{A}(m)} a_{\alpha,\beta}(t,x) \nabla(f \circ X_t)(x) V_{[\beta]}(x) \\ &= \sum_{\beta \in \mathcal{A}(m)} a_{\alpha,\beta}(t,x) V_{[\beta]}(f \circ X_t)(x).\end{aligned}$$

From Lemma 23, $a_{\alpha,\beta} \in \mathcal{K}^{\text{loc}}_{(\|\beta\|-\|\alpha\|)\vee 0}(k-m)$. Hence, it has been shown that:

$$\mathbb{E}\left[\Phi(t,x) V_{[\alpha]} f(X_t^x)\right] = \sum_{\beta \in \mathcal{A}(m)} \mathbb{E}\left[\Phi(t,x) a_{\alpha,\beta}(t,x) V_{[\beta]}(f \circ X_t)(x)\right].$$

The integration by parts formula (52) can then be applied N_m times, after noting that the product $\Phi a_{\alpha,\beta} \in \mathcal{K}^{\text{loc}}_{r+[(\|\beta\|-\|\alpha\|)\vee 0]}((n-1) \wedge (k-m-1))$. And so,

$$\begin{aligned}\mathbb{E}\left[\Phi(t,x) V_{[\alpha]} f(X_t^x)\right] &= \sum_{\beta \in \mathcal{A}(m)} t^{-\frac{\|\beta\|}{2}} \mathbb{E}\left[\Psi_\beta(t,x) f(X_t^x)\right] \\ &= \sum_{\beta \in \mathcal{A}(m)} t^{-\frac{\|\beta\|}{2}} t^{-\frac{\|\alpha\|-\|\beta\|}{2}} \mathbb{E}\left[t^{\frac{\|\alpha\|-\|\beta\|}{2}} \Psi_\beta(t,x) f(X_t^x)\right] \\ &= t^{-\frac{\|\alpha\|}{2}} \mathbb{E}\left[\Phi'_\alpha(t,x) f(X_t^x)\right],\end{aligned}$$

where

$$\Phi'_\alpha = \sum_{\beta \in \mathcal{A}(m)} t^{\frac{\|\alpha\|-\|\beta\|}{2}} \Psi_\beta \in \mathcal{K}^{\text{loc}}_r((n-1) \wedge (k-m-1)).$$

The bounds (57), (58) can be deduced from the previous theorem. □

Corollary 27 (Integration by Parts formula III). *Under the same conditions as Theorem 24, the following integration by parts formula holds:*

$$V_{[\alpha]} \mathbb{E}\left[\Phi(t,x) f(X_t^x)\right] = t^{-\|\alpha\|/2} \mathbb{E}\left[\Phi''_\alpha(t,x) f(X_t^x)\right], \quad t > 0, x \in \mathbb{R}^N, \tag{59}$$

where $\Phi''_\alpha \in \mathcal{K}_r^{loc}((n-1)\wedge(k-m-1))$. *In addition, for any* $q > p$*:*

$$\sup_{t\in(0,T]} \mathbb{E}\left[\left| \Phi''_\alpha(t,x)\right|^p\right] \le C_{p,q}(1+|x|)^p \sup_{t\in(0,T]} \|\Phi(t,x)\|^p_{\mathbb{D}^{2,q}}. \tag{60}$$

Moreover, if $\Phi \in \mathcal{K}_r(n)$ *and the vector fields* V_i, $i = 0,1,\dots,d$ *are uniformly bounded, then* $\Phi''_\alpha \in \mathcal{K}_r((n-1)\wedge(k-m-1))$. *In particular,*

$$\sup_{t\in(0,T]} \sup_{x\in\mathbb{R}^N} \mathbb{E}[\left| \Phi''_\alpha(t,x)\right|^p] < \infty. \tag{61}$$

Proof. Observe that

$$\begin{aligned} V_{[\alpha]}\mathbb{E}[\Phi(t,x)f(X_t^x)] &= \mathbb{E}\left[V_{[\alpha]}(\Phi(t,x))\ f(X_t^x) + \Phi(t,x)V_{[\alpha]}(f\circ X_t)(x)\right] \\ &= \mathbb{E}[V_{[\alpha]}(\Phi(t,x))\ f(X_t^x) + t^{-\|\alpha\|/2}\Phi_\alpha(t,x)f(X_t^x)] \\ &= t^{-\|\alpha\|/2}\mathbb{E}[\Phi''_\alpha(t,x)f(X_t^x)], \end{aligned}$$

where

$$\Phi''_\alpha(t,x) = t^{\|\alpha\|/2}V_{[\alpha]}(\Phi(t,x)) + \Phi_\alpha(t,x) \in \mathcal{K}_r^{\rm loc}((n-1)\wedge(k-m-1)).$$

It is also clear from the previous results that Φ''_α satisfies (60). □

Corollary 28 (Integration by Parts formula IV). *Under the same conditions as Theorem 24, the following integration by parts formula holds for* $m_1+m_2 \le k-m$ *and* $\alpha_1,\dots,\alpha_{m_1+m_2} \in \mathcal{A}(m)$*:*

$$\begin{aligned} &V_{[\alpha_1]}\dots V_{[\alpha_{m_1}]}P_t(V_{[\alpha_{m_1+1}]}\dots V_{[\alpha_{m_1+m_2}]}f)(x) \\ &\quad = t^{-(\|\alpha_1\|+\dots+\|\alpha_{m_1+m_2}\|)/2}\mathbb{E}\left[\Phi_{\alpha_1,\dots,\alpha_{m_1+m_2}}(t,x)f(X_t^x)\right], \end{aligned} \tag{62}$$

where $\Phi_{\alpha_1,\dots,\alpha_{m_1+m_2}} \in \mathcal{K}_0^{loc}((k-m-m_1-m_2))$. *Moreover,*

$$\sup_{t\in(0,T]} \mathbb{E}[\left| \Phi_{\alpha_1,\dots,\alpha_{m_1+m_2}}(t,x)\right|^p] \le C_p(1+|x|)^{(m_1+m_2)p}. \tag{63}$$

If the vector fields V_i, $i = 0,1,\dots,d$ *are uniformly bounded, then* $\Phi_{\alpha_1,\dots,\alpha_{m_1+m_2}} \in \mathcal{K}_0((k-m-m_1-m_2))$. *In particular,*

$$\sup_{t\in(0,T]} \sup_{x\in\mathbb{R}^N} \mathbb{E}[\left| \Phi_{\alpha_1,\dots,\alpha_{m_1+m_2}}(t,x)\right|^p] < \infty. \tag{64}$$

Proof. Once it is noted that constant functions are in $\mathcal{K}_0$, the proof follows from m_2 applications of Theorem 24 followed by m_1 applications of Corollary 26. The bounds (63), (64) follows likewise. □

Remark 29. Observe that we are able to quantify exactly how the derivatives explode (when t tends to 0)—as functions of x-based on an analysis of the integration by parts factors. In the next section, we shall use the above bounds to deduce sharp gradient bounds for the diffusion semigroup P_t.

2.7 Explicit Bounds

We discuss now how the integration by parts formulae allow the acquisition of several explicit gradient bounds. This section is by no means exhaustive, and for a more complete synopsis of obtainable gradient bounds, one should consult Nee [44]. We will use the following norms and semi-norms

$$\|f\|_\infty = \sup_{x\in\mathbb{R}^N} |f(x)|, \quad \|\nabla f\|_\infty = \max_{i\in\{1,\dots,d\}} \left\|\frac{\partial f}{\partial x_i}\right\|_\infty, \quad f \in C_b^\infty(\mathbb{R}^N,\mathbb{R})$$

$$\|f\|_{V,i} = \sum_{u=1}^{i} \sum_{\substack{\alpha_1,\dots,\alpha_u\in\mathcal{A}_0 \\ \|\alpha_1*\dots*\alpha_u\|=i}} \left\|V_{[\alpha_1]}\cdots V_{[\alpha_u]} f\right\|_\infty, \quad i \in \mathbb{N} \quad C_b^{V,i}(\mathbb{R}^N) = \{f : \|f\|_{V,i} < \infty\}$$

$$\|f\|_p = \sum_{i=1}^{p} \left\|\nabla^i f\right\|_\infty, \quad f \in C_b^p(\mathbb{R}^N,\mathbb{R}), \quad p \in \mathbb{N} \tag{65}$$

$$\left\|\nabla^i f\right\|_\infty = \max_{j_1,\dots,j_i\in\{1,\dots,d\}} \left\|\frac{\partial^i f}{\partial x_{j_1}\dots\partial x_{j_i}}\right\|_\infty$$

$$\|f\|_{p,\infty} = \|f\|_\infty + \|f\|_p \quad f \in C_b^\infty(\mathbb{R}^N,\mathbb{R}).$$

Remark 30. Note that $\|V_\alpha f\|_\infty \le C\,\|f\|_{|\alpha|}$ for any $\alpha \in \mathcal{A}$ and $f \in C_b^{|\alpha|}(\mathbb{R}^N,\mathbb{R})$, hence

$$\|\varphi\|_{V,i} \le C\,\|\varphi\|_i \quad i \in \mathbb{N}.$$

Corollary 31. *Let $f \in C_b^\infty(\mathbb{R}^N,\mathbb{R})$ and $\alpha_1,\dots,\alpha_{m_1+m_2} \in \mathcal{A}(m)$ be such that $m_1+m_2 \le k-m$. Then there is a constant C such that, for any $t \in [0,T]$,*

$$\begin{aligned} &|V_{[\alpha_1]}\dots V_{[\alpha_{m_1}]}P_t(V_{[\alpha_{m_1+1}]}\dots V_{[\alpha_{m_1+m_2}]}f)(x)| \\ &\quad = C\|f\|_\infty t^{-(\|\alpha_1\|+\dots+\|\alpha_{m_1+m_2}\|)/2}(1+|x|)^{m_1+m_2}. \end{aligned} \tag{66}$$

Moreover, if the vector fields $V_0, \ldots, V_d$ are uniformly bounded, then there is a constant C such that, for any $t \in [0, T]$,

$$\| V_{[\alpha_1]} \ldots V_{[\alpha_{m_1}]} P_t (V_{[\alpha_{m_1+1}]} \ldots V_{[\alpha_{m_1+m_2}]} f) \|_\infty = C \| f \|_\infty t^{-(\|\alpha_1\| + \ldots + \|\alpha_{m_1+m_2}\|)/2}. \tag{67}$$

Proof. This now follows easily from Corollary 28. □

The following result will be of use in the next section:

Corollary 32. *Assume that $0 < p \leq n \leq k - m$, and let $f \in C_b^\infty(\mathbb{R}^N, \mathbb{R})$. Then there is a constant $C < \infty$ such that for $\alpha_1, \ldots, \alpha_n \in \mathcal{A}(m)$ and any $t \in [0, T]$,*

$$| V_{[\alpha_1]} \ldots V_{[\alpha_n]} P_t f(x) | \leq \frac{C\, t^{p/2}}{t^{(\|\alpha_1\| + \ldots + \|\alpha_n\|)/2}} \| f \|_p (1 + | x |)^n. \tag{68}$$

Moreover, if the vector fields $V_0, \ldots, V_d$ are uniformly bounded, then there is a constant C such that, for any $t \in [0, T]$,

$$\| V_{[\alpha_1]} \ldots V_{[\alpha_n]} P_t f) \|_\infty = \frac{C\, t^{p/2}}{t^{(\|\alpha_1\| + \ldots + \|\alpha_n\|)/2}} \| f \|_p. \tag{69}$$

Proof. We prove the result for $p = 1$ as the general case follows along the same lines. The idea behind this gradient bound is that one can "sacrifice" the derivative along $V_{[\alpha_n]}$ to obtain a new integration by parts formula involving the gradient of f. Observe,

$$\begin{aligned} V_{[\alpha_n]} P_t f(x) &= \sum_{i=1}^N V_{[\alpha_n]}^i (x) \partial_i \mathbb{E} \left[(f \circ X_t)(x) \right] \\ &= \sum_{j=1}^N \mathbb{E} \left[\partial_j f(X_t^x) \sum_{i=1}^N V_{[\alpha_n]}^i (x) (J_t^x)_{j,i} \right] \\ &= \sum_{j=1}^N \mathbb{E} \left[\partial_j f(X_t^x) \Phi^j (t, x) \right], \end{aligned}$$

where $\Phi^j(t, x) = \sum_{i=1}^N V_{[\alpha_k]}^i (x) (J_t^x)_{j,i} \in \mathcal{K}_0^{\text{loc}}(k - m)$. Hence, following $n - 1$ applications of Theorem 24 to the above expression, we see that:

$$V_{[\alpha_1]} \ldots V_{[\alpha_n]} P_t f(x) = t^{-(\|\alpha_1\| + \ldots \|\alpha_{n-1}\|)/2} \sum_{j=1}^N \mathbb{E} \left[\partial_j f(X_t^x) \Phi_{\alpha_1, \ldots, \alpha_{n-1}}^j (t, x) \right].$$

And therefore

$$|V_{[\alpha_1]} \dots V_{[\alpha_n]} P_t f(x)| \leq C t^{-(\|\alpha_1\|+\dots\|\alpha_{n-1}\|)/2} \|\nabla f\|(1 + |x|)^n$$
$$\leq \frac{C t^{1/2}}{t^{(\|\alpha_1\|+\dots+\|\alpha_n\|)/2}} \|\nabla f\|(1 + |x|)^n.$$

The last inequality follows because $t^{(1-\|\alpha_n\|)/2} \geq T^{(1-\|\alpha_n\|)/2}$. □

The gradient bounds presented above play the central role in determining the rates of convergence of the numerical approximations presented in the following chapters. In addition, we can use them to deduce the Hörmander's criterion in the particular case when the vector fields V_i, $i = 0, 1, \dots, d$ satisfy the uniform Hörmander condition.

2.8 Smoothness of the Diffusion Semigroup

In this section, we shall assume for simplicity that the vector fields V_i, $i = 0, 1, \dots, d$ are smooth and uniformly bounded. We prove that, under the assumption of Hörmander's criterion, $x \to P_t f(x)$ is a smooth function. This implies the existence and smoothness of the density of the law of the corresponding diffusion. To show this, we make use of the following proposition, provided by Malliavin in [40]:

Proposition 33. *Let μ be a finite measure defined on the Borel σ-algebra $\mathcal{B}(\mathbb{R}^N)$. Assume that for every multi-index α , there is a constant C_α, such that*

$$\left| \int \partial_\alpha f(x) \mu(dx) \right| \leq C_\alpha \|f\|_\infty$$

for every smooth f with compact support. Then μ has a density with respect to the Lebesgue measure which is smooth on $\mathbb{R}^N$. In particular, if for every multi-index α, there is a constant C_α such that

$$|\mathbb{E}[(\partial_\alpha f)(X_t^x)]| \leq C_\alpha \|f\|_\infty, \tag{70}$$

for every smooth f with compact support, then the law of X_t^x has a density, which is smooth on $\mathbb{R}^N$.

Remark 34. One can "localize" the result in Proposition 33 in the following standard way: Assume that for every $R > 0$ and every multi-index α there is a constant $C_{\alpha,R}$, such that

$$\left| \int \partial_\alpha f(x) \mu(dx) \right| \leq C_{\alpha,R} \|f\|_\infty$$

for every smooth f with compact support in the ball $B(0, R)$. Then μ has a density with respect to the Lebesgue measure which is smooth on $\mathbb{R}^N$. To justify this, one uses Proposition 33 to show that for every $R > 0$, $\mu|_{B(0,R)}$ has a smooth density with respect to the Lebesgue measure.

In particular, if for every $R > 0$ and every multi-index α there is a constant $C_{\alpha,R}$, such that

$$|\,\mathbb{E}[(\partial_\alpha f)(X_t^x)]\,| \leq C_{\alpha,R}\,\|f\|_\infty\,. \tag{71}$$

for every smooth f with compact support in the ball $B(0, R)$ then the law of X_t^x has a smooth density with respect to the Lebesgue measure.

Gradient bounds such as (70), (71) may be deduced from the techniques of Kusuoka, provided some extra assumptions are made.

Theorem 35. *Assume that the following holds for all $x \in \mathbb{R}^N$:*

$$Span\{V_{[\alpha]}(x) : \alpha \in \mathcal{A}(m)\} = \mathbb{R}^N. \tag{72}$$

Then the law of X_t^x has a smooth density with respect to the Lebesgue measure.

Note that we may restate (72), as the property that there exists $\epsilon = \epsilon(x) > 0$ such that

$$\sum_{\alpha\in\mathcal{A}(m)} \left(V_{[\alpha]}(x), \xi\right)^2 \geq \epsilon\,|\,\xi\,|^2, \tag{73}$$

for all $\xi \in \mathbb{R}^N$, or equivalently: the matrix $(VV^T)(x)$ is invertible $\forall x \in \mathbb{R}^N$, where $V(x) := \left(V_{[\alpha]}^j\right)_{\substack{j=1,\dots,N \\ \alpha\in\mathcal{A}(m)}}$. Note: upon taking the infimum over all $|\,\xi\,| = 1$, the LHS of (73) is the minimum eigenvalue of this matrix. The inverse must have smooth entries (by the inverse function theorem) and be bounded on compact sets.

Proof. Showing (71), amounts to deriving an integration by parts formula for the partial derivatives ∂_i. This can easily be iterated to obtain any combination of higher partial derivatives. We claim that there exist smooth functions C_α^i such that:

$$\partial_i = \sum_{\alpha\in\mathcal{A}(m)} C_\alpha^i(x) V_{[\alpha]}(x),$$

for all $x \in \mathbb{R}^m$. This can be re-written in matrix form as $\partial_i = VC^i$, where $V(x) := \left(V_{[\alpha]}^j(x)\right)_{\substack{j=1,\dots,N \\ \alpha\in\mathcal{A}(m)}}$, and $C^i(x) = \left(C_\alpha^i(x)\right)_{\alpha\in\mathcal{A}(m)}$. But it holds that $(VV^T)(x)$ is invertible for all $x \in \mathbb{R}^N$. Therefore, we may choose

$$C^i = V^T(VV^T)^{-1}\partial_i,$$

that is, $C_\alpha^i(x) = (V^T(VV^T)^{-1}\partial_i,)_\alpha(x)$. Clearly, C_α^i is smooth by the inverse function theorem and it is also bounded on compacts. Let φ be a smooth function with compact support in the ball $B(0, R)$. Observe that

$$\mathbb{E}[(\partial_i\varphi)(X_t^x)] = \sum_{\alpha\in\mathcal{A}(m)} \mathbb{E}\left[(C_\alpha^i V_{[\alpha]}\varphi)(X_t^x)\right] = \sum_{\alpha\in\mathcal{A}(m)} \mathbb{E}\left[(\Lambda_R C_\alpha^i V_{[\alpha]}\varphi)(X_t^x)\right],$$

where $\Lambda_R : \mathbb{R}^N \to \mathbb{R}$ is a smooth "truncation" function such that $\Lambda_R(x) = 1$ if $x \in B(0, R)$ and $\Lambda_r(x) = 0$ if $x \notin B(0, 2R)$. We can therefore assume without loss of generality that both C_α^i and $V_{[\alpha]}$ are bounded. By Corollary 26 we deduce that there exists $\Phi'_{\alpha,R}$ such that:

$$\mathbb{E}\left[(\Lambda_R C_\alpha^i V_{[\alpha]}\varphi)(X_t^x)\right] = t^{-\|\alpha\|/2}\mathbb{E}[\Phi'_{\alpha,R}(t,x)\varphi(X_t^x)] \tag{74}$$

and

$$\sup_{t\in(0,T]}\sup_{x\in\mathbb{R}^N} \mathbb{E}[|\,\Phi'_{\alpha,R}(t,x)\,|^p] < \infty. \tag{75}$$

Hence there exists a constant $C_{i,R}$, such that

$$|\,\mathbb{E}[(\partial_i\varphi)(X_t^x)]\,| \leq C_{i,R}\,\|\varphi\|_\infty \tag{76}$$

with

$$C_{i,R} = t^{-\|\alpha\|/2}\sum_{\alpha\in\mathcal{A}(m)}\sup_{x\in\mathbb{R}^N}\mathbb{E}\left[|\,\Phi'_{\alpha,R}(t,x)\,|\right] < \infty.$$

The same argument can be done for any partial derivative and the procedure can be iterated for any multi-index α. The result follows by Remark 34. □

2.9 The V0 Condition

Under the UFG condition alone, one cannot gauge any differentiability properties in the direction V_0. Even if we have differentiability in the direction V_0, the norm $\|V_0 P_t\varphi\|_\infty$ may explode with arbitrary high rate. Kusuoka has given an explicit class of examples where, for arbitrary integers $l \geq 2$, it holds

$$ct^{-\frac{l}{2}}\,\|\varphi\|_\infty \leq \|V_0 P_t\varphi\|_\infty \leq Ct^{-\frac{l}{2}}\,\|\varphi\|_\infty$$

for some constants $c, C > 0$ (see Propositions 14 and 16 in [30]). However the following condition allows us to have a suitable control in the direction V_0.

Definition 36 (The V0 Condition). Let $\{V_i : i = 0, \ldots, d\}$, be a system of vector fields such that $V_1, \ldots, V_d \in C_b^{k+1}(\mathbb{R}^N; \mathbb{R}^N)$ and $V_0 \in C_b^k(\mathbb{R}^N; \mathbb{R}^N)$. We say that $\{V_i : i = 0, \ldots, d\}$ satisfy the V0 condition if, there exist uniformly bounded functions $\varphi_\beta \in C_b^k(\mathbb{R}^N, \mathbb{R})$, with $\beta \in \mathcal{A}(2)$ such that

$$V_0(x) = \sum_{\beta \in \mathcal{A}(2)} \varphi_\beta(x) V_{[\beta]}(x). \tag{77}$$

Condition V0 states that V_0 can be expressed as a linear combination of the vector fields $\{V_1, \ldots V_k\} \cup \{[V_i, V_j], 1 \leq i < j \leq k\}$. This premise is weaker than the ellipticity assumption and has been used, for example, by Jerison and Sánchez–Calle [25] to obtain estimates for the heat kernel. Under the V0 condition all results presented above extend to the differentiability in the direction V_0 as well. For example we have the following equivalent of the corollary 28:

Proposition 37. *Under the same conditions as Theorem 24 and the V0 condition, the following integration by parts formula holds for* $m_1 + m_2 \leq k - m$ *and* $\alpha_1, \ldots, \alpha_{m_1+m_2} \in \mathcal{A}(m) \cup \{(0)\}$*:*

$$\begin{aligned} & V_{[\alpha_1]} \ldots V_{[\alpha_{m_1}]} P_t (V_{[\alpha_{m_1+1}]} \ldots V_{[\alpha_{m_1+m_2}]} f)(x) \\ & \quad = t^{-(\|\alpha_1\| + \ldots + \|\alpha_{m_1+m_2}\|)/2} \mathbb{E}\left[\Phi_{\alpha_1, \ldots, \alpha_{m_1+m_2}}(t, x) f(X_t^x) \right], \end{aligned} \tag{78}$$

where $\Phi_{\alpha_1, \ldots, \alpha_{m_1+m_2}} \in \mathcal{K}_0^{loc}((k - m - m_1 - m_2))$*. Moreover,*

$$\sup_{t \in (0,T]} \mathbb{E}\left[\left| \Phi_{\alpha_1, \ldots, \alpha_{m_1+m_2}}(t, x) \right|^p \right] \leq C_p (1 + |x|)^{(m_1+m_2)p}. \tag{79}$$

Moreover, if the vector fields V_i, $i = 0, 1, \ldots, d$ *are uniformly bounded, then* $\Phi_{\alpha_1, \ldots, \alpha_{m_1+m_2}} \in \mathcal{K}_0((k - m - m_1 - m_2))$*. In particular,*

$$\sup_{t \in (0,T]} \sup_{x \in \mathbb{R}^N} \mathbb{E} \left| \Phi_{\alpha_1, \ldots, \alpha_{m_1+m_2}}(t, x) \right|^p < \infty. \tag{80}$$

From Proposition 37 one can deduce the following corollary similar to Corollary 32

Corollary 38. *Assume* $n \leq k - m$*, and let* $f \in C_b^\infty(\mathbb{R}^N, \mathbb{R})$ *. Then, under the UFG+V0 conditions, there is a constant* $C < \infty$ *such that for* $\alpha_1, \ldots, \alpha_n \in \mathcal{A}(m) \cup \{(0)\}$*:*

$$|V_{[\alpha_1]} \ldots V_{[\alpha_n]} P_t f(x)| \leq \frac{C\, t^{1/2}}{t^{(\|\alpha_1\| + \ldots + \|\alpha_n\|)/2}} \|\nabla f\|_\infty (1 + |x|)^N. \tag{81}$$

Moreover, if the vector fields $V_0, \ldots, V_d$ are uniformly bounded, then there is a constant C such that

$$\|V_{[\alpha_1]} \ldots V_{[\alpha_n]} P_t f)\|_\infty = \frac{C\, t^{1/2}}{t^{(\|\alpha_1\|+\ldots+\|\alpha_n\|)/2}} \|\nabla f\|. \tag{82}$$

and for any integer $p > 0$ there is a constant C_p such that

$$\|V_{[\alpha_1]} \ldots V_{[\alpha_n]} P_t f)\|_\infty = \frac{C_p}{t^{(\|\alpha_1\|+\ldots+\|\alpha_n\|)/2}} \|f\|_p. \tag{83}$$

3 Cubature Methods

3.1 Introduction

In this section we will be concerned with numerical approximations of solutions of stochastic differential equations (SDEs). There are two classes of numerical methods for approximating SDEs. The objective of the first is to produce a pathwise approximation of the solution (strong approximation). The second method involves approximating the distribution of the solution at a particular instance in time (weak approximation). For example when one is only interested in the expectation $\mathbb{E}[\varphi(X_t)]$ for some function φ, it is sufficient to have a good approximation of the distribution of the random variable X_t rather than of its sample paths. This observation was first made by Milstein [42] who showed that pathwise schemes and L^2 estimates of the corresponding errors are irrelevant in this context since the objective is to approximate the law of X_t. This section contains approximations that belong to this second class of methods.

Classical results in this area concentrate on solving numerically SDEs for which the so-called "ellipticity condition", or more generally the "Uniform Hörmander condition" (UH), is satisfied. For a survey of such schemes see, for example, Kloeden and Platen [27] or Burrage, Burrage and Tian [6]. Under this condition, for any bounded measurable function φ, $P_t\varphi$ is smooth for any $t > 0$. It is this property upon which the majority of these schemes rely.

For example, the classical Euler–Maruyama scheme requires $P_t\varphi$ to be four times differentiable in order to obtain the optimal rate of convergence. Talay [57, 58] and, independently, Milstein [43] introduced the appropriate methodology to analyse this scheme. They express the error as a difference including a sum of terms involving $P_t\varphi$. Their analysis also shows the relationship between the smoothness of φ and the corresponding error. Talay and Tubaro [59] prove an even more precise result showing that, under the same conditions, the errors corresponding to the Euler–Maruyama and many other schemes can be expanded in terms of powers of the discretization step. Furthermore, Bally and Talay [2] show the existence of such an expansion under a much weaker hypothesis on φ: that φ need only be measurable and bounded (even the boundedness condition can be relaxed). Higher order schemes require additional smoothness properties of $P_t\varphi$ (see for example, Platen and Wagner [52]).

As explained in the previous chapter, Kusuoka and Stroock [32, 33, 34] studied the properties of $P_t\varphi$ under the UFG condition which is weaker. A number of schemes have recently been developed to work under the UFG conditions rather than the ellipticity condition, their convergence depending intrinsically on the above estimates of $V_{[\alpha_1]} \dots V_{[\alpha_n]} P_t\varphi$. A further advantage of this new generation of schemes is a consequence of the classical result stating that the support of $X(x)$ is the closure of the set $S = \{x^\varphi : [0,\infty) \to \mathbb{R}^d\}$ where x^φ solves the ODE,

$$x_t^\varphi = x + \int_0^t V_0(x_s^\varphi)ds + \sum_{j=1}^d \int_0^t V_j(x_s^\varphi)\varphi(s)\,ds$$

and $\varphi : [0,\infty) \to \mathbb{R}^d$ is an arbitrary smooth function (see Stroock and Varadhan [54–56], Millet and Sanz-Sole[41]). These schemes attempt to keep the support of the approximating process on the set S. In this way, stability problems that are known to affect classical schemes can be avoided. For example, Ninomyia and Victoir [49] give an explicit example where the Euler–Maruyama approximation fails whilst their algorithm succeeds (see Example 43 below for their algorithm). Their example involves an SDE related to the Heston stochastic volatility model in finance.

In this chapter we give a general criterion for the convergence of a class of weak approximations incorporating this new category of schemes. The criterion is based upon the stochastic Stratonovich–Taylor expansion of $\varphi(X_t)$ and demonstrates how the rate of convergence depends on the smoothness of the test function φ.

For smooth test functions, an equidistant partition of the time interval on which the approximation is sought is optimal. For less smooth functions, this is no longer true. We emphasize that the UFG+V0 conditions are not required for smooth test functions.

3.2 M-Perfect Families

In this section we introduce the concept of an *m-perfect* family. Such families correspond to various weak approximations of SDEs, including the Lyons–Victoir and Ninomiya–Victoir schemes. The main result appears in Theorem 46 and Corollary 47.

For $\alpha = (i_1, \dots, i_r) \in \mathcal{A}$ and $\varphi \in C_b^r(\mathbb{R}^N)$, let $f_{\alpha,\phi}$ be defined as $f_{(i_1,\dots,i_r),\varphi} := V_{i_1} \dots V_{i_r}\varphi$ and $I_{f_{\alpha,\varphi}}(t)$ be the iterated Stratonovich integral

$$I_{f_{\alpha,\varphi}}(t) := \int_0^t \int_0^{s_0} \cdots \left(\int_0^{s_{r-2}} f_{\alpha,\varphi}(X_{s_{r-1}}) \circ dW_{s_{r-1}}^{i_1} \right) \circ \cdots \circ dW_{s_1}^{i_{r-1}} \circ dW_{s_0}^{i_r},$$

for $t \geq 0$. If $i_1 = 0$ then $I_{f_{\alpha,\varphi}}(t)$ is well defined for $\varphi \in C_b^r(\mathbb{R}^N)$. However, if $i_1 \neq 0$ then $I_{f_{\alpha,\varphi}}(t)$ is well defined provided $\varphi \in C_b^{r+2}(\mathbb{R}^N)$, since the semimartingale property of $f_{\alpha,\varphi}(X)$ is required in the definition of the first Stratonovich integral $\int_0^{s_{r-2}} f_{\alpha,\varphi}(X_{s_{r-1}}) \circ dW_{s_{r-1}}^{i_1}$. Note that the Stratonovich integrals are evaluated innermost first. Finally let

$$I_\alpha(t) := \int_0^t \int_0^{s_0} \cdots \left(\int_0^{s_{r-2}} 1 \circ dW_{s_{r-1}}^{i_1} \right) \circ \cdots \circ dW_{s_1}^{i_{r-1}} \circ dW_{s_0}^{i_r}.$$

Let $\alpha = (i_1, \ldots, i_r) \in \mathcal{A}_0$ be an arbitrary multi-index such that $\|\alpha\| = m \in \mathbb{N}$ (and $|\alpha| = r \in \mathbb{N}$). If m is odd, then $\mathbb{E}[I_\alpha(t)] = 0$ and if m is even then

$$\mathbb{E}[I_\alpha(t)] = \begin{cases} \frac{t^{\frac{m}{2}}}{2^{r-\frac{m}{2}} (\frac{m}{2})!} & \text{if } \alpha \in \mathcal{A}_0^{m,r} \\ 0 & \text{otherwise} \end{cases}, \tag{84}$$

where $\mathcal{A}_0^{m,r}$ is the set of multi-indices $\alpha = \alpha_1 * \cdots * \alpha_{\frac{m}{2}} \in \mathcal{A}_0(m)$ such that each $\alpha_i = (0)$ or (j, j) for some $j \in \{1, \ldots, k\}$. Note that $r - \frac{m}{2}$ is equal to the number of pairs of indices (j, j) occurring in α. A proof of this result can be found in [19].

We state three further results in (85), (86) and (88). The proofs are all elementary and can be found in [19]. The first two give an upper bound on the L^2 norm of $I_{f_{\alpha,\varphi}}(t)$ for smooth φ. The third provides an explicit form for the remainder of $\varphi(X_t)$ when expanded in terms of iterated integrals.

For $\varphi \in C_b^{\|\alpha\|+2}(\mathbb{R}^N)$ and any multi-index $\alpha = (i_1, \ldots, i_r) \in \mathcal{A}_0$ such that $i_1 \neq 0$, we have[9]

$$\left\| I_{f_{\alpha,\varphi}}(t) \right\|_2 \leq c \left\| f_{\alpha,\varphi} \right\|_\infty t^{\frac{\|\alpha\|}{2}} + c \sum_{i=1}^k \left\| V_i f_{\alpha,\varphi} \right\|_\infty t^{\frac{\|\alpha\|+1}{2}} \tag{85}$$

for some constant $c = c(\alpha) > 0$. For $\varphi \in C_b^{\|\alpha\|}(\mathbb{R}^N)$ and any multi-index $\alpha = (i_1, \ldots, i_r) \in \mathcal{A}_0$ such that $i_1 = 0$ we have

$$\left\| I_{f_{\alpha,\varphi}}(t) \right\|_2 \leq c \left\| f_{\alpha,\varphi} \right\|_\infty t^{\frac{\|\alpha\|}{2}}. \tag{86}$$

For $m \in \mathbb{N}$, $\varphi \in C_b^{m+3}(\mathbb{R}^N)$ and $x \in \mathbb{R}^N$, we define the truncation,

$$\varphi_t^m(x) := \varphi(x) + \sum_{\alpha \in \mathcal{A}_0(m)} f_{\alpha,\varphi}(x) I_\alpha(t). \tag{87}$$

[9]In the following, we allow for the constant c to take different values from one line to another.

Then for $t \geq 0$ the remainder is

$$R_{m,t,\varphi}(x) := \varphi(X_t) - \varphi_t^m(x) = \left(\sum_{\|\alpha\|=m+1} + \sum_{\|\alpha\|=m+2, \alpha=0*\beta, \|\beta\|=m} \right) I_{f_{\alpha,\varphi}}(t). \tag{88}$$

In the following, we define a class of approximations of X expressed in terms of certain families of stochastic processes, $\bar{X}(x) = \{\bar{X}_t(x)\}_{t\in[0,\infty)}$ for $x \in \mathbb{R}^N$, which are explicitly solvable. In particular, we can explicitly compute the operator,

$$(Q_t\varphi)(x) = \mathbb{E}[\varphi(\bar{X}_t(x))]. \tag{89}$$

The semigroup P_T will then be approximated by $Q_{h_n}^m Q_{h_{n-1}}^m \cdots Q_{h_1}^m$ where $\{h_j := t_j - t_{j-1}\}_{j=1}^n$ and $\pi_n = \{t_j := (\frac{j}{n})^\gamma T\}_{j=0}^n$ for $n \in \mathbb{N}$, is a sufficiently fine partition of the interval $[0, T]$. In particular $h_j \in [0, 1)$ for $j = 1, \ldots, n$. The underlying idea is that $Q_t\varphi$ will have the same truncation as $P_t\varphi$.

So let $\bar{X}(x) = \{\bar{X}_t(x)\}_{t\in[0,\infty)}$, where $x \in \mathbb{R}^N$, be a family of progressively measurable stochastic processes such that, $\lim_{y\to x_0} \bar{X}_t(y) = \bar{X}_t(x_0)$ $\mathbb{P}$-almost surely, for any $t \geq 0$ and $x_0 \in \mathbb{R}^N$. As a result, the operator Q_t defined in (89) has the property that $Q_t\varphi \in C_b(\mathbb{R}^N)$ for any $\varphi \in C_b(\mathbb{R}^N)$. In particular, $Q_t : C_b(\mathbb{R}^N) \to C_b(\mathbb{R}^N)$ is a Markov operator.

Definition 39. For $m \in \mathbb{N}$, the family $\bar{X}(x) = \{\bar{X}_t(x)\}_{t\in[0,\infty)}$ where $x \in \mathbb{R}^N$, is said to be **m-perfect** for the process X if there exist a constant $c > 0$ and an integer $M \geq m + 1$ such that for $\varphi \in C_b^{V,M}(\mathbb{R}^N)$,

$$\sup_{x\in\mathbb{R}^N} |Q_t\varphi(x) - \mathbb{E}[\varphi_t^m(x)]| \leq c \sum_{i=m+1}^{M} t^{i/2} \|\varphi\|_{V,i} \,. \tag{90}$$

As we can see from (90), the quantity $\mathbb{E}[\varphi_t^m(x)]$ plays the same role as the classical truncation in the standard Taylor expansion of a function. Using (84) we deduce that,

$$\begin{aligned}
\mathbb{E}[\varphi_t^0(x)] &= \varphi(x) \\
\mathbb{E}[\varphi_t^2(x)] &= \varphi(x) + L\varphi(x)t \\
\mathbb{E}[\varphi_t^4(x)] &= \varphi(x) + L\varphi(x)t + L^2\varphi(x)\frac{t^2}{2},
\end{aligned}$$

where $L = V_0 + \frac{1}{2}\sum_{i=1}^d V_i^2$. Furthermore, since $\mathbb{E}[I_\alpha(t)] = 0$ for odd $\|\alpha\|$, it follows that $\mathbb{E}[\varphi_t^1(x)] = \mathbb{E}[\varphi_t^0(x)], \mathbb{E}[\varphi_t^3(x)] = \mathbb{E}[\varphi_t^2(x)]$ and $\mathbb{E}[\varphi_t^5(x)] = \mathbb{E}[\varphi_t^4(x)]$.

3.3 Examples

There now follow some examples of m-perfect families corresponding to the semigroup $\{P_t\}_{t\in[0,\infty)}$, the Lyons–Victoir method and the Ninomiya–Victoir algorithm.

Example 40. The family of stochastic processes $\{X_t(x)\}_{t\in[0,\infty)}$, where $x \in \mathbb{R}^N$, is m-perfect. More precisely, there exists a constant $c > 0$ such that for $\varphi \in \mathcal{C}_b^{V,m+2}(\mathbb{R}^N)$,

$$\sup_{x\in\mathbb{R}^N} |P_t\varphi(x) - \mathbb{E}[\varphi_t^m(x)]| \leq c \sum_{i=m+1}^{m+2} t^{i/2} \|\varphi\|_{V,i}\,, \tag{91}$$

Proof. For $\varphi \in \mathcal{C}_b^{V,m+3}(\mathbb{R}^N)$,

$$\left|P_t\varphi(x) - \mathbb{E}[\varphi_t^m(x)]\right| = \left|\mathbb{E}[R_{m,t,\varphi}(x)]\right| = \left|\mathbb{E}[(\sum_{\|\alpha\|=m+1} + \sum_{\|\alpha\|=m+2,\alpha=0*\beta,\|\beta\|=m})I_{f_{\alpha,\varphi}}(t)]\right|$$

Applying inequality (85) to the first sum,

$$\begin{aligned}\sum_{\|\alpha\|=m+1} \left\|I_{f_{\alpha,\varphi}}(t)\right\|_2 &\leq \sum_{\|\alpha\|=m+1} \{c\left\|f_{\alpha,\varphi}\right\|_\infty t^{\frac{m+1}{2}} + c\sum_{i=1}^{k}\left\|V_i f_{\alpha,\varphi}\right\|_\infty t^{\frac{m+2}{2}}\} \\ &\leq c \sum_{i=m+1}^{m+2} t^{i/2} \|\varphi\|_{V,i}\end{aligned} \tag{92}$$

for some constant $\bar{c} > 0$. Applying result (86) to the second sum,

$$\begin{aligned}\sum_{\|\alpha\|=m+2,\alpha=0*\beta,\|\beta\|=m} \left\|I_{f_{\alpha,\varphi}}(t)\right\|_2 &\leq \sum_{\|\alpha\|=m+2,\alpha=0*\beta,\|\beta\|=m} c\left\|f_{\alpha,\psi}\right\|_\infty t \\ &\leq c\,\|\varphi\|_{V,m+2}\, t^{\frac{m+2}{2}}.\end{aligned} \tag{93}$$

The result for $\varphi \in \mathcal{C}_b^{V,m+3}(\mathbb{R}^N)$ follows from combining (92) and (93). Since none of the terms in (91) depend on partial derivatives of order $m+3$, the inequality is also valid for any $\varphi \in \mathcal{C}_b^{V,m+2}(\mathbb{R}^N)$ (a standard approximation method can be used). □

In the following example, the family of processes $\bar{X}(x) = \{\bar{X}_t(x)\}_{t\in[0,1]}$, where $x \in \mathbb{R}^N$, corresponds to the Lyons–Victoir approximation (see [36]). The example involves a set of l finite variation paths, $\omega_1, \ldots, \omega_l \in \mathcal{C}_0^0([0,1],\mathbb{R}^d)$, for some $l \in \mathbb{N}$, together with some weights $\lambda_1, \ldots, \lambda_l \in \mathbb{R}^+$ such that $\sum_{j=1}^{l} \lambda_j = 1$. These paths are said to define a **cubature formula on Wiener Space of degree m** if, for any $\alpha \in \mathcal{A}_0(m)$,

$$\mathbb{E}[I_\alpha(1)] = \sum_{j=1}^{l} \lambda_j I_\alpha^{\omega_j}(1),$$

where,

$$I_{(i_1,\ldots,i_r)}^{\omega_j}(1) := \int_0^1 \int_0^{s_0} \cdots \left(\int_0^{s_{r-2}} d\omega_j^{i_1}(s_{r-1}) \right) \cdots d\omega_j^{i_{r-1}}(s_1) d\omega_j^{i_r}(s_0).$$

From the scaling properties of the Brownian motion we can deduce, for $t \geq 0$,

$$\mathbb{E}[I_\alpha(t)] = \sum_{j=1}^{l} \lambda_j I_\alpha^{\omega_{t,j}}(t),$$

where $\omega_{t,1}, \ldots, \omega_{t,l} \in C_0^0([0,t], \mathbb{R}^d)$ is defined by $\omega_{t,j}(s) = \sqrt{t}\omega_j\left(\frac{s}{t}\right)$, $s \in [0,t]$. In other words, the expectation of the iterated Stratonovich integrals $I_\alpha(t)$ is the same under the Wiener measure as it is under the measure,

$$\mathbb{Q}_t := \sum_{j=1}^{l} \lambda_j \delta_{\omega_{t,j}}.$$

Example 41. If we choose $\bar{X}$ to be the stochastic flow defined in (1), but with the driving Brownian motion replaced by the paths $\omega_{t,1}, \ldots, \omega_{t,l}$ defined above then the family of processes, $\{\overline{X}_t(x)\}_{t \in [0,1]}$, with corresponding operator $(\overline{Q}_t \varphi)(x) := \mathbb{E}_{\mathbb{Q}_t}[\varphi(\overline{X}_t(x))]$, is m-perfect. More precisely, there exists a constant $c > 0$ such that for $\varphi \in C_b^{V,m+2}(\mathbb{R}^N)$,

$$\sup_x \left| \overline{Q}_t \varphi(x) - \mathbb{E}[\varphi_t^m(x)] \right| \leq c \sum_{i=m+1}^{m+2} t^{i/2} \|\varphi\|_{V,i}$$

For example, if $(\lambda_j, \omega_{t,j})$ are chosen such that for $l = 2^d$ the paths are $\omega_{t,j} : t \mapsto t(1, z_j^1, \ldots, z_j^d)$ for $j = 1, \ldots, 2^d$ with points $z_j \in \{-1, 1\}^d$ and weights $\lambda_j = 2^{-d}$, we obtain a cubature formula of degree 3 and a corresponding 3-perfect family.

Proof. Let us first observe that $I_\alpha^{\omega_{t,j}}(t) = t^{\frac{|\alpha|}{2}} I_\alpha^{\omega_j}(1)$ Hence, for $\varphi \in C_b^{V,m+2}(\mathbb{R}^N)$,

$$\begin{aligned}
\left|\overline{Q}_t\varphi(x) - \mathbb{E}[\varphi_t^m(x)]\right| &= \left|\mathbb{E}_{\mathbb{Q}_t}[R_{m,t,\varphi}(x)]\right| \\
&\leq \Big(\sum_{\|\alpha\|=m+1} + \sum_{\|\alpha\|=m+2,\alpha=0*\beta,\|\beta\|=m} \Big) \left\| f_{\alpha,\varphi} \right\|_\infty \left\| \mathbb{E}_{\mathbb{Q}_t}[I_\alpha(t)] \right\|_2 \\
&\leq \Big(\sum_{\|\alpha\|=m+1} + \sum_{\|\alpha\|=m+2,\alpha=0*\beta,\|\beta\|=m} \Big) \left\| f_{\alpha,\varphi} \right\|_\infty \sum_{j=1}^{l} \lambda_j \left\| I_\alpha^{\omega_{t,j}}(t)] \right\|_2 \\
&\leq \Big(\sum_{\|\alpha\|=m+1} + \sum_{\|\alpha\|=m+2,\alpha=0*\beta,\|\beta\|=m} \Big) k_\alpha \left\| f_{\alpha,\varphi} \right\|_\infty t^{\frac{|\alpha|}{2}}.
\end{aligned}$$

where $k_\alpha = \sum_{j=1}^{l} \lambda_j \left\| I_\alpha^{\omega_j}(1) \right\|_2$. □

Remark 42. (i) There has been no change to the underlying measure in the example above. Merely a representation in terms of the measure $\mathbb{Q}_t$ has been introduced to ease the computation of $\overline{Q}_t$. More precisely, the family of processes $\left\{\overline{X}_t(x)\right\}_{t\in[0,1]}$ where $x \in \mathbb{R}^N$ is constructed as follows. We take,

$$\overline{X}_0(x) = x$$

and then randomly choose a path $\omega_{t,r}$ from the set $\{\omega_{t,1}, \ldots, \omega_{t,l}\}$ with corresponding probabilities $(\lambda_1, \ldots, \lambda_l)$. Each process then follows a deterministic trajectory driven by the solution of the ordinary differential equation,

$$d\overline{X}_t = V_0(\overline{X}_t)dt + \sum_{j=1}^{d} V_j(\overline{X}_t)d\omega_{t,k}^j.$$

We can therefore compute the expected value of a functional of $X_t(x)$ as integrals on the path space with respect to the Radon measure $\mathbb{Q}_t$. Hence the identities,

$$\overline{Q}_t\varphi(x) = \mathbb{E}\left[\varphi(\overline{X}_t(x))\right] = \mathbb{E}_{\mathbb{Q}_t}\left[\varphi(\overline{X}_t(x))\right]$$

(ii) The approach adopted by Lyons and Victoir to construct the above approximation resembles the ideas developed by Clark and Newton in a series of papers [10, 11, 45, 46]. Heuristically, Clark and Newton constructed strong approximations of SDEs using flows driven by vector fields which were measurable with respect to the filtration generated by the driving Wiener process. In a similar vein, Castell and Gaines [8] provide a method of strongly approximating the solution of an SDE by means of exponential Lie series.

(iii) The family of processes $\bar{X}(x) = \{\bar{X}_t(x)\}_{\{t\in[0,1],x\in\mathbb{R}^N\}}$ corresponding to the Lyons–Victoir approximation (see [36]) have the fundamental property that they match the expectation of the truncated signature as sketched in the Introduction. In [36], Lyons and Victoir constructed degree 3 and degree 5 approximations in general dimensions. More recently, Gyurko and Lyons developed in [22] higher degree approximation (degree 7, 9 and 11) in low dimensions and show how to extend the cubature method to piece-wise smooth test functions.

For the following example, we will denote by $\exp(Vt)f$ the value at time t of the solution of the ODE $y' = V(y)$, $y(0) = f$ where $V \in C_b^\infty(\mathbb{R}^N, \mathbb{R}^N)$. In particular, $\exp(Vt)(x)$ is $\exp(Vt)f$ for f being the identity function. The family of processes $Y(x) = \{Y_t(x)\}_{t\in[0,1]}$ below corresponds to the Ninomiya–Victoir approximation (see [49]).

Example 43. Let Λ and Z be two independent random variables such that Λ is Bernoulli distributed $\mathbb{P}(\Lambda = 1) = \mathbb{P}(\Lambda = -1) = \frac{1}{2}$ and $Z = (Z^i)_{i=1}^k$ is a standard normal k-dimensional random variable. Consider the family of processes $Y(x) = \{Y_t(x)\}_{t\in[0,1]}$ defined by

$$Y_t(x) = \begin{cases} \exp(\frac{V_0}{2}t)\prod_{i=1}^{k}\exp(Z^i V_i t^{1/2})\exp(\frac{V_0}{2}t)(x) & \text{if } \Lambda = 1 \\ \exp(\frac{V_0}{2}t)\prod_{i=1}^{k}\exp(Z^{k+1-i} V_{k+1-i} t^{1/2})\exp(\frac{V_0}{2}t)(x) & \text{if } \Lambda = -1 \end{cases}$$

with the corresponding operator $(Q_t\varphi)(x) := \mathbb{E}[\varphi(Y_t(x))]$. Then there exists a constant $c > 0$ such that for $\varphi \in C_b^{V,8}(\mathbb{R}^N)$

$$\sup_x \left|\overline{Q}_t\varphi(x) - \mathbb{E}[\varphi_t^5(x)]\right| \le ct^3 \|\varphi\|_{V,6}$$

Hence $\{Y_t(x)\}_{t\in[0,1]}$ is 5-perfect.

Proof. See [13]. □

The following lemma is required to prove the main theorem below.

Lemma 44. *For $0 < s \le t \le 1$ and any m-perfect family $\{\overline{X}_t(x)\}_{t\in(0,1]}$ with corresponding operator $Q = \{Q_t\}_{t\in(0,1]}$ we have,*

$$\|P_t(P_s\varphi) - Q_t(P_s\varphi)\|_\infty \le c\|\varphi\|_p \sum_{j=m+1}^{M} \frac{t^{j/2}}{s^{\frac{j-p}{2}}}, \tag{94}$$

where $\varphi \in C_b^p(\mathbb{R}^N)$ for $0 \le p < \infty$ and some constant $c > 0$. In particular, for $\varphi \in C_b^M(\mathbb{R}^d)$,

$$\|P_t(P_s\varphi) - Q_t(P_s\varphi)\|_\infty \le c\|\varphi\|_p t^{\frac{m+1}{2}}. \tag{95}$$

Proof. Since $\mathcal{C}_b^\infty(\mathbb{R}^N)$ is dense in $\mathcal{C}_b^p(\mathbb{R}^d)$ in the topology generated by the norm $||\cdot||_{p,\infty}$ it suffices to prove (94) and (95) only for a function $\varphi \in \mathcal{C}_b^\infty(\mathbb{R}^N)$. By Corollary 32, we have

$$\|P_t\varphi\|_{V,j} = \sum_{i=1}^{j} \sum_{\substack{\alpha_1,\ldots,\alpha_i \in \mathcal{A}_0 \\ \|\alpha_1 * \ldots * \alpha_i\| = j}} \left\| V_{[\alpha_1]} \cdots V_{[\alpha_i]} P_t \varphi \right\|_\infty$$

$$\leq \sum_{i=1}^{j} \sum_{\substack{\alpha_1,\ldots,\alpha_i \in \mathcal{A}_0 \\ \|\alpha_1 * \ldots * \alpha_i\| = j}} \frac{c\,\|\varphi\|_p}{t^{(\|\alpha_1 * \ldots * \alpha_i\| - p)/2}} \leq \frac{c\,\|\varphi\|_p}{t^{\frac{j-p}{2}}}$$

Then (94) and (95) follow from the definition of an m-perfect family. □

The family of processes $\bar{X}(x) = \{\bar{X}_t(x)\}_{t\in[0,\infty)}$ below corresponds to the Kusuoka approximation. We recall that Kusuoka's result requires only the UFG condition.

Example 45. A family of random variables $\{Z_\alpha : \alpha \in \mathcal{A}_0\}$ is said to be **m-moment** similar if $\mathbb{E}[\,|Z_\alpha|^r] < \infty$ for any $r \in \mathbb{N}$, $\alpha \in \mathcal{A}_0$ and $Z_{(0)} = 1$ with,

$$\mathbb{E}[Z_{\alpha_1} \ldots Z_{\alpha_j}] = \mathbb{E}[I_{\alpha_1} \ldots I_{\alpha_j}]$$

for any $j = 1, \ldots, m$ and $\alpha_1, \ldots, \alpha_j \in \mathcal{A}_0$ such that $\|\alpha_1\| + \cdots + \|\alpha_j\| \leq m$ where I_α is defined as above.

Let $\{Z_\alpha : \alpha \in \mathcal{A}_0\}$ be a family of m-moment similar random variables and let $\bar{X}(x) = \{\bar{X}_t(x)\}_{t\in[0,\infty)}$ be the family of processes,

$$\bar{X}_t(x) = \sum_{j=0}^{m} \frac{1}{j!} \sum_{\substack{\alpha_1,\ldots,\alpha_j \in \mathcal{A}_0, \\ \|\alpha_1\| + \cdots + \|\alpha_j\| \leq m}} t^{\frac{\|\alpha_1\| + \cdots + \|\alpha_j\|}{2}} (P^0_{\alpha_1} \ldots P^0_{\alpha_j})(V_{[\alpha_1]} \ldots V_{[\alpha_j]} H)(x) \tag{96}$$

where $H : \mathbb{R}^N \to \mathbb{R}^N$ is defined $H(x) = x$ and

$$P^0_\alpha := |\alpha|^{-1} \sum_{j=0}^{|\alpha|} \frac{(-1)^{j+1}}{j} \sum_{\beta_1 * \ldots * \beta_j = \alpha} Z_{\beta_1} \ldots Z_{\beta_j}$$

with the corresponding operator $Q = \{Q_t\}_{t\in(0,1]}$ defined by,

$$Q_t\varphi(x) = \mathbb{E}[\varphi(\bar{X}_t(x))]$$

for $\varphi \in \mathcal{C}_b(\mathbb{R}^N)$ Then,

$$\|P_{t+s}\varphi - Q_t P_s\varphi\|_\infty \le c\,\|\nabla\varphi\|_\infty \sum_{j=m+1}^{m^{m+1}} \frac{t^{j/2}}{s^{\frac{j-1}{2}}} \tag{97}$$

for some constant $c > 0$.

Proof. See Definition 1, Theorem 3 and Lemma 18 in Kusuoka[29] for (97). □

The family $\bar{X}(x),\ x \in \mathbb{R}^N$ as defined in (96) is not m-perfect. However, inequality (97) is a particular case of (94) where $p = 1$ and $M = m^{m+1}$. Since (94) is the only result required to obtain (98), we deduce from the proof of Theorem 46 that (98), with $p = 1$, holds for Kusuoka's method as well. Similarly part (ii) of Corollary 47 holds for Kusuoka's method. For numerical algorithms related to the family $\bar{X}(x),\ x \in \mathbb{R}^N$ as defined in (96) see [31, 47, 48]. In particular, paper [48] uses a control on the computational effort based on the same algorithm (the TBBA) as the one employed in Sect. 3.5.

The set of vector fields appearing in (96) belong to the Lie algebra generated by the original vector fields $\{V_0, V_1, \ldots, V_d\}$. Ben Arous [1] and Burrage and Burrage [5] employ the same set of vector fields to produce *strong* approximations of solutions of SDEs. Notably, the same ideas appear much earlier in Magnus [39], in the context of approximations of the solution of linear (deterministic) differential equations. Castell [7] also gives an explicit formula for the solution of an SDE in terms of Lie brackets and iterated Stratonovich integrals.

3.4 Rates of Convergence

We now prove our main result on m-perfect families, the gist of which can be conveyed by the concept of local and global order of an approximation. Local order measures how close an approximation is to the exact solution on a sub-interval of the integration, given an exact initial condition at the start of that subinterval. The global order of an approximation looks at the build up of errors over the entire integration range. The theorem below states that, in the best possible case, the global order of an approximation obtained using an m-perfect family is one less than the local order. More precisely, for a suitable partition, the global error is of order $\frac{m-1}{2}$ whilst the local error is of order $\frac{m+1}{2}$.

Let us define the function,

$$\Upsilon^p(n) = \begin{cases} n^{-\frac{1}{2}\min(\gamma p,(m-1))} & \text{if } \gamma p \neq m-1 \\ n^{-(m-1)/2}\ln n & \text{for } \gamma p = m-1 \end{cases}.$$

In the following,

$$\mathcal{E}^{\gamma,n}(\varphi) := \left\| P_T\varphi - Q^m_{h_n} Q^m_{h_{n-1}} \cdots Q^m_{h_1}\varphi \right\|_\infty$$

for $\gamma \in \mathbb{R}$, $n \in \mathbb{N}$.

Theorem 46. *Let $T, \gamma > 0$ and $\pi_n = \{t_j = (\frac{j}{n})^\gamma T\}_{j=0}^n$ be a partition of the interval $[0,T]$ where $n \in \mathbb{N}$ is such that $\{h_j = t_j - t_{j-1}\}_{j=1}^n \subseteq (0,1]$. Then for any m-perfect family $\{\overline{X}_t(x)\}_{t\in[0,T]}$ with corresponding operator $Q = \{Q_t\}_{t\in(0,1]}$ we have, for $\varphi \in C_b^p(\mathbb{R}^N)$ where $p = 1,\ldots,m$,*

$$\mathcal{E}^{\gamma,n}(\varphi) \leq c\Upsilon^p(n)\|\varphi\|_p + \left\| P_{h_1}\varphi - Q_{h_1}^m\varphi \right\|_\infty \tag{98}$$

for some constant $c \equiv c(\gamma, M, T) > 0$ where $M \geq m+1$, as in Definition 39. In particular, if $\gamma \geq \frac{m-1}{p}$ then,

$$\mathcal{E}^{\gamma,n}(\varphi) \leq \frac{c}{n^{\frac{m-1}{2}}}\|\varphi\|_p + \left\| P_{h_1}\varphi - Q_{h_1}^m\varphi \right\|_\infty .$$

Proof. We have,

$$\begin{aligned}
\mathcal{E}^{\gamma,n}(\varphi) &= P_{h_n}(P_{T-h_n}\varphi) - Q_{h_n}^m(P_{T-h_n}\varphi) \\
&\quad + \sum_{j=1}^{n-1} Q_{h_n}^m \cdots Q_{h_{j+1}}^m (P_{T-h_{j+1}-\cdots-h_n}\varphi - Q_{h_j}^m P_{T-h_j-\cdots-h_n}\varphi) \\
&= P_{h_n}(P_{t_{n-1}}\varphi) - Q_{h_n}^m(P_{t_{n-1}}\varphi) \\
&\quad + \sum_{j=1}^{n-1} Q_{h_n}^m \cdots Q_{h_{j+1}}^m (P_{h_j}(P_{t_{j-1}}\varphi) - Q_{h_j}^m(P_{t_{j-1}}\varphi)).
\end{aligned}$$

By Lemma 44, there exists a constant $c > 0$ such that,

$$\left\| P_{h_n}(P_{t_{n-1}}\varphi) - Q_{h_n}^m(P_{t_{n-1}}\varphi) \right\|_\infty \leq c\|\varphi\|_p \sum_{l=m+1}^{M} \frac{h_n^{l/2}}{t_{n-1}^{\frac{l-p}{2}}}$$

Since P is a semigroup and $Q_{h_j}^m$ is a Markov operator for $j = 2,\ldots,n-1$,

$$\begin{aligned}
\left\| Q_{h_n}^m \cdots Q_{h_{j+1}}^m (P_{h_j}(P_{t_{j-1}}\varphi) - Q_{h_j}^m(P_{t_{j-1}}\varphi)) \right\|_\infty &\leq \left\| P_{h_j}(P_{t_{j-1}}\varphi) - Q_{h_j}^m(P_{t_{j-1}}\varphi) \right\|_\infty \\
&\leq c\|\varphi\|_p \sum_{l=m+1}^{M} \frac{h_j^{l/2}}{t_{j-1}^{\frac{l-p}{2}}}
\end{aligned}$$

for some $c > 0$. Finally, since $Q_{h_j}^m$ is a Markov operator, it follows from (102) that,

$$\left\| Q_{h_n}^m \cdots Q_{h_2}^m (P_{h_1}\varphi - Q_{h_1}^m\varphi) \right\|_\infty \leq \left\| P_{h_1}\varphi - Q_{h_1}^m\varphi \right\|_\infty$$

Combining these last four results gives,

$$\mathcal{E}^{\gamma,n}(\varphi) = \left\| P_T\varphi - Q^m_{h_n}\cdots Q^m_{h_1}\varphi \right\|_\infty \le \left\| P_{h_1}\varphi - Q^m_{h_1}\varphi \right\|_\infty + c\,||\varphi||_p \sum_{j=2}^{n}\sum_{l=m+1}^{M} \frac{h_j^{l/2}}{t_{j-1}^{\frac{l-p}{2}}}.$$

It follows, almost immediately from the definition of h_j that,

$$h_j = \frac{\gamma T(j-1)^{\gamma-1}}{n^\gamma}\int_{j-1}^{j}\left(\frac{u}{j-1}\right)^{\gamma-1} du,$$

but for $j \in \{2,\ldots,n\}$, $(\frac{u}{j-1})^{\gamma-1} \le \max[(\frac{j}{j-1})^{\gamma-1}, 1] \le \max[2^{\gamma-1}, 1]$. Hence for $l = m+1,\ldots,M$,

$$\frac{h_j^{l/2}}{t_{j-1}^{(l-p)/2}} \le \frac{(\frac{\gamma T(j-1)^{\gamma-1}}{n^\gamma}\max[2^{\gamma-1},1])^{l/2}}{\left(\left(\frac{j-1}{n}\right)^\gamma T\right)^{(l-p)/2}}$$
$$\le c(\frac{T}{n^\gamma})^{\frac{l}{2}-\frac{(l-p)}{2}}(j-1)^{\frac{(\gamma-1)l}{2}-\frac{\gamma(l-p)}{2}} = c(\frac{T}{n^\gamma})^{\frac{p}{2}}(j-1)^{\frac{\gamma p-l}{2}}$$

where $c = \max[1, (\gamma\max[2^{\gamma-1},1])^{M/2}]$. It follows that,

$$\sum_{l=m+1}^{M}\frac{h_j^{l/2}}{t_{j-1}^{(l-p)/2}} \le c\left(\frac{1}{n}\right)^{\frac{\gamma p}{2}}\sum_{l=m+1}^{M}(j-1)^{\frac{\gamma p-l}{2}}.$$

Since $\sum_{l=m+1}^{M}(j-1)^{\frac{\gamma p-l}{2}} = (j-1)^{\frac{\gamma p-(m+1)}{2}}\sum_{l=0}^{M-(m+1)}(j-1)^{-\frac{l}{2}} \le (j-1)^{\frac{\gamma p-(m+1)}{2}} M$ we have,

$$\sum_{l=m+1}^{M}\frac{h_j^{l/2}}{t_{j-1}^{(l-p)/2}} \le M\left(\frac{1}{n}\right)^{\frac{\gamma p}{2}}(j-1)^{\frac{\gamma p-(m+1)}{2}} \tag{99}$$

We now consider (99) for three different ranges of γ.
For $\gamma \in \left(0, \frac{m-1}{p}\right)$, $\sum_{j=2}^{n}(j-1)^{\frac{\gamma p-(m+1)}{2}} \le \sum_{j=2}^{\infty}(j-1)^{\frac{\gamma p-(m+1)}{2}}$ and since the series on the right hand side is convergent, we have,

$$n^{-\frac{\gamma p}{2}}\sum_{j=2}^{n}(j-1)^{\frac{\gamma p-(m+1)}{2}} \le cn^{-\frac{\gamma p}{2}}$$

for some constant $c = c(\gamma, M) > 0$.

For $\gamma = \frac{m-1}{p}$, $\sum_{j=2}^{n}(j-1)^{-1} \leq c \ln n$ for some constant $c \equiv c(\gamma, M) > 0$ so we have,

$$n^{-\frac{\gamma p}{2}} \sum_{j=2}^{n} (j-1)^{\frac{\gamma p-(m+1)}{2}} \leq c n^{-\frac{(m-1)}{2}} \ln n.$$

For $\gamma > \frac{m-1}{p}$, we have

$$\sum_{j=2}^{n} (\frac{j-1}{n})^{\frac{\gamma p-(m+1)}{2}} \frac{1}{n} \leq c \int_0^1 x^{\frac{\gamma p-(m+1)}{2}} dx = c \int_0^1 x^{-1+\frac{\gamma p-(m-1)}{2}} dx < \infty$$

so,

$$n^{-\frac{\gamma p}{2}} \sum_{j=2}^{n} (j-1)^{\frac{\gamma p-(m+1)}{2}} = n^{-\frac{m-1}{2}} \sum_{j=2}^{n} \left(\frac{j-1}{n}\right)^{\frac{\gamma p-(m+1)}{2}} \frac{1}{n} \leq c n^{-\frac{m-1}{2}}. \qquad \square$$

We observe that the rate of convergence is the controlled by the maximum between $\Upsilon(n)$ and the rate at which $\left\| P_{h_1}\varphi - Q_{h_1}^m \varphi \right\|_\infty$ converges to 0. We define $\bar{\Upsilon}^{k_1,k_2}(n) := \Upsilon^{k_1}(n) + n^{-\frac{\gamma k_2}{2}}$. We have the following corollary:

Corollary 47. *(i) For any $\varphi \in C_b^M(\mathbb{R}^N)$,*

$$\mathcal{E}^{\gamma,n}(\varphi) \leq c\bar{\Upsilon}^{m+1,m+1}(n) \|\varphi\|_M.$$

for some constant $c > 0$. In particular, if $\gamma \geq 1$, then $\mathcal{E}^{\gamma,n}(\varphi) \leq \frac{c}{n^{\frac{m-1}{2}}} \|\varphi\|_M$.

(ii) If there exists a constant $c > 0$ independent of t such that,

$$\sup_{x\in\mathbb{R}^N} \left|\bar{X}_t(x) - x\right| \leq c\sqrt{t}, \tag{100}$$

then, for any $\varphi \in C_b^1(\mathbb{R}^N)$,

$$\mathcal{E}^{\gamma,n}(\varphi) \leq c\bar{\Upsilon}^{1,1}(n) \|\varphi\|_1$$

for some constant $c > 0$. In particular, if $\gamma \geq m-1$, then $\mathcal{E}^{\gamma,n}(\varphi) \leq \frac{c}{n^{\frac{m-1}{2}}} \|\varphi\|_1$.

(iii) if there exist constants $c, \bar{c} > 0$ independent of t such that,

$$\|P_t\varphi - Q_t^m\varphi\|_\infty \leq ct^{\frac{\bar{c}}{2}} \|\varphi\|_l, \tag{101}$$

then, for any $\varphi \in C_b^l(\mathbb{R}^N)$ *where* $1 < l < M$, *we have*

$$\mathcal{E}^{\gamma,n}(\varphi) \leq c\bar{\Upsilon}^{l,\bar{c}}(n)\,\|\varphi\|_l$$

for some constant $c > 0$, *In particular, if* $\gamma \geq m-1$, *then* $\mathcal{E}^{\gamma,n}(\varphi) \leq \frac{c}{n^{\frac{m-1}{2}}}\|\varphi\|_l$.

Proof. (i) The result follows from Theorem 46 and the definition of an m-perfect family.

(ii) If $\varphi \in C_b(\mathbb{R}^N)$ is Lipschitz then,

$$|Q_t\varphi(x) - \varphi(x)| \leq c\,||\nabla\varphi||_\infty\sqrt{t} \tag{102}$$

hence,

$$\left\| P_{h_1}\varphi - Q_{h_1}^m\varphi\right\|_\infty \leq c\,||\varphi||_1\sqrt{t}.$$

(iii) The result follows from Theorem 46 and (101). □

Finally we define μ_t to be the law of X_t, that is $\mu_t(\varphi) = E[\varphi(X_t)]$ for $\varphi \in C_b(\mathbb{R}^N)$. We also define μ_t^N to be the probability measure defined by,

$$\mu_t^N(\varphi) = \mathbb{E}\left[Q_{h_n}^m Q_{h_{n-1}}^m \cdots Q_{h_1}^m\varphi(X_0)\right] = \int_{\mathbb{R}^N} Q_{h_n}^m Q_{h_{n-1}}^m \cdots Q_{h_1}^m\varphi(x)\,\mu_0(dx)$$

for $\varphi \in C_b(\mathbb{R}^N)$ and introduce the family of norms on the set of signed measures:

$$|\mu|_l = \sup\left\{|\mu(\varphi)|, \varphi \in C_b^l(\mathbb{R}^N), \|\varphi\|_{l,\infty} \leq 1\right\}, \quad l \geq 1.$$

Obviously, $|\mu|_l \leq |\mu|_{l'}$ if $l \leq l'$. In other words, the higher the value of l, the coarser the norm. We have the following:

Corollary 48. *(i) For* $l \geq M$, *we have* $\left|\mu_t - \mu_t^N\right|_l \leq c\bar{\Upsilon}^{m+1,m+1}(n)$. *In particular, if* $\gamma \geq 1$, *then* $\left|\mu_t - \mu_t^N\right|_l \leq \frac{c}{n^{\frac{m-1}{2}}}$.

(ii) If (100) is satisfied then $\left|\mu_t - \mu_t^N\right|_l \leq c\bar{\Upsilon}^{1,1}(n)$. *In particular, if* $\gamma \geq m-1$, *then* $\left|\mu_t - \mu_t^N\right|_l \leq \frac{c}{n^{\frac{m-1}{2}}}$.

(iii) If (101) is satisfied then $\left|\mu_t - \mu_t^N\right|_l \leq c\bar{\Upsilon}^{l,c_{23}}(n)$. *In particular, if* $\gamma \geq m-1$, *then* $\left|\mu_t - \mu_t^N\right|_l \leq \frac{c}{n^{\frac{m-1}{2}}}$.

Remark 49. We deduce that there is a payoff between the rate of convergence and the coarseness of the norm employed: the finer the norm the slower the rate of convergence. Hence intermediate results such as part *(iii)* of Corollaries 47 and 48 may prove useful in subsequent applications. The additional constraint (101) holds, for example, for the Lyons–Victoir method, as a cubature formula of degree m is also a cubature formula of degree m' for $m' \leq m$. Similarly, it holds for Kusuoka's approximation since an m-similar family is also m'-similar for any $m' \leq m$.

3.5 Cubature and TBBA

In this section we discuss an algorithm that is used to control the computational effort required for the implementation of the Lyons–Victoir cubature method. This method suffers from the usual drawback of any tree based method, namely an exponentially increasing support. This is not an issue in low dimensional problems or when only a sparse partition is used. However, the exponential growth is a major hurdle in more complex and/or high-dimensional problems. To the best of our knowledge, currently, there exist two methods that may be applied to control this growth: The recombination method of Litterer and Lyons [35] and the tree based branching algorithm (TBBA) of Crisan and Lyons [14]. The application of the former to the cubature method has been extensively discussed in [35], where as here, we focus on the TBBA.[10]

The idea behind the TBBA is to construct a finite random measure with a support of size less than a pre-determined value that is an unbiased, minimal variance estimator of the original measure. The method insures that every point in the support of the original measure remains in the support of the resulting measure with a probability (approximately) proportional to its original weight. To fix ideas, let us consider the cubature measure $\mathbb{Q}_1^m$ of degree $m \geq 3$ supported on the paths $\omega_1, \ldots, \omega_{c_d^m}$ with corresponding weights $\lambda_1, \ldots, \lambda_{c_d^m}$, $c_d^m \in \mathbb{N}_+$. As usual we may consider by scaling, cubature measures $\mathbb{Q}_t^m$ on any interval $[0, t]$. Let $\Xi_{t,x}(\omega)$, $\omega \in C_{0,bv}\left([0,t];\mathbb{R}^d\right)$ denote the solution at time t of the ODE

$$\begin{cases} dy_{t,x} = \sum_{j=0}^{d} V_j(y_{t,x})d\omega^j(t) \\ y_{0,x} = x \end{cases}. \tag{103}$$

Consider also a partition $\pi := \{0 = t_0 < t_1 < \ldots < t_n = t\}$ of $[0, t]$. By iterating the cubature measure along this partition and solving the successive ODEs (see Remark 42), we generate a collection of discrete measures $\{\mathbb{Q}_{t_k}^m\}_{k \leq n}$, where the cardinality of the support of the measure $\mathbb{Q}_{t_k}^m$ is $\left(c_d^m\right)^k$.

We wish to replace the measure $\mathbb{Q}_{t_k}^m$ by a random measure $\tilde{\mathbb{Q}}_{t_k}^m$ whose support is included in the support of the measure $\mathbb{Q}_{t_k}^m$ and whose cardinality is at most N (with $N < \left(c_d^m\right)^k$). Moreover we want $\tilde{\mathbb{Q}}_{t_k}^m$ to be an unbiased minimal variance estimator of $\mathbb{Q}_{t_k}^m$ in a sense that we will make explicit below. To handle the additional

[10] The TBBA has also been used to control on the computational effort for a class of numerical algorithm using the family $\bar{X}(x)$, $x \in \mathbb{R}^d$ as defined in (96), see [47] for details.

randomness we introduce an additional probability space $\left(\tilde{\Omega}, \tilde{\mathcal{F}}, \tilde{\mathbb{P}}\right)$ which supports the random probability measure $\tilde{\mathbb{Q}}^m_{t_k}$. We will require that $\tilde{\mathbb{E}}\left[\tilde{\mathbb{Q}}^m_{t_k}\right] = \mathbb{Q}^m_{t_k}$, where $\tilde{\mathbb{E}}$ denotes integration with respect to $\tilde{\mathbb{P}}$. Let

$$\mathbb{Q}^m_{t_k} = \sum_{j=1}^{c_d^m} \lambda_j \delta_{\gamma_j}, \quad \gamma_j = \omega_{\delta_1, i_1} \otimes \ldots \otimes \omega_{\delta_k, i_k}, \text{ for some } i_1, \ldots, i_k = 1, \ldots, c_d^m,$$

where $\omega_i \otimes \omega_j$ denotes the concatenation of two paths. We will construct a random *probability* measure $\tilde{\mathbb{Q}}^m_{t_k}$ such that

$$\tilde{\mathbb{Q}}^m_{t_k}(\gamma) = \begin{cases} \frac{\lfloor N\mathbb{Q}^m_{t_k}(\gamma)\rfloor}{N} & \text{with probability } 1 - \{N\mathbb{Q}^m_k(\gamma)\} \\ \frac{\lfloor N\mathbb{Q}^m_{t_k}(\gamma)\rfloor + 1}{N} & \text{with probability } \{N\mathbb{Q}^m_{t_k}(\gamma)\} \end{cases}, \gamma \in \operatorname{supp}(\mathbb{Q}^m_{t_k}), \tag{104}$$

where for any real number y, $\lfloor y \rfloor$ denotes the lower integer part and $\{y\}$ the fractional part, $\{y\} = y - \lfloor y \rfloor$. As a result each point in the support of $\mathbb{Q}^m_k(\gamma)$ has either mass 0 (i.e. it does not appear in the support of $\tilde{\mathbb{Q}}^m_{t_k}$ or its mass is an integer multiple of $1/N$. Since $\tilde{\mathbb{Q}}^m_{t_k}$ is a probability measure, its support cannot therefore have cardinality larger than N and is included in the support of $\mathbb{Q}^m_{t_k}$. If $\tilde{\mathbb{Q}}^m_{t_k}(\gamma)$ has distribution described by (104) for any $\gamma \in \operatorname{supp}(\mathbb{Q}^m_{t_k})$, it is clearly an unbiased estimator of $\mathbb{Q}^m_{t_k}$, that is $\tilde{\mathbb{E}}\left[\tilde{\mathbb{Q}}^m_{t_k}\right] = \mathbb{Q}^m_{t_k}$. Moreover it has minimal variance amongst all unbiased estimators of $\mathbb{Q}^m_{t_k}$ for which the mass associated to any element in the support of the original measure takes values in the set $\{0, \frac{1}{N}, \frac{2}{N}, \ldots, 1\}$. See [14] pg. 344–345 for further optimality properties of $\tilde{\mathbb{Q}}^m_{t_k}$.

The algorithm that produces the random probability measure $\hat{\mathbb{Q}}^m_k$ from $\mathbb{Q}^m_k$ such that (104) is satisfied for every element in the support of the cubature measure is the subject of Theorem 2.6 of [14]. The idea is to embed the support of the cubature measure into a binary tree and distribute the weights recursively, targeting distribution (104) at every nod/leaf.

Each element in the support of $\mathbb{Q}^m_{t_k}$ is associated to the end nodes or leaves of the binary tree. We associate a weight to each leaf equal to the mass of the corresponding element in the support of $\mathbb{Q}^m_{t_k}$. We then recursively associate a weight to each of the (intermediate) nodes, equal to the sum of the weights of its offspring nodes. Eventually we associate weight 1 to the root node (the sum of the masses of all the elements in the support of $\mathbb{Q}^m_{t_k}$). We note that any tree (not necessarily a binary one) can be embedded into a binary tree as follows: For each intermediate node, we separate the set of its offsprings nodes in two sets. On the left, we take a singleton consisting of the first of the offsprings nodes and on the right we add a new intermediate node with offsprings corresponding to the rest of the offsprings of the original node. We then apply the same procedure to the intermediate node and repeat the process until we are left with only two offspring nodes which we keep as part of the new tree. The next example explain this procedure further through a concrete example.

Example 50. Let us consider the cubature method of order 3 in dimension 2. Starting from x_0 (the initial condition for the forward diffusion) we take on step forward say at time 1. This produces a measure with four elements in its support (see Example 41), as there are four paths that define the cubature formula,[11] all carrying equal weight. Schematically, $\mathbb{Q}_1^m$ looks as in the figure below

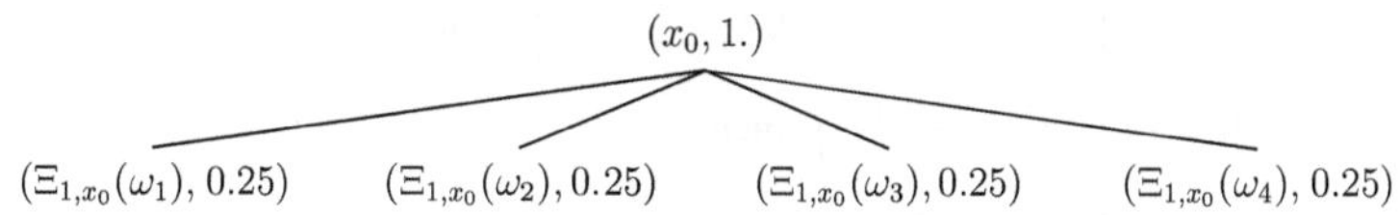

We embed the above tree into the following (by no means unique) binary tree:

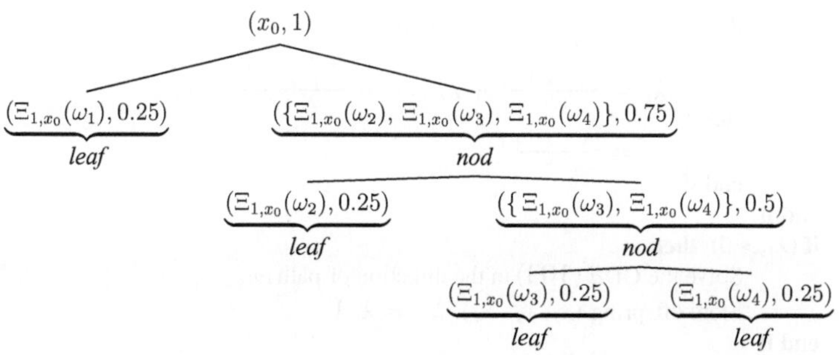

Notice how every node carries the total weight of all of its offspring leaves.

We will next describe how one distributes the mass according to TBBA *per family*, hence achieving the distribution (104) at every point in the support of the measure. The reasoning relies of course on the structure of the binary tree.

Any path $\gamma \in \operatorname{supp}\left(\mathbb{Q}_k^m\right)$ carries the weight $\lambda_x = \mathbb{Q}_k^m(\gamma)$ which is the product of the cubature weights that correspond to the ODEs we solve to arrive at $x \equiv \Xi_{t_k,x_0}(\gamma)$ along the path γ. Assume that we have assigned to x the random weight $\hat{\lambda}_x = \tilde{\mathbb{Q}}_{t_k}^m(\gamma)$ distributed according to (104). The following algorithm shows how one assigns the corresponding weights to any of the offsprings of x:

[11] These are the straight lines connecting the origin with $(-1,-1)$, $(1,-1)$, $(-1,1)$, $(1,1)$.

Algorithm 1 TBBA(x, λ_x, $\hat{\lambda}_x$)

Require: $\lambda_1, \ldots, \lambda_{c_d^m}$ {The cubature weights and c_d^m is the cubature dimension.}
  Define $\lambda_{i:n} = \sum_{j=i}^{c_d^m} \lambda_j$
  Declare $\hat{\lambda}_1, \ldots, \hat{\lambda}_{c_d^m}$ and $\hat{\lambda}_{1:c_d^m}, \ldots, \hat{\lambda}_{c_d-1:c_d^m}$
  {$\hat{\lambda}_1, \ldots, \hat{\lambda}_{c_d^m}$ store the TBBA weights at every leaf}
  { whereas the $\hat{\lambda}_{1:c_d^m}, \ldots, \hat{\lambda}_{c_d^m-1:c_d^m}$ store the TBBA weights at every nod.}
  Set $\hat{\lambda}_{1:c_d^m} = \hat{\lambda}_x$.
  for $i = 1$ **to** $N_m - 1$ **do**
    $u_i(x) \sim U[0,1]$, {Draw uniform}
    if $\left(\{N\lambda_x\lambda_{i:c_d^m}\} = \{N\lambda_x\lambda_i\} + \{N\lambda_x\lambda_{i+1:c_d^m}\}\right)$ **then**
      if $\left(u_i(x) < \frac{\{N\lambda_x\lambda_i\}}{\{N\lambda_x\lambda_{i:c_d^m}\}}\right)$ **then**
        $\hat{\lambda}_i = \frac{\lfloor N\lambda_x\lambda_i\rfloor}{N} + \hat{\lambda}_{i:c_d^m} - \frac{\lfloor N\lambda_x\lambda_{i:c_d^m}\rfloor}{N}$
      else
        $\hat{\lambda}_i = \frac{\lfloor N\lambda_x\lambda_i\rfloor}{N}$
      end if
    else
      if $\left(u_i(x) < \frac{1-\{N\lambda_x\lambda_i\}}{1-\{N\lambda_x\lambda_{i:c_d^m}\}}\right)$ **then**
        $\hat{\lambda}_i = \frac{\lfloor N\lambda_x\lambda_i\rfloor+1}{N} + \hat{\lambda}_{i:c_d^m} - \frac{\lfloor N\lambda_x\lambda_{i:c_d^m}\rfloor+1}{N}$
      else
        $\hat{\lambda}_i = \frac{\lfloor N\lambda_x\lambda_i\rfloor+1}{N}$
      end if
    end if
    if $(\hat{\lambda}_i > 0)$ **then**
      Solve the ODE (103) in the direction of path ω_i
      Store offspring $(x_i, \lambda_{x_i}, \hat{\lambda}_i)$, $\lambda_{x_i} = \lambda_x\lambda_i$
    end if
    Set $\hat{\lambda}_{i+1:c_d^m} = \hat{\lambda}_{i:c_d^m} - \hat{\lambda}_i$.
  end for

Remark 51. All uniform random variables used in **Algorithm 1** are drawn independent of each other.

We apply **Algorithm 1** recursively until all nodes in the support of the cubature measure are assigned their corresponding random weights $\tilde{\lambda}_x$. We continue in this way until we reach the leaves of the tree. Recall that there can be at most N elements in the support of the original measure that get assigned a positive weight by the TBBA, hence indeed the new measure $\hat{\mathbb{Q}}_k^m$ will have a support of cardinality at most N.

We denote the set of all nodes corresponding to the (original) cubature tree at time t_k by

$$\mathcal{C}_k \equiv \bigcup_{i_1,\ldots,i_k=1}^{c_d^m} \left\{\Xi_{x_0,t_k}\left(\omega_{i_1} \otimes \ldots \otimes \omega_{i_k}\right)\right\}, \quad k = 1\ldots,n.$$

We denote by $\hat{\mathcal{C}}_k$ the set of remaining nodes after the TBBA is applied

$$\hat{\mathcal{C}}_k := \{x \in \mathcal{C}_k, \quad \hat{\lambda}_x > 0\},$$

where $\hat{\lambda}_x$ is the random weight computed by **Algorithm 1**. In other words, $\hat{\mathcal{C}}_k$ is the set of all nodes to which the TBBA assigns a positive weight. Finally, we shall also use the notation

$$\hat{\mathcal{C}}_k^x := \hat{\mathcal{C}}_k \bigcap \{\text{ children of x}\}, \; x \in \hat{\mathcal{C}}_{k-1}, \quad k = 1, \ldots, n.$$

We collect in the following Lemma some properties of the random weights constructed via the **Algorithm 1**. In particular, the algorithm produces an unbiased estimator of the pure cubature measure and the random weights are sampled with minimal variance.

Lemma 52. *For any point $x \in \bigcup_{i=1}^n \hat{\mathcal{C}}_i$, algorithm 1 produces a random weight $\hat{\lambda}_x$, that is distributed according to* (104)*, i.e.,*

$$\hat{\lambda}_x = \begin{cases} \frac{\lfloor N\lambda_x \rfloor}{N} & \text{with probability } 1 - \{N\lambda_x\} \\ \frac{\lfloor N\lambda_x \rfloor + 1}{N} & \text{with probability } \{N\lambda_x\} \end{cases}, \tag{105}$$

where λ_x is the original cubature weight. Moreover

$$\tilde{\mathbb{E}}\left[\hat{\lambda}_x\right] = \lambda_x, \quad \tilde{\mathbb{E}}\left[\left(\hat{\lambda}_x - \lambda_x\right)^2\right] = \frac{\{N\lambda_x\}\left(1 - \{N\lambda_x\}\right)}{N^2}.$$

Finally, the random weights that correspond to different leaves are negatively correlated, i.e.

$$\tilde{\mathbb{E}}\left[\left(\hat{\lambda}_x - \lambda_x\right)\left(\hat{\lambda}_y - \lambda_y\right)\right] \le 0, \quad x \neq y, \quad x, y \in \hat{\mathcal{C}}_i, \; i = 1, \ldots, n.$$

A proof of the previous Lemma can be found in the appendix of [15].

Theorem 53. *Let $\pi := \{0 = t_0 < t_1 < \ldots < t_n = T\}$ be a partition of the time interval $[0, T]$ on which we use a cubature formula of degree m to construct cubature measures along the partition π, $\{\mathbb{Q}_{h_i}^m\}_{i=1}^n$, $h_i = t_i - t_{i-1}$. Let $N \in \mathbb{N}_+$ be a given parameter which we use to define the cubature+TBBA measures $\{\tilde{\mathbb{Q}}_{h_i}^m\}_{i=1}^n$ supported on an additional probability space $\left(\tilde{\Omega}, \tilde{\mathcal{F}}, \tilde{\mathbb{P}}\right)$. Then for any function $\phi \in C_b^1\left(\mathbb{R}^d\right)$ we have*

$$\tilde{\mathbb{E}}\left[\left|P_T\phi - \tilde{\mathbb{Q}}_T^m\phi\right|^2\right]^{1/2} \le C\left(\frac{1}{n^{\frac{m-1}{2}}} + \frac{n}{\sqrt{N}}\right).$$

Proof. The first term in the control of the error is explained via Corollary 47 as the error between the diffusion semigroup operator and the cubature measure.

$$\tilde{\mathbb{E}}\left[\left|P_T\phi - \tilde{\mathbb{Q}}_T^m\phi\right|^2\right]^{1/2} \leq 2\tilde{\mathbb{E}}\left[\left|P_T\phi - \mathbb{Q}_T^m\phi\right|^2\right]^{1/2} + 2\tilde{\mathbb{E}}\left[\left|\mathbb{Q}_T^m\phi - \tilde{\mathbb{Q}}_T^m\phi\right|^2\right]^{1/2}$$

and hence we can focus on the second term. We proceed with the usual telescopic sum expansion

$$\tilde{\mathbb{Q}}_T^m\phi - \mathbb{Q}_T^m\phi = \sum_{i=1}^{n-1} \tilde{\mathbb{Q}}_{h_1}^m \dots \tilde{\mathbb{Q}}_{h_i}^m \mathbb{Q}_{h_{i+1}}^m \dots \mathbb{Q}_{h_n}^m\phi - \tilde{\mathbb{Q}}_{h_1}^m \dots \tilde{\mathbb{Q}}_{h_{i+1}}^m \mathbb{Q}_{h_{i+2}}^m \dots \mathbb{Q}_{h_n}^m\phi$$

From the Markov property of the cubature method and TBBA algorithm we understand that taking expectations under the family $\{\tilde{\mathbb{Q}}_{h_i}^m\}$ composes in the obvious manner, i.e.

$$\tilde{\mathbb{Q}}_{h_1}^m \dots \tilde{\mathbb{Q}}_{h_i}^m\phi = \sum_{x_1\in\hat{C}_1} \hat{\lambda}_{x_1} \sum_{x_2\in C_2^{x_1}} \frac{\hat{\lambda}_{x_2}}{\hat{\lambda}_{x_1}} \dots \sum_{x_i\in C_i^{x_{i-1}}} \frac{\hat{\lambda}_{x_i}}{\hat{\lambda}_{x_{i-1}}}\phi(x_i) = \sum_{x_i\in\hat{C}_i} \hat{\lambda}_{x_i}\phi(x_i)$$

In this way, we see that every term in the telescopic sum, may be written as

$$\begin{aligned}
&\tilde{\mathbb{Q}}_{h_1}^m \dots \tilde{\mathbb{Q}}_{h_i}^m \mathbb{Q}_{h_{i+1}}^m \dots \mathbb{Q}_{h_n}^m\phi - \tilde{\mathbb{Q}}_{h_1}^m \dots \tilde{\mathbb{Q}}_{h_{i+1}}^m \mathbb{Q}_{h_{i+2}}^m \dots \mathbb{Q}_{h_n}^m\phi \\
&= \tilde{\mathbb{Q}}_{h_1}^m \dots \tilde{\mathbb{Q}}_{h_i}^m \left(\tilde{\mathbb{Q}}_{h_{i+1}}^m - \mathbb{Q}_{h_{i+1}}^m\right) \mathbb{Q}_{T-t_{i+1}}^m\phi \\
&= \sum_{x_i\in\hat{C}_i} \hat{\lambda}_{x_i} \sum_{x_{i+1}\in\hat{C}_{i+1}^{x_i}} \left(\frac{\hat{\lambda}_{x_{i+1}}}{\hat{\lambda}_{x_i}} - \frac{\lambda_{x_{i+1}}}{\lambda_{x_i}}\right) \mathbb{Q}_{T-t_{i+1}}^m\phi(x_{i+1})
\end{aligned}$$

Next, by using the identity

$$\frac{\hat{a}}{\hat{b}} - \frac{a}{b} = \frac{\hat{a}-a}{\hat{b}} + \frac{a}{b\hat{b}}(b-\hat{b}),$$

we can re-write the above generic term of the telescopic sum as

$$\begin{aligned}
&\sum_{x_{i+1}\in\hat{C}_{i+1}} (\hat{\lambda}_{x_{i+1}} - \lambda_{x_{i+1}})\mathbb{Q}_{T-t_{i+1}}^m\phi(x_{i+1}) \\
&\quad + \sum_{x_i\in\hat{C}_i} (\hat{\lambda}_{x_i} - \lambda_{x_i}) \sum_{x_{i+1}\in\hat{C}_{i+1}^{x_i}} \frac{\lambda_{x_{i+1}}}{\lambda_{x_i}} \mathbb{Q}_{T-t_{i+1}}^m\phi(x_{i+1})
\end{aligned}$$

We can then take squares in the above, and using the fact that the random TBBA weights are negatively correlated as well as the expression on the variance of the error (see Lemma 52) and the fact that $\phi(X_T)$ has a finite second moment under the pure cubature measure (this is quite trivial to show), we have that

$$\tilde{\mathbb{E}}\left[\left|\tilde{\mathbb{Q}}^m_{h_1}\cdots\tilde{\mathbb{Q}}^m_{h_i}\mathbb{Q}^m_{h_{i+1}}\cdots\mathbb{Q}^m_{h_n}\phi-\tilde{\mathbb{Q}}^m_{h_1}\cdots\tilde{\mathbb{Q}}^m_{h_{i+1}}\mathbb{Q}^m_{h_{i+2}}\cdots\mathbb{Q}^m_{h_n}\phi\right|^2\right]^{1/2}\leq C/\sqrt{N}$$

and the result follows. □

3.6 Numerical Simulations Under the Heston Model

In this section we present the application of the cubature and TBBA method for the approximation of a call option on a Heston model price process. This is a favorable set up since the Heston model is well known for capturing the volatility dynamics in various asset classes and hence has received a lot of attention by practioners and academics. On the other hand, pricing call options under the Heston model admits semi closed solutions (see [23]) against which we can compare the efficiency of our method. Let us recall briefly the Heston model. In the following, we consider $X=\{(X^1_t(x),X^1_t(x)),t\geq 0,x\in\mathbb{R}^2\}$ satisfying

$$X^1_t(x)=x^1+\int_0^t rX^1_s(x)\,ds+\int_0^t X^1_s(x)\sqrt{X^2_s(x)}dB^1_s \tag{106}$$

$$X^2_t(x)=x^2+\int_0^t \alpha\left(\theta-X^2_s(x)\right)ds+\int_0^t \beta\sqrt{X^2_s(x)}\left(\rho dB^1_s+\sqrt{1-\rho^2}dB^2_s\right), \tag{107}$$

where $x^1,x^2>0$ are positive values, $\left(B^1,B^2\right)$ is a standard two dimensional Brownian motion and α,θ,μ are positive constants satisfying

$$2\alpha\theta-\beta^2>0$$

to ensure the existence and uniqueness of a solution of the SDE (107) which never hits 0. This is a two factor stochastic volatility model with ρ being the correlation between the two random noises, $|\rho|\leq 1$. The payoff of a vanilla call option with maturity $T>0$ and strike price $K>0$ is given

$$C(T,K)=E\left[\left(X^1_T-K\right)_+\right].$$

In the numerical example below we consider the following values for the various parameters

X_0	r	α	θ	β	ρ	T
100.	0.05	1	0.4	0.2	0	1.

We price a call option with strikes varying between $K = 80, 90, 100$ and 110. We keep the (maximal) number of particles that the TBBA allows to survive fixed at $N = 200000$. For every strike price and any number of steps, we launch the algorithm 10 times and average out the results. In other words if $\hat{c}(K, X_0, N, n)$ denotes the value computed by our algorithm when N particles and n steps for the discretization of time are used for a call option with strike K and spot at X_0 at time 0, we report on

$$\sum_{i=1}^{10} |(\hat{c}_i(K, X_0, N, n) - c(K, X_0)) / c(K, X_0)|,$$

where $\hat{c}_i(K, X_0, N, n)$ is the result of the i-th run of our algorithm and $c(K, X_0)$ is the value of the call option in the Heston model. We plot the results for the various strikes and varying number of steps in the figure below:

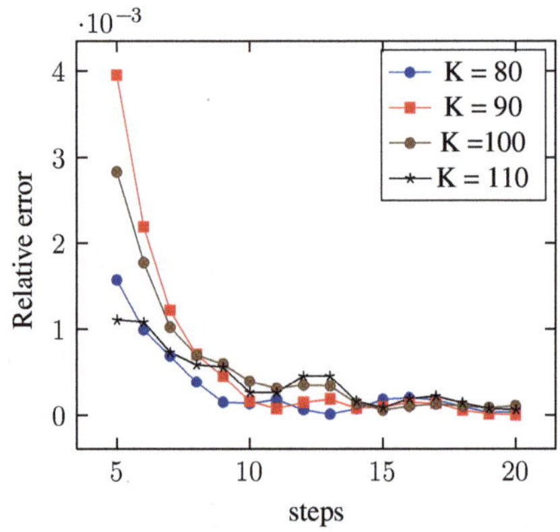

In all different strikes, we see that the algorithm behaves satisfactorily. It achieves an accuracy between 10^{-3} and 10^{-4} in the relevant error when 15 or more steps are used to discretize time. Recall that an at-the-money call is in general more difficult to approximate than in or out of the money calls, as its derivatives oscillate more as we approach maturity. However our algorithm does not seem affected by this.

4 Backward SDEs

In this section we present a brief overview to the theory of backward stochastic differential equations (BSDEs). These objects have received considerable attention over the last 20 years as they are intrinsically connected with three areas of stochastic analysis where research is very active: Non linear pricing, stochastic control and probabilistic representations of (viscosity) solutions of nonlinear PDEs and associated numerical methods. We will not go deep into the subject of BSDEs. Rather, we present some key points, mostly for ready reference, as in the following section we discuss an algorithm designed for the numerical solution of a BSDE (equivalently of a non linear PDE) based on the cubature and TBBA method.

4.1 The General Framework for Backward SDEs

Let $(\Omega, \mathcal{F}, \mathbb{P})$ be complete probability space endowed with a filtration that satisfies the usual conditions $\{\mathcal{F}_t\}_{t\geq 0}$. Let W be a d-dimensional, $\{\mathcal{F}_t\}$-adapted Brownian motion and let $(X, Y, Z) = \{(X_t, Y_t, Z_t), t \in [0, T]\}$ be the solution of the (decoupled) system, called a Forward–Backward SDE:

$$X_t = X_0 + \int_0^t V_0(X_s)ds + \sum_{i=1}^d \int_0^t V_i(X_s) \circ dW_s^i, \quad \text{forward component} \tag{108}$$

$$Y_t = \Phi(X_T) + \int_t^T f(s, X_s, Y_s, Z_s)ds - \sum_{i=1}^d \int_t^T Z_s^i dW_s^i, \quad \text{backward component.} \tag{109}$$

In (108)+(109), the process X is d-dimensional, Y is one dimensional and Z is d-dimensional. The coefficients $V_i : \mathbb{R}^d \to \mathbb{R}$ are smooth vector fields with $V_i \in C_b^\infty(\mathbb{R}^d)$, $i = 0, 1, \ldots, d$. The stochastic integrals in (108) are *Stratonovitch* type integral. The quantity $\Phi(X_T)$ is called *the final condition*, whilst $f : [0, T] \times \mathbb{R}^d \times \mathbb{R} \times \mathbb{R}^d \to \mathbb{R}$ is a Lipschitz function called "the driver".

Initially, existence and uniqueness for solution of equations of the form (108)+(109) was shown under a general Lipschitz assumption on the coefficients. This has since been relaxed considerably but here, we will only consider systems whose coefficients satisfy at least the following Lipschitz assumptions:

(A) The coefficients of the forward SDE $V_i : \mathbb{R}^d \to \mathbb{R}^d$, $i = 0, 1, \ldots, d$ and the driver f are globally Lipschitz with respect to the spatial variables. Further on, the driver is $1/2$-Hölder continuous with respect to t.

(B) The coefficients of the forward SDE $V_i : \mathbb{R}^d \to \mathbb{R}^d$, $i = 0, 1, \ldots, d$ have all entries belonging to $C_b^m(\mathbb{R}^d)$, the space of bounded m times differentiable functions with all partial derivatives bounded. The value of the parameter m shall be determined further on.

(C) The final condition Φ is Lipschitz continuous.

We denote by K the bound associated with all assumptions **(A), (B), (C)**.

Theorem 54 (Pardoux and Peng (1990)). *Under assumptions (A),(C) there exists a unique $\mathcal{F}_t$-adapted solution (X, Y, Z) of the system* (108) + (109).

Let us consider the simplest form of BSDE

$$Y_t = \Phi(X_T) + \int_t^T f(s, X_s)ds - \sum_{i=1}^{d} \int_t^T Z_s^i dW_s^i. \tag{110}$$

By the Martingale Representation Theorem, for

$$\xi \equiv \Phi(X_T) + \int_0^T f(s, X_s)ds$$

there exists a unique $\mathcal{F}_t$-adapted process Z such that the martingale $M = \{M_t, t \in [0, T]\}$ defined as $M_t = \mathbb{E}[\xi|\mathcal{F}_t], t \in [0, T]$ has the following representation

$$M_t = \mathbb{E}[\xi] + \sum_{i=1}^{d} \int_0^t Z_s^i dW_s^i$$

Define $Y = (Y_t, t \in [0, T])$ to be the $\mathcal{F}_t$-adapted process

$$Y_t \equiv M_t - \int_0^t f(s, X_s)ds. \tag{111}$$

It is the straightforward to show that the pair (Y, Z) are the unique solution of (110). Indeed from (111) we deduce that

$$\begin{aligned} Y_T &= M_T - \int_0^T f(s, X_s)ds \\ &= \mathbb{E}\left[\overbrace{\Phi(X_T) + \int_0^T f(s, X_s)ds}^{\xi} |\mathcal{F}_T\right] - \int_0^T f(s, X_s)ds = \Phi(X_T), \end{aligned}$$

hence

$$\begin{aligned} Y_t - \overbrace{\Phi(X_T)}^{Y_T} &= \overbrace{(M_t - \int_0^t f(s, X_s)ds)}^{Y_t} - \overbrace{(M_T - \int_0^T f(s, X_s)ds)}^{Y_T} \\ &= -\sum_{i=1}^{d} \int_t^T Z_s^i dW_s + \int_t^T f(s, X_s)ds. \end{aligned}$$

Thus (Y, Z) satisfies (110). The martingale representation theorem, as applied above, lies at the heart of the Picard iteration style argument for the proof of Theorem 54.

A celebrated result in the theory of BSDEs, is a theorem due to Pardoux and Peng that links their solution to the (viscosity) solution of semilinear PDEs. This is achieved by a Feynman–Kac type representation and it has since been extended to obstacle problems [16], quasi-linear PDEs [38] and indeed recently to fully non-linear PDEs [9,53]. Here we restrict ourselves to the simplest possible case, namely the one corresponding to semilinear PDEs (equivalently decoupled FBSDEs). Let us consider the following semilinear PDE,

$$\begin{cases} (\partial_t + L)u = -f\left(t, x, u, (\nabla u V)(x)\right), & t \in [0,T),\ x \in \mathbb{R}^d \\ u(T,x) = \Phi(x), & x \in \mathbb{R}^d \end{cases} . \tag{112}$$

In (112), L is the second order differential operator

$$Lv = V_0 + \frac{1}{2}\sum_{i=1}^{d} V_i^2, \tag{113}$$

V is the matrix valued function with columns $V_i(x)$, $i = 1,\dots,d$, $V^*(x)$ is the transpose of $V(x)$ and u has final condition $u(T,x) = \Phi(x)$.

Theorem 55 (Pardoux and Peng 1992). *Under additional smoothness assumptions on its coefficients, the unique solution of the Cauchy problem* (112) *admits the following* ***Feynman–Kac representation***

$$u(t,x) = Y_t^{t,x} = \mathbb{E}\left[\Phi(X_T^{t,x}) + \int_t^T f(s, X_s^{t,x}, Y_s^{t,x}, Z_s^{t,x})\right], \tag{114}$$

where $(X^{t,x}, Y^{t,x}, Z^{t,x})$ *is the stochastic flow associated FBSDE* (108) + (109)*, i.e.,*

$$X_s^{t,x} = x + \int_t^s V_0(X_u^{t,x})du + \sum_{i=1}^{d}\int_t^s V_i(X_u^{t,x}) \circ dW_u^i, \quad s \in [t,T], \tag{115}$$

$$Y_s^{t,x} = \Phi(X_T^{t,x}) + \int_s^T f(u, X_u^{t,x}, Y_u^{t,x}, Z_u^{t,x})du - \sum_{i=1}^{d}\int_s^T (Z_u^{t,x})^i dW_u^i. \tag{116}$$

Moreover $Z_s^{t,x} = \nabla u(s, X_s^{t,x})V(X_s^{t,x})$ *for* $s \in [t,T]$.

The representation for Y is true even if u exists only in the viscosity sense. Given such a viscosity solution, Ma and Zhang [37] show that the representation for Z holds as well, provided that the driver and the terminal condition are continuously differentiable. Numerical algorithms that are designed for the approximation of solutions of BSDEs are, in effect, probabilistic methods for solving semilinear PDEs.

4.2 Discretization of Backward SDEs

The Feynman–Kac representation (114) is instructive as it implies that the solution to a BSDE can be expressed as an integral against the law of the forward diffusion. Indeed taking expectations in a BSDE and substituting for Z, we have,

$$Y_t^{t,x} = \mathbb{E}\left[\Phi(X_T^{t,x}) + \int_t^T f(s, X_s^{t,x}, u(s, X_s^{t,x}), \nabla u(s, X_s^{t,x})V(X_s^{t,x}))ds\right]$$

As $Y_s^{t,x}$ is adapted to $\{\mathcal{F}_s^{t,x}\}_{t\leq s\leq T}$, the filtration associated to $X^{t,x}$, it is almost surely deterministic and there exists a functional $\Lambda_t : C\,[t,T] \to \mathbb{R}$ such that

$$Y_t^{t,x} = \mathbb{E}[\Lambda_t\,(X_\cdot^{t,x})],$$

where $C\,[t,T]$ is the space of continuous functions $\alpha : [t,T] \to \mathbb{R}^d$ and $X_\cdot^{t,x}$ is the path valued random map

$$\omega \in \Omega \longrightarrow \left\{X_s^{t,x}(\omega), s \in [t,T]\right\}.$$

Obviously the functional Λ_t is only implicitly defined by the dynamics of the backward equation. Hence, a numerical method for the approximation of $Y_t^{t,x}$ should rely on two components : A method that substitutes Λ_t with an explicitly computable functional and an approximation of the law of the forward diffusion to integrate against.

We approximate Λ_t in the following manner: Consider a partition $\pi = \{0 = t_0 < \ldots < t_{n-1} < t_n = T\}$ of $[0,T]$ with $h_i := t_i - t_{i-1}, i = 1,\ldots,n$. Assume that we know the values of Y, Z at time t_{i+1}, Y_{i+1}, Z_{i+1}. Consider the BSDE between times t_i, t_{i+1}

$$Y_{t_i} = Y_{t_{i+1}} + \int_{t_i}^{t_{i+1}} f(X_s, Y_s, Z_s)ds - \int_{t_i}^{t_{i+1}} Z_s \cdot dB_s$$

and discretize the Riemann integral using the left hand side point (the so called implicit Euler scheme of Bouchard–Touzi [3]), thus leading to an implicit equation for Y_{t_i} and the stochastic part in the usual way, to obtain

$$Y_{t_i} \simeq Y_{t_{i+1}} + h_{i+1} f(X_{t_i}, Y_{t_i}, Z_{t_i}) - Z_{t_i} \cdot \Delta W_{i+1}. \tag{117}$$

By conditioning (117) with respect to $\mathcal{F}_{t_i}$ we obtain a first order approximation for Y_{t_i}

$$Y_{t_i} \simeq \mathbb{E}\left[Y_{t_{i+1}} \middle| \mathcal{F}_{t_i}\right] + h_{i+1} f(X_{t_i}, Y_{t_i}, Z_{t_i}), \tag{118}$$

but for the presence of Z_{t_i}. To treat the Z_{t_i}, we can multiply both sides of (117) by ΔW^l_{i+1}, $l = 1, \ldots, d$ and condition with respect to $\mathcal{F}_{t_i}$, to obtain

$$Z^l_{t_i} \simeq \mathbb{E}\left[Y_{t_{i+1}} \frac{\Delta W^l_{i+1}}{h_{i+1}} \Big| \mathcal{F}_{t_i} \right], \quad l = 1, \ldots, d. \tag{119}$$

Inspired by (118), (119) we define the family $R_i : C_{Lip}\left(\mathbb{R}^d\right) \to C_{Lip}\left(\mathbb{R}^d\right) i = 0, 1, \ldots, n-1$ of operators defined on the set of Lipschitz continuous functions $C_{Lip}\left(\mathbb{R}^d\right)$:

$$R_i g(x) = \mathbb{E}\left[g\left(X^{t_i,x}_{t_{i+1}}\right)\right] \\ +h_{i+1} f\left(t_i, x, R_i g(x), \frac{1}{h_{i+1}} \mathbb{E}\left[g\left(X^{t_i,x}_{t_{i+1}}\right)\left(W_{t_{i+1}} - W_{t_i}\right)\right]\right). \tag{120}$$

The iteration of this family of operators $R_{i:n-1} := R_i \ldots R_{n-1}$ gives rise to an explicitly defined functional $\Lambda^\pi_{t_i}$, $i = 0, \ldots, n-1$,

$$\mathbb{E}[\Lambda^\pi_{t_i}(X^{t_i,x}_\cdot)] = R_{i:n-1}\Phi(x).$$

The operator $R_{i:n-1}$ applied to the boundary data $\Phi(\cdot)$ and evaluated at $x = X^{t_i,x}_{t_i}$, can be viewed as a discretized version (corresponding to the partition π) of $Y^{t_i,x}_{t_i}$. In fact the above discretization is merely the Euler scheme for BSDEs (it should be clear that the Riemann integral is discretized in an Euler fashion). Relative to this, we have the following convergence result due to Bouchard and Touzi [3] and independently to Zhang [60]. This results were further refined by Gobet and Labart [20], where an error expansion, under additional smoothness assumptions, was obtained.

Theorem 56 (Bouchard and Touzi, Zhang). *Set $Y^{\pi,x}_0 = R_{0:n-1}\Phi(x)$. Under assumptions **(A)**, **(C)***

$$|Y^{\pi,x}_0 - Y^x_0| \leq C\sqrt{\|\pi\|},$$

where $\|\pi\|$ is the size of the partition mesh.

Remark 57. Originally, the proof of the convergence of the Euler scheme for BSDEs required an ellipticity assumption on the diffusion matrix of the forward component. However, the proof can be redone without this at least in the Markovian case. All one needs to show is that the value functions describing Y_t, Z_t as functions of time and X_t are smooth enough for the relevant stochastic Taylor expansions to be applied.

Theorem 58 (Gobet and Labart). *Let assumption **(B)** hold true with $m \geq 3$ and assume also that the partial derivatives of the driver with respect to space are*

Hölder continuous. Assume also that the terminal condition is twice continuously differentiable with bounded partial derivatives. Then

$$|Y_0^{\pi,x} - Y_0^x| \leq C\,\|\pi\|.$$

To obtain a fully implementable scheme, a method of computation for the expectations appearing in (120) involved needs to be introduced. We will present next an algorithm that uses the cubature method to approximate the law of the forward diffusion and the TBBA algorithm to control the computational effort. Both of these when combined with the Euler style discretization (118), (119) provide a fully implementable scheme for BSDEs.

4.3 Cubature on BSDEs

We will use a cubature formula of degree m, supported on paths $\omega_1, \ldots, \omega_{c_d^m} : [0,1] \to \mathbb{R}^d$. We also fix throughout a parameter N to be used in the application of the TBBA. Using this cubature formula and TBBA we build (see Sect. 3.5) the sequence of explicit measures $\{\tilde{\mathbb{Q}}_{t_i}^m\}_{i=1}^n$. Substituting integration against the Wiener measure, with integration against the explicit measures $\{\tilde{\mathbb{Q}}_{t_i}^m\}_{i=1}^n$ in (120), we can define the following family of operators:

$$\begin{aligned}\hat{R}_i g\,(x) &= \mathbb{E}_{\tilde{\mathbb{Q}}_{t_i}^m}[g(X_{t_{i+1}}^{t_i,x})] \\ &\quad + h_{i+1} f\left(t_i, x, \hat{R}_i g\,(x)\,, \frac{1}{h_{i+1}} \mathbb{E}_{\tilde{\mathbb{Q}}_{t_i}^m}\left[g(X_{t_{i+1}}^{t_i,x})(W_{t_{i+1}} - W_{t_i})\right]\right)\end{aligned} \tag{121}$$

where $g : \mathbb{R}^d \to \mathbb{R}$. Computations of the involved expectations in (121) are done in the obvious way, namely we work our way backwards along the cubature+TBBA tree.

Recall from the Sect. 3.5 the sets $\hat{C}_i$, $i = 1, \ldots, n$ and for every $x \in C_i$ the subset of its children $\hat{C}^x$. Given that we are standing at depth i (equivalently, at time t_i), we need to evaluate the operator $\hat{R}_i$, when applied to $\hat{R}_{i+1:n-1}\Phi$, at all points $x \in \hat{C}_i$. We have

$$\begin{aligned}\mathbb{E}_{\tilde{\mathbb{Q}}_{t_i}^m}[g(X_{t_{i+1}}^{t_i,x})] \equiv \mathbb{E}_{\tilde{\mathbb{Q}}_{t_i}^m}\left[g(X_{t_{i+1}})|X_{t_i} = x\right] &:= \sum_{\bar{x} \in C_{i+1}^x} \frac{\hat{\lambda}_{\bar{x}}}{\lambda_{\bar{x}}} g(\bar{x}), \quad x \in \hat{C}_i \\ \mathbb{E}_{\tilde{\mathbb{Q}}_{t_i}^m}\left[g(X_{t_{i+1}})\Delta W_{i+1}^l | X_{t_i} = x\right] &:= \sum_{\bar{x} \in C_{i+1}^x} \frac{\hat{\lambda}_{\bar{x}}}{\lambda_{\bar{x}}} g(\bar{x}) \delta\omega_{h_{i+1},\bar{x}}^l, \quad x \in \hat{C}_i,\, l = 1, \ldots, d,\end{aligned} \tag{122}$$

where $\omega^l_{h_k,\bar{x}}$ is the l-th coordinate of the path $\omega_{h_{i+1},\bar{x}}$ in the cubature formula, that was used in the ODE that lead to the point $\bar{x} \in \hat{\mathcal{C}}^x_{i+1}$, scaled over the time interval $[t_i, t_{i+1})$. It should then be clear how one computes $\hat{R}_{i:n-1}\Phi(x)$ for $x \in \hat{\mathcal{C}}_i$.

Estimating the global error $\hat{R}_{0:n-1}\Phi(x_0) - Y_0^{0,x_0}$ requires standard numerical analysis arguments as well as some knowledge of the behavior of the solution to PDE (112). As estimating the errors of cubature formulas is done with the help of Taylor expansions, the derivatives of the involved functions need to be estimated. In other words, we need gradient bounds, in the spirit of Sect. 2 but here for the semilinear PDEs. For elliptic PDEs, such bounds are of course well known for a long time. But when one wishes to step into the realm of degenerate PDEs/SDEs the subject becomes quite technical and difficult. Recently, these issues were addressed in Crisan and Delarue [12] and we are able to report here on this gradient bounds for semi linear PDEs without discussing its proof.

Theorem 59 (Crisan and Delarue [12]). *Let assumption (B) hold true and consider an $m \geq 3$. Assume further that the vector fields $\{V_i : i = 0, \ldots, d\}$ satisfy the UFG condition. Assume also $\Phi \in C_b^m\left(\mathbb{R}^d\right)$. Define $u(t,x) = Y_t^{t,x}$. Then u is differentiable in all the direction that appear in* (112). *Moreover, for any multi-index $\alpha \in \mathcal{A}^1_m$, there exist increasing function $c_\alpha, \bar{c}_\alpha : [0,\infty) \to [0,\infty)$ such that for any $\Phi \in C_b^m\left(\mathbb{R}^d\right)$, we have*

$$\|V_\alpha u(t,\cdot)\|_\infty \leq c_\alpha \left(\sum_{\alpha \in \mathcal{A}_m} \|V_\alpha \Phi\| \right), \tag{123}$$

$$\|V_\alpha u(t,\cdot)\|_\infty \leq \frac{\bar{c}_\alpha(\|\Phi\|_{Lip})}{(T-t)^{(\|\alpha\|-1)/2}}, \quad t \in [0,T), \tag{124}$$

In analyzing the error we split it into two parts: The error between the solution of the BSDE and the Euler scheme and the error between the Euler scheme and its cubature and TBBA realization. The first part of the error is treated by Theorem 56. The second part of the error is split to the error due to cubature method and the error due to TBBA. Let us define the family of intermediate operators

$$\begin{aligned} \bar{R}_i g(x) &= \mathbb{E}_{\mathbb{Q}^m_{t_i}}[g(X^{t_i,x}_{t_{i+1}})] \\ &\quad + h_{i+1} f\left(t_i, x, \bar{R}_i g(x), \frac{1}{h_{i+1}} \mathbb{E}_{\mathbb{Q}^m_{t_i}}\left[g(X^{t_i,x}_{t_{i+1}})(W_{t_{i+1}} - W_{t_i})\right]\right), \end{aligned} \tag{125}$$

which is merely the equivalent definition to the family $\{\hat{R}_i\}_{1\leq i\leq n}$ but using the pure cubature measures. It is obvious that in quantifying the error between $R_{i:n-1}\Phi$, $\bar{R}_{i:n-1}\Phi$, $i = 0, \ldots, n-1$ we need to quantify the errors

$$\mathbb{E}_{\mathbb{Q}^m_{t_i}}[g(X^{t_i,x}_{t_{i+1}})] - \mathbb{E}[g(X^{t_i,x}_{t_{i+1}})],\ \mathbb{E}_{\mathbb{Q}^m_{t_i}}[g(X^{t_i,x}_{t_{i+1}})\Delta W_{i+1}] - \mathbb{E}[g(X^{t_i,x}_{t_{i+1}})\Delta W_{i+1}],\ i = 0, \ldots, n-1.$$

We have already seen in Sect. 3 that

$$\sup_x \left|\mathbb{E}\left[g(X^{t_i,x}_{t_{i+1}})\right] - \mathbb{E}_{\mathbb{Q}^m_{t_{i+1}}}\left[g(X^{t_i,x}_{t_{i+1}}))\right]\right| \leq C \sum_{j=m+1}^{m+2} h_{i+1}^{j/2} \sup_{I\in\mathcal{A}(j)\setminus\mathcal{A}(j-1)} \|V_I g\|_\infty. \tag{126}$$

For the second term, we also have

$$\begin{aligned}&\sup_x \left|\mathbb{E}\left[g(X^{t_i,x}_{t_{i+1}})\Delta W^l_{i+1}\right] - \mathbb{E}_{\mathbb{Q}^m_{h_{i+1}}}\left[g(X^{t_i,x}_{t_{i+1}})\Delta W^l_{i+1}\right]\right| \\ &\leq C \sum_{j=m}^{m+2} h_{i+1}^{(j+1)/2} \sup_{I\in\mathcal{A}(j)\setminus\mathcal{A}(j-1)} \|V_I g\|_\infty.\end{aligned} \tag{127}$$

Proof of (127) Let us fix a value $l \in \{1, \ldots, d\}$. Since the function g is smooth it admits the Stratonovich–Taylor expansion. An easy application of Itô's formula, shows that the product of an iterated Stratonovich integral and a Brownian motion can be expressed as a sum of higher order iterated integrals (see for example Proposition 5.2.10 of [27]).

$$\left(\int_{\Delta^k[0,t]} \circ dW^I\right) W^l_t = \sum_{j=0}^{k} \int_{\Delta^{k+1}[0,t]} \circ dW^{(i_1,\ldots,i_j,l,i_{j+1},\ldots,i_k)},$$

where for any multi index $\alpha = (i_1, \ldots, i_k)$ we denote

$$\int_{\Delta^k[0,t]} \circ dW^\alpha := \int_{0<t_1<\ldots<t_k<t} \circ d\, W^{i_1}_{t_1} \ldots \circ dW^{i_k}_{t_k}.$$

Hence, we have that

$$\begin{aligned}g(X_t(0,x))W^l_t = &\sum_{(i_1,\ldots,i_k)\in\mathcal{A}_m} V_{(i_1,\ldots,i_k)}g(x) \sum_{j=0}^{k} \int_{\Delta^{k+1}[0,t]} \circ dW^{(i_1,\ldots,i_j,l,i_{j+1},\ldots,i_k)} \\ &+ R_m(t,x,g)W^l_t.\end{aligned}$$

Using this formula the error is

$$\begin{aligned}&\left|\mathbb{E}\left[g(X_t(0,x))W^l_t\right] - \mathbb{E}_{\mathbb{Q}^m_t}\left[g(X_t(0,x))W^l_t\right]\right| \\ &\leq \left|\mathbb{E}\left[R_m(t,x,g)W^l_t\right]\right| + \left|\mathbb{E}_{\mathbb{Q}^m_t}\left[R_m(t,x,g)W^l_t\right]\right| \\ &\quad + \left|\left(\mathbb{E} - \mathbb{E}_{\mathbb{Q}^m_t}\right)\left[\sum_{(i_1,\ldots,i_k)\in\mathcal{A}_m} \sum_{j=0}^{k} V_{(i_1,\ldots,i_k)}g(x) \int_{\Delta^{k+1}[0,t]} \circ dW^{(i_1,\ldots,l,\ldots,i_k)}\right]\right|.\end{aligned} \tag{128}$$

According to estimates of Lemma 8 in [36] and (88), we have that

$$\left.\begin{array}{l}\sup_x \mathbb{E}\left[R_m(t,x,g)^2\right]^{1/2}\\ \sup_x \mathbb{E}_{\mathbb{Q}_t^m}\left[|R_{m,t,g}|^2\right]^{1/2}\end{array}\right\} \leq C \sum_{j=m+1}^{m+2} t^{j/2} \sup_{I\in\mathcal{A}(j)\setminus\mathcal{A}(j-1)} \|V_\alpha g\|_\infty .$$

An application of Hölder's inequality gives us

$$\sup_x \left|\mathbb{E}\left[R_{m,t,g}W_t\right]\right| \leq \sum_{j=m+1}^{m+2} t^{(j+1)/2} \sup_{\alpha\in\mathcal{A}(j)\setminus\mathcal{A}(j-1)} \|V_\alpha g\|_\infty .$$

To estimate the term $\mathbb{E}_{\mathbb{Q}_t^m}\left[R_{m,t,g}W_t\right]$ observe that

$$R_{m,t,g} = \sum_{\substack{(i_2,\ldots,i_k)\in\mathcal{A}_m\\ (i_1,\ldots,i_k)\notin\mathcal{A}_m}} \int_{\Delta^k[0,t]} V_{i_1}\ldots V_{i_k} g(X_{t_1}(0,x)) \circ dW_{t_1}^{i_1} \circ \ldots \circ dW_{t_k}^{i_k}.$$

So that, with $l \in \{1,\ldots,d\}$ fixed,

$$\begin{aligned}&\left|\mathbb{E}_{\mathbb{Q}_t^m}\left[R_{m,t,g}W_t^l\right]\right|\\ &\leq \sum_{j=1}^{N} \lambda_j \times \sum_{\substack{(i_2,\ldots,i_k)\in\mathcal{A}_m\\ (i_1,\ldots,i_k)\notin\mathcal{A}_m}} \left|\int_{\Delta^k[0,t]} V_{i_1}\ldots V_{i_k} g\left(X_{t_1}(0,x)(\omega_{t_1,j})\right)\right.\\ &\qquad \left. d\omega_{t,j}^{i_1}(t_1)\ldots d\omega_{t,j}^{i_k}(t_k)\omega_{t,j}^{l}(t)\right| .\end{aligned}$$

Performing a change of variables to the paths $\omega_{t,j}$ to pass back to the paths that define the cubature formula on [0, 1] we obtain the estimate

$$\sup_x \left|\mathbb{E}_{\mathbb{Q}_t^m}\left[R_{m,t,g}W_t\right]\right| \leq C \sum_{j=m+1}^{m+2} t^{(j+1)/2} \sup_{\alpha\in\mathcal{A}(j)\setminus\mathcal{A}(j-1)} \|V_\alpha g\|_\infty , \tag{129}$$

where the constant C depends on the bounds on the total variation of the paths $\omega_1,\ldots,\omega_N$. We now focus on the last term of (128).

$$\begin{aligned}&|(\mathbb{E}-\mathbb{E}_{\mathbb{Q}_t^m})[\sum_{\alpha\in\mathcal{A}(m)} V_\alpha g(x) \int_{\Delta^k[0,t]} \circ dW^\alpha\, W_t^l]|\\ &= \left|\sum_{\alpha\in\mathcal{A}(m)} V_\alpha\, g(x)\left(\mathbb{E}-\mathbb{E}_{\mathbb{Q}_t^m}\right)\left[\sum_{j=0}^{k}\int_{\Delta^{k+1}[0,t]} \circ dW^{(i_1,\ldots,l,\ldots,i_k)}\right]\right|\\ &= \left|\sum_{\alpha\in\mathcal{A}(m)\setminus\mathcal{A}(m-1)} V_\alpha g(x)\left(\mathbb{E}-\mathbb{E}_{\mathbb{Q}_t^m}\right)\left[\sum_{j=0}^{k}\int_{\Delta^{k+1}[0,t]} \circ dW^{(i_1,\ldots,l,\ldots,i_k)}\right]\right|\end{aligned}$$

since the terms corresponding to $\alpha \in \mathcal{A}(m-1)$ are 0 by definition of the measure $\mathbb{Q}_t^m$.

Hence, to obtain the estimate, observe that, for any $\alpha \in \mathcal{A}(m)\backslash\mathcal{A}(m-1)$ the terms under the cubature measure satisfy

$$\left|\mathbb{E}_{\mathbb{Q}_t^m}\left[\int_{\Delta^{k+1}[0,t]} \circ dW^{(i_1,\dots,l,\dots,i_k)}\right]\right| \leq C t^{(m+1)/2}$$

since they are iterated integrals along paths of bounded variation and hence, with similar arguments to the ones we used to derive (129), we may show that they are of order $t^{(m+1)/2}$. As for the ones under the Wiener measure, they are either 0 or of order $t^{(m+1)/2}$ according to (84). The bounds on the derivatives of the vector fields complete the proof. □

We can now report on the main cubature for BSDEs error estimate

Theorem 60. *Consider a fixed $m \geq 3$ and assume that the system* (115) + (116) *satisfies assumption* **(B)** *and* **(C)**. *Given a partition π we consider the family of operators $\{\bar{R}_i\}_{0\leq i\leq n-1}$ along it and consider a $p > 1$. Then, there exists a constant C independent of the partition, such that*

$$\begin{aligned}|Y_0 - \bar{Y}_0^\pi| \leq C \sum_{i=0}^{n-2} \Bigg(&\sum_{j=3}^{4} h_{i+1}^{(j+1)/2} \sup_{\|I\|=j} \|V_I u(t_i,\cdot)\|_\infty \\ &+ \sum_{j=m+1}^{m+2} h_{i+1}^{j/2} \sup_{\|I\|=j,j-1} \|V_I u(t_i,\cdot)\|_\infty \Bigg) \\ &+ \mathbb{E}_{\mathbb{Q}_{t_{n-1}}^m}\left[\left| Y_{t_{n-1}} - \bar{R}_{n-1}\Phi\left(X_{t_{n-1}}\right)\right|^p\right]^{1/p} \end{aligned} \tag{130}$$

where $\bar{Y}_0^\pi = \bar{R}_{0:n-1}\Phi(x_0)$, $X_0 = x_0$.

The proof of the theorem requires the following lemma:

Lemma 61. *Consider two measurable functions $g_1, g_2 : \mathbb{R}^d \to \mathbb{R}$. The operators $\{\bar{R}_i\}_{i=1}^n$ $i = 0,\dots,n$ enjoy the following property*

$$|\bar{R}_i g_1 - \bar{R}_i g_2|(x) \leq \frac{1 + C h_{i+1}}{1 - K h_{i+1}} \mathbb{E}_{\mathbb{Q}^m}[|g_1 - g_2|^p (g(X_{t_{i+1}}^{t_i,x}))]^{1/p} \tag{131}$$

for any $p > 1$, where C is a constant which depends on the bounded variation constants of the paths ω_j, $j = 1,\dots,N$ that define cubature on the Wiener space and K is the Lipschitz constant of the driver f.

Proof. The Lipschitz property of f tells us that there exists bounded deterministic functions $\nu(x)$, $\zeta(x)$ such that

$$(1 - h_{i+1}\nu(x))(\bar{R}_i g_1(x) - \bar{R}_i g_2(x))$$
$$= \mathbb{E}_{\mathbb{Q}^m}\left[(g_1 - g_2)(X^{t_i,x}_{t_{i+1}})\right] + \zeta(x) \cdot \mathbb{E}_{\mathbb{Q}^m}\left[(g_1 - g_2)(X^{t_i,x}_{t_{i+1}})\Delta W_{i+1}\right].$$

Hence, for h_{i+1} small enough,

$$(1 - Kh_{i+1})|\bar{R}_i g_1(x) - \bar{R}_i g_2(x)|$$
$$\leq \mathbb{E}_{\mathbb{Q}^m}\left[|g_1 - g_2|\,(X^{t_i,x}_{t_{i+1}})|\Delta W_{i+1} \cdot \zeta(x) + 1|\right]$$
$$\leq \mathbb{E}_{\mathbb{Q}^m}\left[(|g_1 - g_2|^p\,(X^{t_i,x}_{t_{i+1}})\right]^{1/p} \mathbb{E}_{\mathbb{Q}^m}\left[(\Delta W_{i+1} \cdot \zeta(x) + 1)^{2k}\right]^{1/2k},$$

where $k > q/2$ and q is the conjugate of p. Observe that $\mathbb{E}_{\mathbb{Q}^m}[\Delta W_{i+1}] = \mathbb{E}[\Delta W_{i+1}] = 0$, since ΔW_{i+1} can be written as a stochastic integral of length 1. For any higher powers of the Brownian increment, it holds that

$$\mathbb{E}_{\mathbb{Q}^m}\left[\left(\Delta W^l_{i+1}\right)^r\right] \leq C h^{r/2}_{i+1},\ \forall l = 1, \ldots, d.$$

To see this, observe that for any $r \leq m$ we may express the increment $(\Delta W^l_{i+1})^r$ as a linear combination of iterated integrals of length less than m. The estimate then follows from the definition of the measure $\mathbb{Q}^m$. Hence,

$$\mathbb{E}_{\mathbb{Q}^m}\left[(\Delta W_{i+1} \cdot \zeta(x) + 1)^{2k}\right]^{1/2k} \leq (1 + C h_{i+1})$$

and this completes the proof. □

Proof of Theorem 60. To begin with, set

$$\epsilon_i = \sum_{j=3}^{4} h^{(j+1)/2}_{i+1} \sup_{\|I\|=j} \|V_I u(t_i, \cdot)\|_\infty, \quad i = 0, \ldots, n-1.$$

We expand the error as a telescopic sum

$$Y_0 - \bar{R}_{0:n-1}\Phi(x) = \sum_{i=0}^{n-1} \bar{R}_{0:i-1} Y^{0,x}_{t_i} - \bar{R}_{0:i} Y^{0,x}_{t_{i+1}}. \tag{132}$$

The size of each of the terms $\bar{R}_{0:i-1} Y^{0,x}_{t_i} - \bar{R}_{0:i} Y^{0,x}_{t_{i+1}}$ is then controlled using Lemma 61. We have, with for $p > 1$,

$$|Y_0 - \bar{Y}^\pi_0| \leq C \sum_{i=0}^{n-1} \mathbb{E}_{\mathbb{Q}^m_{t_i}}\left[\left|Y^{t_i,X_{t_i}}_{t_i} - \bar{R}_i Y^{t_i,X_{t_i}}_{t_{i+1}}\right|^p\right]^{1/p}. \tag{133}$$

Observe that for any $i \in \{0, \ldots, n-1\}$ and $x \in \mathbb{R}^d$, by taking expectations on the backward part of (109), we have

$$Y_{t_i}^{t_i,x} = \mathbb{E}\left[Y_{t_{i+1}}^{t_i,x} + \int_{t_i}^{t_{i+1}} f(s, X_s^{t_i,x}, Y_s^{t_i,x}, Z_s^{t_i,x}) ds \right].$$

The above together with the definition of $\bar{R}_i$ tells us

$$Y_{t_i}^{t_i,x} - \bar{R}_i Y_{t_{i+1}}^{t_i,x} = (\mathbb{E} - \mathbb{E}_{\mathbb{Q}_{t_{i+1}}^m})[Y_{t_{i+1}}^{t_i,x}] + \int_{t_i}^{t_{i+1}} \mathbb{E} f(s, X_s^{t_i,x}, Y_s^{t_i,x}, Z_s^{t_i,x}) ds$$
$$-h_{i+1} f\left(t_i, x, \bar{R}_i Y_{t_{i+1}}^{t_i,x}, \frac{1}{h_{i+1}} \mathbb{E}_{\mathbb{Q}_{t_{i+1}}^m} \left[\bar{R}_i Y_{t_{i+1}}^{t_i,x} \Delta W_{i+1} \right] \right)$$

We now fix a value for $i = 0, \ldots, n-2$. To compare the drivers we need to add and subtract the right terms:

$$\begin{aligned} Y_{t_i}^{t_i,x} - \bar{R}_i Y_{t_{i+1}}^{t_i,x} &= (\mathbb{E} - \mathbb{E}_{\mathbb{Q}_{t_{i+1}}^m})[Y_{t_{i+1}}^{t_i,x}] \\ &\quad + \int_{t_i}^{t_{i+1}} \mathbb{E}\left[f(s, X_s^{t_i,x}, Y_s^{t_i,x}, Z_s^{t_i,x}) - f(t_i, x, Y_{t_i}^{t_i,x}, Z_{t_i}^{t_i,x}) \right] ds \\ &\quad + h_{i+1} \Big(f(t_i, x, Y_{t_i}^{t_i,x}, Z_{t_i}^{t_i,x}) \\ &\qquad - f(t_i, x, \bar{R}_i Y_{t_{i+1}}^{t_i,x}, \frac{1}{h_{i+1}} \mathbb{E}_{\mathbb{Q}_{t_{i+1}}^m} \left[Y_{t_{i+1}}^{t_i,x} \Delta W_{i+1} \right] \Big) \\ &=: I_1^{t_i,x} + I_2^{t_i,x} + I_3^{t_i,x} \end{aligned} \tag{134}$$

with the obvious definition for the $I_k^{t_i,x}$'s. To estimate each of these terms the non linear Feynman–Kac formula for BSDE's plays a central role.

Since (112) has a classical solution on $[0, T) \times \mathbb{R}^d$, it holds that

$$Y_s^{t,x} = u(s, X_s^{t,x}), \qquad Z_s^{t,x} = \nabla u(s, X_s^{t,x}) V(X_s^{t,x}).$$

We can apply Itô's formula to the function $\bar{f}:(t,x) \to f(t, x, u(t,x), \nabla u(t,x) V(x))$ to control I_2,

$$\begin{aligned} |I_2^{t_i,x}| = \Bigg| \mathbb{E}\Bigg[\int_{t_i}^{t_{i+1}} & \int_{t_i}^{s} \partial_t \bar{f}(r, X_r^{t_i,x}) + V_0 \bar{f}(r, X_r^{t_i,x}) dr ds \\ &+ \int_{t_i}^{s} \left(\frac{1}{2} \sum_{i=1}^{d} V_i^2 \bar{f}(r, X_r^{t_i,x}) dr + \sum_{i=1}^{d} V_i \bar{f}(r, X_r^{t_i,x}) dW_r^i \right) ds \Bigg] \Bigg| \end{aligned} \tag{135}$$

Hence

$$\sup_{x\in\mathbb{R}^d} | I_2^{t_i,x} | \leq C h_{i+1}^2 \left(\| V_0 \bar{f} + \partial_t \bar{f} \|_\infty + \max_{i,=1,\ldots,d} \| V_i^2 \bar{f} \|_\infty \right)$$
$$\leq C h_{i+1}^2 \sup_{\|I\|=3} \| V_I u \|_\infty , \tag{136}$$

where, the latter estimate follows by the chain rule. To estimate $I_3^{t_i,x}$ we use the mean value theorem, so that we can find two points $\theta_1 \in \mathbb{R}$, $\theta_2 \in \mathbb{R}^d$ such that

$$I_3^{t_i,x} = h_{i+1} \Big(f_y(t_i, x, \theta_1, \theta_2)(Y_{t_i}^{t_i,x} - \bar{R}_i Y_{t_{i+1}}^{t_i,x})$$
$$+ f_z(t_i, x, \theta_1, \theta_2) \cdot \Big(Z_{t_i}^{t_i,x} - \frac{1}{h_{i+1}} \mathbb{E}_{\mathbb{Q}^m_{t_{i+1}}} \Big[Y_{t_{i+1}}^{t_i,x} \Delta W_{i+1} \Big] \Big) \Big)$$
$$|I_3^{t_i,x}| \leq K h_{i+1} \Big(\Big| Y_{t_i}^{t_i,x} - \bar{R}_i Y_{t_{i+1}}^{t_i,x} \Big| + \Big| Z_{t_i}^{t_i,x} - \frac{1}{h_{i+1}} \mathbb{E}_{\mathbb{Q}^m_{t_{i+1}}} \Big[Y_{t_{i+1}}^{t_i,x} \Delta W_{i+1} \Big] \Big| \Big), \tag{137}$$

since the partial derivatives of f are bounded by K. As a next step observe that

$$h_{i+1} \Big| Z_{t_i}^{t_i,x} - \frac{1}{h_{i+1}} \mathbb{E}_{\mathbb{Q}^m_{t_{i+1}}} \Big[Y_{t_{i+1}}^{t_i,x} \Delta W_{i+1} \Big] \Big| \tag{138}$$
$$\leq \Big| h_{i+1} Z_{t_i}^{t_i,x} - \mathbb{E} \Big[Y_{t_{i+1}}^{t_i,x} \Delta W_{i+1} \Big] \Big| + \Big| \mathbb{E} \Big[Y_{t_{i+1}}^{t_i,x} \Delta W_{i+1} \Big] - \mathbb{E}_{\mathbb{Q}^m_{t_{i+1}}} \Big[Y_{t_{i+1}}^{t_i,x} \Delta W_{i+1} \Big] \Big|$$

As before, $Y_{t_{i+1}}^{t_i,x} = u(t_{i+1}, X_{t_{i+1}}^{t_i,x})$ and we may apply the stochastic Taylor expansion to the latter, to treat the first term above. In particular, we do so using the hierarchical set $\mathcal{A}_2$. Let us fix an integer value $l = 1, \ldots, d$ and denote by $Z_{t_i}^{t_i,x,l}$ the l-th entry of the vector $Z_{t_i}^{t_i,x}$. We then have,

$$\Big| h_{i+1} Z_{t_i}^{t_i,x,l} - \mathbb{E} \Big[u(t_{i+1}, X_{t_{i+1}}^{t_i,x}) \Delta W_{i+1}^l \Big] \Big|$$
$$= \Big| h_{i+1} Z_{t_i}^{t_i,x,l} - \mathbb{E} \Big[\Big(u(t_i, x) + \sum_{i=0}^{d} V_i u(t_i, x) \int_{t_i}^{t_{i+1}} \circ dW_s^i \tag{139}$$
$$+ \sum_{i,j=1}^{d} V_i V_j u(t_i, x) \int_{t_i}^{t_{i+1}} \int_{t_i}^{t} \circ dW_s^i \circ dW_t^j + R_2(h_{i+1}, x, u) \Big) \Delta W_{i+1}^l \Big] \Big| .$$

Observe that

$$\mathbb{E} \Big[\sum_{i=1}^{d} V_i u(t_i, x) \int_{t_i}^{t_{i+1}} \circ dW_s^i \Delta W_{i+1}^l \Big] = h_{i+1} V_l u(t_i, x).$$

Moreover, according to Proposition 5.2.10 of Kloeden and Platen [27] we have that for any $k, r = 1, \ldots, d$,

$$\int_{t_i}^{t_i+1} \int_{t_i}^{t} \circ dW_s^k \circ dW_t^r \Delta W_{i+1}^l = J_{(k,r,l)}[1]_{t_i,t_i+1} + J_{(k,r,j)}[1]_{t_i,t_i+1} + J_{(l,k,r)}[1]_{t_i,t_i+1}$$

and the three terms on the right hand side will have expectation 0 according to (84). Due to the non linear Feynman- Kac formula we have for $i = 0, \ldots, n-2$ that $Z_{t_i}^{t_i,x,l} = \nabla u(t_i, x)^* \cdot V_l(x) = V_l u(t_i, x)$. Hence, (139) together with (138) and the estimate on the remainder process, give us

$$\begin{aligned} &h_{i+1} \left| Z_{t_i}^{t_i,x} - \frac{1}{h_{i+1}} \mathbb{E}_{\mathbb{Q}_{t_i+1}^m} \left[Y_{t_i+1}^{t_i,x} \Delta W_{i+1} \right] \right| \\ &\leq C\mathbb{E}\left[|R_2(t_i,x,u)\Delta W_{i+1}| \right] + \frac{1}{h_{i+1}} \left| \left(\mathbb{E} - \mathbb{E}_{\mathbb{Q}_{t_i+1}^m} \right) \left[u(t_{i+1}, X_{t_i+1}^{t_i,x}) \Delta W_{i+1} \right] \right| \end{aligned} \tag{140}$$

Equations (140) and (139) are plugged in (138) and the resulting estimate (137). The latter together with (136) and (134) gives us

$$\begin{aligned} &(1 - h_{i+1}K)\, \mathbb{E}_{\mathbb{Q}_{t_i}^m} \left[\left| Y_{t_i}^{t_i,X_{t_i}} - \bar{R}_i Y_{t_i+1}^{t_i,X_{t_i}} \right|^p \right]^{1/p} \\ &\leq \mathbb{E}_{\mathbb{Q}_{t_i}^m} \left[\left| \left(\mathbb{E} - \mathbb{E}_{\mathbb{Q}_{t_i+1}^m} \right) \left[Y_{t_i+1}^{t_i,X_{t_i}} \right] \right|^p \right]^{1/p} \\ &\quad + \mathbb{E}_{\mathbb{Q}_{t_i}^m} \left[\left| \left(\mathbb{E} - \mathbb{E}_{\mathbb{Q}_{t_i+1}^m} \right) \left[Y_{t_i+1}^{t_i,X_{t_i}} \Delta W_{i+1} \right] \right|^p \right]^{1/p} + \epsilon_i \\ &\leq \epsilon_i + C \sum_{j=m+1}^{m+2} h_{i+1}^{j/2} \sup_{I \in \mathcal{A}_j \setminus \mathcal{A}_{j-1}} \|V_I g\|_\infty, \quad i = 0, \ldots, n-2. \end{aligned} \tag{141}$$

where we have used the estimates (126) and (127). This completes the proof. □

We have already discussed how a non even partition can compensate for the explosion in the gradient bounds, in the linear case. In view of Theorem 59, we have a similar result in the semilinear case. In the more interesting case where the terminal condition is only Lipschitz continuous, we have to appeal to the derivative bounds (124). In this case the control on the derivatives of u explodes as t approaches T. To compensate for this negative impact of the derivative bounds on the error estimate we shall use a non equidistant partition that becomes denser as we approach T.

Corollary 62. *Let (A) and (B) hold true, fix and $m \geq 3$ and assume further that the vector fields $\{V_i : i = 0, \ldots, d\}$ satisfy the UFG condition and that the final condition Φ is Lipschitz. We consider the family $\{\bar{R}_i\}_{0 \leq i \leq n-1}$ along the partition π:*

$$t_i = T\left(1 - \left(1 - \frac{i}{n}\right)^{\beta}\right), \quad i = 0, \ldots, n, \quad \beta \geq 2.$$

Then, there exists an increasing function $c : [0, \infty) \to [0, \infty)$ independent of the partition such that

$$\left| Y_0 - \bar{R}_{0:n-1}\Phi(x_0) \right| \leq \frac{c(\|\Phi\|_{Lip})}{n}$$

Proof. Let us assume first that $\Phi \in C_b^m\left(\mathbb{R}^d\right)$. In the following, the functions $c_i : [0, \infty) \to [0, \infty)$ are all strictly increasing. Given the estimates of Theorems 130, 59, it is straightforward to see that the dominating term in our error bound is $h_i^2 \sup_{\|\alpha\|=3} \|V_\alpha u\|_\infty$. On the above partition we have, for a given multi index α with $\|\alpha\| = 3$,

$$\begin{aligned}(t_i - t_{i-1})^2 \|V_\alpha u\|_\infty &\leq c_1(\|\Phi\|_{Lip}) T^2 \left(\int_{1-\frac{i}{n}}^{1-\frac{i-1}{n}} \beta s^{\beta-1} ds \right)^2 \frac{1}{T\,(1 - i/n)^{\beta}} \\ &\leq \frac{c_2(\|\Phi\|_{Lip})}{n^2}\end{aligned}$$

On the other hand, for the term corresponding to t_{n-1} we may argue, using the mean value theorem, that,

$$\begin{aligned}&\mathbb{E}_{\mathbb{Q}^m_{t_{n-1}}}\left[\left| Y_{t_{n-1}} - \bar{R}_{n-1}\Phi(X_{t_{n-1}}) | X_{t_{n-1}} \right|^p \right]^{1/p} \\ &\leq C \sum_{l=0}^{d} \mathbb{E}_{\mathbb{Q}^m_{t_{n-1}}}\left[\left| \left(\mathbb{E} - \mathbb{E}_{\mathbb{Q}^m_{t_n}}\right)\left[\Phi(X_{t_n}) \Delta W_n^l | X_{t_{n-1}} \right] \right|^p \right]^{1/p}\end{aligned}$$

from elementary properties of the Wiener and cubature measure, it is clear that

$$\left(\mathbb{E} - \mathbb{E}_{\mathbb{Q}^m_{t_n}}\right)\left[\Phi(X_{t_n}) \Delta W_n^l | X_{t_{n-1}} \right] = \left(\mathbb{E} - \mathbb{E}_{\mathbb{Q}^m_{t_n}}\right)\left[\left(\Phi(X_{t_n}) - \Phi(X_{t_{n-1}})\right) \Delta W_n^l | X_{t_{n-1}} \right]$$

and hence, standard estimates on the increments of the forward diffusion together with the Lipschitz property of Φ, lead to

$$\mathbb{E}_{\mathbb{Q}^m_{t_{n-1}}}\left[\left| Y_{t_{n-1}} - \bar{R}_{n-1}\Phi(X_{t_{n-1}}) | X_{t_{n-1}} \right|^p \right]^{1/p} \leq \frac{c_3(\|\Phi\|_{Lip})}{n^{\beta/2}}$$

which concludes the proof for the case of smooth terminal conditions. Assume next that Φ is Lipschitz. Via a standard mollification result, one can construct a sequence of smooth functions $\{\Phi_m\}_{m \geq 0}$ that converge uniformly to Φ and such that $\|\Phi_m\|_{Lip} \leq \|\Phi\|_{Lip}$ for all $m \geq 0$. Using the continuity properties of both Y_0 and $\bar{R}_{0:n}$ as functions of the final condition, it follows that

$$\left| Y_0 - \bar{R}_{0:n-1}\Phi(x_0) \right| = \lim_{m\to\infty} \left| Y_0^m - \bar{R}_{0:n-1}\Phi^m(x_0) \right| \le \frac{c(\|\Phi\|_{Lip})}{n},$$

where Y_0^m is the solution of the BSDE corresponding to the final condition Φ_m. Crucially in the above inequality the function c is independent of m. The proof is complete. □

It remains to estimate the error $\bar{R}_{0:n-1}\Phi(x_0) - \bar{R}_{0:n-1}\Phi(x_0)$, i.e. the error due to the application of the TBBA. In this case one needs only to combine the arguments of the previous proof with the arguments that were presented in the proof of Theorem 53. Such analysis can be found in [15]. We report here on the this estimate.

Theorem 63. *Let assumptions (A) and (B) hold true and assume that Φ is Lipschitz continuous. Consider the family $\{\hat{R}_i\}_{0\le i\le n}$ defined with N particles. With the usual notation, on the iteration of operators, there exists a constant C independent of the partition, such that*

$$\tilde{\mathbb{E}}\left[\left| \bar{R}_{0:n}\Phi(x_0) - \hat{R}_{0:n}\Phi(x_0) \right|^2 \right]^{1/2} \le \frac{Cn}{\sqrt{N}}. \tag{142}$$

4.4 Numerical Simulations

In this section, we apply our numerical scheme for BSDEs in one and multidimensional problems where the involved coefficients can be smooth or non smooth. This empirical study helps us to validate the method described above.

One Dimensional Numerical Examples

Firstly, we consider the following popular non-linear example from finance, the problem of pricing with differential interest rates. In this set up, one is able to invest money in the money account at an interest rate r and borrow at an interest rate R with $R > r$. The underlying asset price evolves as a geometric Brownian motion under the objective probability measure:

$$X_t^{0,x_0} = \int_0^t \mu\, X_s^{0,x_0} ds + \int_0^t \sigma\, X_s^{0,x_0} dW_s.$$

It is shown in El Karoui et al. [17] that a self-financing trading strategy of portfolio Z and wealth process Y solves a BSDE with driver

$$f(t,x,y,z) = -\left(ry + z\theta - (R-r)(y - z/\sigma)_-\right)$$

where $(x)_-$ denotes the negative part of x and $\theta = (\mu - r)/\sigma$. The problem of pricing a call option corresponds to a terminal condition of the form $\Phi(x) = (x - K)_+$.

We test our algorithm with parameters

μ	r	R	σ	X_0	K
0.03	0.06	0.08	0.2	10	10

As explained in Gobet et al. [21], in such an economy the issuer of the call option keeps borrowing money to hedge the call option so that the price of the option is the Black–Scholes with interest rate R. Hence we have the favorable set up of a non linear driver, but yet we know Y_0. Moreover we see that, even though the driver is *not differentiable* our algorithm still produces very good estimates. In the figure below, we plot the ratio of the computed value over the Black Scholes price against the number of steps.

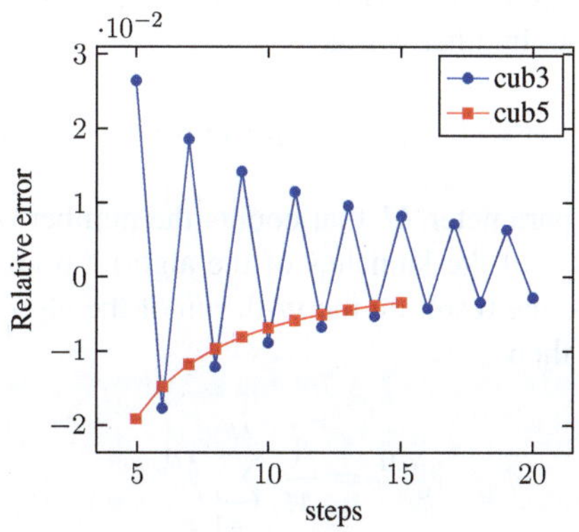

Since this is only a one dimensional set up, we manage to achieve an accuracy of 10^{-3} with only a few time discretization steps and hence the application of TBBA to control the computational effort is not necessary here.

Since pure cubature can be applied successfully in one dimensional examples, we can next try to monitor the effect that TBBA has on the overall error. We do so in a smooth example. We consider a FBSDE system with smooth coefficients and a non linear driver for the backward part:

$$X_t^{0,x_0} = x_0 + \int_0^t \mu X_s ds + \int_0^t \sqrt{1 + X_t^2} dW_t, \quad 0 \le t \le T$$

$$Y_t^{0,x_0} = \arctan(X_T^{0,x_0}) - \int_t^T rY_s + e^{r(T-s)} (\mu - 1) X_s^{0,x_0} \left(Z_s^{0,x_0}\right)^2 ds$$

$$- \int_t^T Z_s^{0,x_0} dW_s. \tag{143}$$

It is easy to check , by means of Itô's lemma, that the solution to the above system is given by

$$Y_t^{0,x_0} = e^{-r(T-t)} \arctan(X_t^{0,x_0}), \quad Z_t^{0,x_0} = \frac{e^{-r(T-t)}}{\sqrt{1+\left(X_t^{0,x_0}\right)^2}}.$$

We test our example with parameters

T	μ	r	x_0
1.	0.02	0.1	2

We denote by N the (maximal) number of paths that the support of the "pruned" cubature measure is allowed to hold, at every point on the partition. Let $Y_0 = e^{-rT}\arctan(X_0)$ denote the solution of (143) at time 0. We denote by $\hat{y}_0^N \equiv \hat{y}_0^N(\omega)$ the result we get at time 0 by solving the BSDE along the tree produced by one launch of the algorithm. In other words

$$\hat{y}_0^N = \hat{R}_{0:n-1}\Phi(x_0).$$

We also fix a further parameter M that counts the number of times the algorithm is launched. Obviously all the launches of the algorithm are independent of each other. Let $\hat{y}_0^{N,m}$ denote the result on the m-th run of the algorithm, $m = 1, \dots, M$. Our approximation is then

$$\hat{y}_0^{N,M} = \frac{1}{M}\sum_{m=1}^{M} \hat{y}_0^{N,m}.$$

The figure below, monitors the error (we plot $\frac{\hat{y}_0^{N,M} - \bar{y}_0}{\bar{y}_0}$) on example (143), when using cubature of order 3, 5 with and without sampling, against the number of steps. In this case the parameters N, M are fixed as $N = 100000$, $M = 10$.

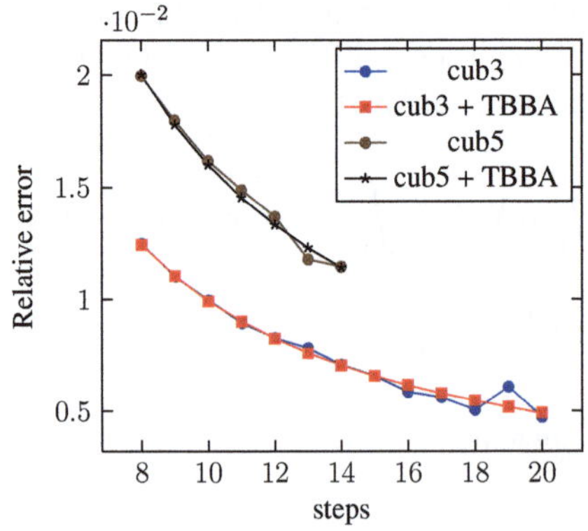

In particular we see that no accuracy is lost when applying the TBBA. Next we turn to a multidimensional example. The goal here is to show that the method produces good estimates but also to compare its performance with existing methods for solving BSDE. In a recent publication of Bouchard and Warin [4] the authors study the application of three other numerical methods (quantization, Malliavin integration by parts and regression on function basis) for BSDEs on the pricing of American/Bermudan options. In particular, we consider the case where the underlying is a Geometric Brownian motion and the payoff is a call or a put written on geometric/arithmetic averages. Here we shall consider the equivalent European pricing problem. In terms of computational complexity (on which the authors of [4] report), there is no significant difference. Indeed, the pricing of the Bermudan counterpart amounts to checking for optimal exercise on every point in the support of the underlying measure which would be negligible given the overall complexity of the algorithm.

We look at a five-dimensional example:

$$
\begin{aligned}
X_t^i &= x_0^i + \int_0^t \mu_i X_s^i ds + \sigma_i X_s^i dW_s^i, \quad i = 1, \ldots, 5 \\
Y_t &= \left(\prod_{i=1}^{5} X_1^i - K \right)_+ - \int_t^1 rY_s + \theta Z_s ds - \int_t^1 Z_s \cdot dW_s
\end{aligned}
\tag{144}
$$

where $\theta_i = (\mu_i - r)/\sigma_i$, $i = 1, \ldots, 5$ is the market price of risk. The theoretical value for Y_0 can be produced with the usual Black Scholes methodology. Again we fix the number of steps to 10 and we do a plot the usual relative error. Of course we normalize against the Black Scholes price.

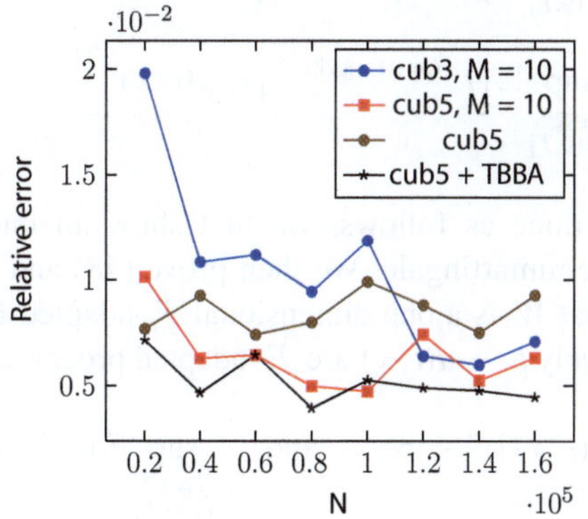

As far as the computational time is concerned, we report on the following values (the computational time is measured here in seconds)[12]:

N	40000	100000	160000
cub3(M=10)	3.8	8.6	13.2
cub5(M=10)	6.9	16.9	26.3
cub3(M=20)	7.4	17.4	26.5
cub5(M=20)	13.7	33.5	53

Comparing these performance results, in conjunction with the information on the errors, with Fig. 7(e), Fig. 8 of [4] we see that the cubature+TBBA algorithm can achieve similar accuracy in lesser time. On the other hand, we see that there is a small bias (relative error of order 0.5 %) that the algorithm does not treat with the increase in N. This bias is due to the discretization error (recall that we are normalizing against the theoretical Black Scholes value).

Appendix

In this section we provide various proofs of results left outstanding from the main body.

Proposition 20. *For any $T > 0$, $p \in [1, \infty)$, $\alpha, \beta \in \mathcal{A}(m)$ and $\gamma \in \mathcal{A}$, the following hold*

$$\sup_{t\in(0,T]} \mathbb{E}\left[t^{-\|\gamma\|/2} \left| \hat{B}_t^{\circ\gamma} \right|\right]^p < \infty, \tag{48}$$

$$\sup_{\substack{x\in\mathbb{R}^N \\ t\in(0,T]}} \mathbb{E}\left[t^{-(m+1-\|\alpha\|)/2} \left| r_{\alpha,\beta}(t,x) \right|\right]^p < \infty. \tag{49}$$

Proof. The proof is done as follows: we first show an intermediate result that holds for a general semimartingale. We then prove (48) and (49) via an inductive argument. Assume that W is a one dimensional $\mathcal{F}_t$-adapted Brownian motion and $t \to u(t,x)$, respectively $t \to u(t,x)$ are $\mathcal{F}_t$-adapted processes such that

$$\sup_{\substack{x\in\mathbb{R}^N \\ t\in[0,T]}} \mathbb{E}\left(t^{-r_u} \left| u(t,x) \right|\right)^p < \infty, \quad \text{and} \quad \sup_{\substack{x\in\mathbb{R}^N \\ t\in[0,T]}} \mathbb{E}\left(t^{-r_v} \left| v(t,x) \right|\right)^p < \infty,$$

[12]Numerical experiments were performed with single-threaded code on a Intel i7 processor at 2.8 GHz.

for some constants, $r_u, r_v \in [0, \infty)$. Next let ξ^x be the process defined as

$$\xi_t^x = \int_0^t u(s,x)dW_s + \int_0^t v(s,x)ds,$$

Then, for $p \geq 1$:

$$\begin{aligned}
\mathbb{E}\left[\mid \xi_t^x \mid^p \right] &= \mathbb{E}\left[\left| \int_0^t u(s,x)dW_s + \int_0^t v(s,x)ds \right|^p \right] \\
&\overset{(1)}{\leq} 2^{p-1}\left\{ \mathbb{E}\left[\left| \int_0^t u(s,x)dW_s \right|^p \right] + \mathbb{E}\left[\left| \int_0^t v(s,x)ds \right|^p \right] \right\} \\
&\overset{(2)}{\leq} 2^{p-1}\left\{ C_p \mathbb{E}\left[\left(\int_0^t \mid u(s,x) \mid^2 ds \right)^{\frac{p}{2}} \right] + t^{p-1}\mathbb{E}\left[\int_0^t \mid v(s,x) \mid^p ds \right] \right\} \\
&\overset{(3)}{\leq} 2^{p-1}\left\{ C_p \, t^{\frac{p}{2}-1}\mathbb{E}\left[\int_0^t \mid u(s,x) \mid^p ds \right] + t^{p-1}\mathbb{E}\left[\int_0^t \mid v(s,x) \mid^p ds \right] \right\} \\
&\overset{(4)}{\leq} 2^{p-1}\left\{ C_p t^{\frac{1}{2}(p-1)} \int_0^t \mathbb{E}\left[\mid u(s,x) \mid^p \right] ds + t^{p-1} \int_0^t \mathbb{E}\left[\mid v(s,x) \mid^p \right] ds \right\},
\end{aligned}$$

where we used the following: Hölder's inequality for finite sums for (1), Burkholder's inequality, Jensen's inequality respectively, for (2), Jensen's inequality for definite integrals for (3), Fubini's theorem for (4).

Now we observe that

$$\mathbb{E} \mid u(s,x) \mid^p \leq \left(\sup_{\substack{x \in \mathbb{R}^N \\ s \in [0,T]}} \mathbb{E}\left[s^{-r_u} \mid u(s,x) \mid \right]^p \right) s^{pr_u}$$

$$\mathbb{E} \mid v(s,x) \mid^p \leq \left(\sup_{\substack{x \in \mathbb{R}^N \\ s \in [0,T]}} \mathbb{E}\left[s^{-r_v} \mid v(s,x) \mid \right]^p \right) s^{pr_v}.$$

And so,

$$\begin{aligned}
\mathbb{E} \mid \xi_t^x \mid^p \leq \tilde{C}_p \Bigg\{ & t^{\frac{1}{2}p-1} \left(\sup_{\substack{x \in \mathbb{R}^N \\ s \in [0,T]}} \mathbb{E}\left[s^{-r_u} \mid u(s,x) \mid \right]^p \right) \left(\int_0^t s^{pr_u} ds \right) \\
& + t^{p-1} \left(\sup_{\substack{x \in \mathbb{R}^N \\ s \in [0,T]}} \mathbb{E}\left[s^{-r_v} \mid v(s,x) \mid \right]^p \right) \left(\int_0^t s^{pr_v} ds \right) \Bigg\}
\end{aligned}$$

$$\leq \hat{C}_p \Bigg\{ t^{\frac{1}{2}p-1} t^{pr_u+1} \Bigg(\sup_{\substack{x\in\mathbb{R}^N \\ s\in[0,T]}} \mathbb{E}\,[s^{-r_u} \mid u(s,x) \mid]^p \Bigg)$$

$$+ t^{p-1} t^{pr_v+1} \Bigg(\sup_{\substack{x\in\mathbb{R}^N \\ s\in[0,T]}} \mathbb{E}\big(s^{-r_v} \mid v(s,x) \mid\big)^p \Bigg) \Bigg\}$$

$$\leq \hat{C}_{p,u,v} \Big\{ t^{p(r_u+\frac{1}{2})} + t^{p(r_v+1)} \Big\}.$$

That is, if we take $r_\xi = \min\{r_u + \frac{1}{2}, r_v + 1\}$, then for all $p \in [1, \infty)$,

$$\sup_{\substack{x\in\mathbb{R}^N \\ t\in(0,T]}} \mathbb{E}\left[\left(t^{-r_\xi} \mid \xi_t^x \mid\right)^p\right] < \infty. \tag{145}$$

Proof of (48): We prove by induction on $\mid \gamma \mid$. Observe that for $\mid \gamma \mid = 1$, we have that:

$$\hat{B}_t^{\circ\gamma} = \begin{cases} B_t^\gamma & \text{if } \gamma \in \{1, \ldots, d\} \\ t & \text{if } \gamma = 0 \end{cases}, \tag{146}$$

in which case, we split $\mid \gamma \mid = 1$ into $\gamma \in \{1, \ldots, d\}$ and $\gamma = 0$. For the former we apply the inductive step with $u \equiv 1$ and $v \equiv 0$. Then we may choose $0 = r_u \ll r_v$ to obtain:

$$\sup_{t\in[0,T]} \mathbb{E}\left[\left(t^{-1/2} \mid B_t^\gamma \mid\right)^p\right] < \infty.$$

In the latter case we obviously have:

$$\sup_{t\in[0,T]} \mathbb{E}\left[t^{-1} \mid B_t^\gamma \mid\right]^p < \infty.$$

We now assume that the result holds for some $k \in \mathbb{N}$, i.e. we have the following for all $\gamma \in \mathcal{A}$ satisfying $\mid \gamma \mid = k$:

$$\sup_{t\in(0,T]} \mathbb{E}\left[t^{-\|\gamma\|/2} \left| \hat{B}_t^{\circ\gamma} \right|\right]^p. \tag{147}$$

Observe, that for $i \in \{1, \ldots, d\}$

$$\hat{B}_t^{\circ(\gamma*i)} = \int_0^t \hat{B}_s^{\circ\gamma} \circ dB_s^i \tag{148}$$

$$= \int_0^t \hat{B}_s^{\circ\gamma} dB_s^i + \frac{1}{2}\left\langle \hat{B}^{\circ\gamma}, B^i \right\rangle_t, \tag{149}$$

and noting that

$$\hat{B}_t^{\gamma} = \int_0^t \hat{B}_s^{\circ\gamma'} dB_s^{\gamma_k} + \frac{1}{2}\left\langle \hat{B}^{\circ\gamma'}, B^{\gamma_k}\right\rangle_t .$$

It is clear that

$$\hat{B}_t^{\circ(\gamma*i)} = \int_0^t \hat{B}_s^{\circ\gamma} dB_s^{i} + \frac{1}{2}\delta_{\gamma_k,i}\hat{B}^{\circ(\gamma'*0)}.$$

Now $|\gamma' * 0| = k$, so $\hat{B}^{\circ(\gamma'*0)}$ satisfies (147) with $\|\gamma' * 0\| = \|\gamma'\| + 2$. Moreover, we can control $\int_0^t \hat{B}_s^{\circ\gamma} dB_s^i$ by using the inductive step with $u(t,x) = \hat{B}_t^{\circ\gamma}$ and $v \equiv 0$, so that $\frac{1}{2}\|\gamma\| = r_u \ll r_v$, by the inductive hypothesis, and we have:

$$\sup_{t\in(0,T]} \mathbb{E}\left[t^{-r_{\gamma*i}}\left|\hat{B}_t^{\circ(\gamma*i)}\right|\right]^p < \infty,$$

where $r_{\gamma*i} = \min\{(\|\gamma\|+1)/2, (\|\gamma'\|+2)/2\} = \|\gamma * i\|/2$.

If $i = 0$, then we may apply the inductive step with $u \equiv 0$ and $v(t,x) = \hat{B}_t^{\circ\gamma}$, so that $\frac{1}{2}\|\gamma\| = r_v \ll r_u$, by the inductive hypothesis. In this case,

$$\sup_{t\in(0,T]} \mathbb{E}\left[t^{-r_{\gamma*i}}\left|\hat{B}_t^{\circ(\gamma*i)}\right|\right]^p < \infty,$$

with, again, $r_{\gamma*i} = \|\gamma * i\|/2$. Hence the result is proved.

Proof of (49): The proof of this result is similar to the induction carried out above. We notice that the remainder term, as defined, is the sum of numerous iterated Stratonovich integrals. We prove that the result holds for each element of the sum. This may then be easily extended to the sum of multiple such objects. We have already seen (cf. Proposition 42) that, for any $\alpha, \beta \in \mathcal{A}(m)$, $p \in [1,\infty)$, $T > 0$:

$$\sup_{\substack{x\in\mathbb{R}^N\\ t\in[0,T]}} \mathbb{E}\left|a_{\alpha,\beta}(t,x)\right|^p < \infty. \tag{150}$$

Moreover, since $c_{\alpha,\beta}^i \in C_b^{k+1-|\alpha|}(\mathbb{R}^N)$ is uniformly bounded, it follows that

$$\sup_{\substack{x\in\mathbb{R}^N\\ t\in[0,T]}} \mathbb{E}\left|c_{\alpha*\gamma,\beta}^j(X_t^x)\right|^p < \infty. \tag{151}$$

We again prove the result by induction on $|\gamma|$. Assume $|\gamma| = 1$. Using the fact that $c_{\alpha*\gamma,\delta}^j(X_t^x)$ and $a_{\delta,\beta}(t,x)$ satisfy (150) and (151) respectively, the product must satisfy an analogous inequality (by Hörmander's inequality). Note that this semimartingale will be comprised of integrands which are sums and products of objects like those in (150), (151), and hence if $\gamma \in \{1, \dots, d\}$ it has been demonstrated already that (cf. the first part of the proof, i.e. $r_u = r_v = 0$),

$$\mathbb{E}\left[t^{-r_\gamma} \int_0^t c^j_{\alpha*\gamma,\delta}(X^x_s) a_{\delta,\beta}(s,x) \circ dB^\gamma_t\right]^p < \infty, \tag{152}$$

where $r_\gamma = \min\{\frac{1}{2}, 1\} = \frac{1}{2}$. Now if $\gamma = 0$, then we apply the step with $u \equiv 0$ and $v(t,x) = c^j_{\alpha*\gamma,\delta}(X^x_t) a_{\delta,\beta}(t,x)$. That is, $0 = r_v \ll r_u$, to obtain

$$\mathbb{E}\left[t^{-r_\gamma} \int_0^t c^j_{\alpha*\gamma,\delta}(X^x_s) a_{\delta,\beta}(s,x) ds\right]^p < \infty, \tag{153}$$

where $r_\gamma = 1$. We now assume the result holds for some $k \in \mathbb{N}$. i.e. we have the following for all $\gamma \in \mathcal{A}$ satisfying $|\gamma| = k$:

$$\sup_{\substack{x \in \mathbb{R}^N \\ t \in (0,T]}} \mathbb{E}\left[t^{-\|\gamma\|/2} \left| \int_0^t \int_0^{s_k} \dots \int_0^{s_2} (-1)^{|\gamma|} c^j_{\alpha*\gamma,\delta}(X^x_{s_1}) a_{\delta,\beta}(s_1,x) \circ dB^{\gamma_1}_{s_1} \dots \circ dB^{\gamma_k}_{s_k} \right|\right]^p < \infty. \tag{154}$$

To ease the notational burden, we write,

$$Z(t,x,\gamma) := \int_0^t \int_0^{s_k} \dots \int_0^{s_2} (-1)^{|\gamma|} c^j_{\alpha*\gamma,\delta}(X^x_{s_1}) a_{\delta,\beta}(s_1,x) \circ dB^{\gamma_1}_{s_1} \dots \circ dB^{\gamma_k}_{s_k},$$

for $\gamma = (\gamma_1, \dots, \gamma_k)$. Observe, that for $i \in \{1, \dots, d\}$

$$\begin{aligned} Z(t,x,\gamma * i) &= \int_0^t Z(s,x,\gamma) \circ dB^i_s \\ &= \int_0^t Z(s,x,\gamma) dB^i_s + \frac{1}{2}\langle Z(.,x,\gamma), B^i\rangle_t \\ &= \int_0^t Z(s,x,\gamma) dB^i_s + \frac{1}{2}\delta_{\gamma_{k-1},\gamma_k} \int_0^t Z(s,x,\gamma') dt \\ &= \int_0^t Z(s,x,\gamma) dB^i_s + \frac{1}{2}\delta_{\gamma_{k-1},\gamma_k} Z(t,x,\gamma' * 0). \end{aligned}$$

By the inductive hypothesis, $Z(t,x,\gamma' * 0)$ satisfies (147) with $r_{\gamma'*0} = (\|\gamma'\| + 2)/2$, and we also use the inductive step on the right-hand term with $u(t,x) = Z(t,x,\gamma)$ and $v \equiv 0$, so that $r_v \gg r_u = \|\gamma\|/2$, with

$$\sup_{\substack{x \in \mathbb{R}^N \\ t \in (0,T]}} \mathbb{E}\left[t^{-r_{\gamma*i}} |Z(t,x,\gamma * i)|\right]^p < \infty,$$

where $r_{\gamma*i} = \min\left\{\frac{\|\gamma\|+1}{2}, \frac{\|\gamma'\|+2}{2}\right\} = \frac{\|\gamma\|+1}{2}$. If $i = 0$ then we may apply the inductive step with $u \equiv 0$ and $v(t,x) = Z(t,x,\gamma)$, so that with $\|\gamma\|/2 = r_v \ll r_u$ we get

$$\sup_{\substack{x\in\mathbb{R}^N \\ t\in[0,T]}} \mathbb{E}\,[t^{-r_{\gamma*0}} \mid Z(t,x,\gamma*0)\mid]^p < \infty,$$

where $r_{\gamma*0} = \frac{\|\gamma\|+2}{2}$. Hence the result is proved.

Finally, note that a finite sum of these would also satisfy a similar inequality with $r_{sum} = \min\{r_k; r_k$ is optimal (i.e. (145) holds) for k-th sum member$\}$. i.e.

$$\sup_{\substack{x\in\mathbb{R}^N \\ t\in(0,T]}} \mathbb{E}\left[t^{-(m+1-\|\alpha\|)/2} \mid r_{\alpha,\beta}(t,x)\mid\right]^p < \infty,$$

as required. □

A.1 *Invertibility of the Malliavin Covariance Matrix*

The aim of this section is to prove the following proposition from the main body. The proof is demanding, but fundamental to the results, and so it is given its own subsection.

Proposition 21. *$M(t,x)$ is $\mathbb{P}$-a.s. invertible. Moreover, for $p \in [1,\infty)$, $\alpha, \beta \in \mathcal{A}(m)$,*

$$\sup_{t\in(0,1],\, x\in\mathbb{R}^N} \mathbb{E}\left[M^{-1}_{\alpha,\beta}(t,x)\right]^p < \infty. \tag{155}$$

For real-symmetric matrices such as $M(t,x)$ there is an elegant representation of the minimal eigenvalue. The following lemma utilises this to simplify the requirements for invertibility.

Lemma 64. *The statement of the previous proposition holds, providing the following can be shown for each $p \in [1,\infty)$: there exists $C > 0$ s.t.*

$$\mathbb{P}\left(\inf_{|\xi|=1}(\xi, M(t,x)\xi) < \frac{1}{n}\right) < Cn^{-p},$$

for all $n \geq 1,\ t \in (0,1],\ and\ x \in \mathbb{R}^N$.

Proof. We sketch this proof. It is obvious from what has gone before that elements of the matrix M (rather than those of the inverse) satisfy (155). As the inverse matrix is comprised of the inverse of the determinant multiplied with multilinear combinations of elements of M, it suffices to show that the inverse of the determinant satisfies (155). The element $\inf_{|\xi|=1}(\xi, M(t,x)\xi)$ represents the smallest eigenvalue of M and hence its $-N$th power (where N is $\dim(M)$) provides an upper bound for the inverse of the determinant. Finally the expression in Lemma 64 may be used to deduce the L^p integrability (uniform over $t \in (0,1]$, $x \in \mathbb{R}$) of this upper bound, as it provides the required tail decay. □

In view of these results, consider $(\xi, M(t,x)\xi)$. The determinant of $M(t,x)$ is non-negative and increasing with t. This means that if $M(t,x)$ is a.s. invertible for some $t > 0$, then it must be invertible thereafter. Let $y \geq 1$.

$$\begin{aligned}
(\xi, M(t,x)\xi) &= \sum_{\alpha,\beta\in\mathcal{A}(m)} \xi_\alpha \xi_\beta M_{\alpha,\beta}(t,x) \\
&= \sum_{\alpha,\beta\in\mathcal{A}(m)} \xi_\alpha \xi_\beta t^{-(\frac{\|\alpha\|}{2}+\frac{\|\beta\|}{2})} \left\langle k_\alpha(t,x), k_\beta(t,x)\right\rangle_H \\
&= \left\| \sum_{\alpha\in\mathcal{A}(m)} \xi_\alpha t^{-\frac{\|\alpha\|}{2}} k_\alpha(t,x) \right\|_H^2 \\
&= \left\| \sum_{\alpha\in\mathcal{A}(m)} \xi_\alpha t^{-\frac{\|\alpha\|}{2}} \int_0^{t\wedge\cdot} (a^0_{\cdot,\alpha} + r_{\cdot,\alpha})(u,x)du \right\|_H^2 \\
&\geq \left\| \sum_{\alpha\in\mathcal{A}(m)} \xi_\alpha t^{-\frac{\|\alpha\|}{2}} \int_0^{t/y\wedge\cdot} (a^0_{\cdot,\alpha} + r_{\cdot,\alpha})(u,x)du \right\|_H^2 .
\end{aligned} \tag{156}$$

Observe that, since $y \geq 1$, using the notation: $S^n := \{x \in \mathbb{R}^{n+1} : \|x\| = 1\}$,

$$\begin{aligned}
&\inf_{\xi\in S^{N_m-1}} \left\| \sum_{\alpha\in\mathcal{A}(m)} \xi_\alpha t^{-\frac{\|\alpha\|}{2}} \int_0^{t/y\wedge\cdot} a_{\cdot,\alpha}(u,x)du \right\|_H^2 \\
&\geq \inf_{\xi\in S^{N_m-1}} \left\| \sum_{\alpha\in\mathcal{A}(m)} \xi_\alpha \left[\frac{t}{y}\right]^{-\frac{\|\alpha\|}{2}} \int_0^{t/y\wedge\cdot} a_{\cdot,\alpha}(u,x)du \right\|_H^2 .
\end{aligned}$$

Now focus on the term appearing on the RHS:

$$\begin{aligned}
&\left\| \sum_{\alpha \in \mathcal{A}(m)} \xi_\alpha \left[\frac{t}{y}\right]^{-\frac{\|\alpha\|}{2}} \int_0^{t/y \wedge .} a_{.,\alpha}(u,x)du \right\|_H^2 \\
&\geq \frac{1}{2} \left\| \sum_{\alpha \in \mathcal{A}(m)} \xi_\alpha \left[\frac{t}{y}\right]^{-\frac{\|\alpha\|}{2}} \int_0^{t/y \wedge .} a^0_{.,\alpha}(u,x)du \right\|_H^2 \\
&\quad - \left\| \sum_{\alpha \in \mathcal{A}(m)} \xi_\alpha \left[\frac{t}{y}\right]^{-\frac{\|\alpha\|}{2}} \int_0^{t/y \wedge .} r_{.,\alpha}(u,x)du \right\|_H^2 \\
&\geq \frac{1}{2} \left\| \sum_{\alpha \in \mathcal{A}(m)} \xi_\alpha \left[\frac{t}{y}\right]^{-\frac{\|\alpha\|}{2}} \int_0^{t/y \wedge .} a^0_{.,\alpha}(u,x)du \right\|_H^2 \\
&\quad - \left(\sum_{\alpha \in \mathcal{A}(m)} \xi_\alpha^2 \right) \left(\int_0^{t/y} \sum_{\alpha \in \mathcal{A}(m)} \sum_{i=1}^{d} \left[\frac{t}{y}\right]^{-\|\alpha\|} r_{i,\alpha}(u,x)^2 du \right).
\end{aligned}$$

Recall that $a^0_{i,\alpha}(u,x) = 0$ whenever $\alpha \neq i * \gamma$ for all multiindices γ. Moreover, $a^0_{i,\alpha}(u,x) = \hat{B}_u^{\circ\gamma}$ when $i * \gamma = \alpha$. That is, as each multindex $\alpha \in \mathcal{A}(m)$ satisfies $\alpha_1 \in \{1, \ldots, d\}$:

$$a^0_{.,\alpha}(u,x) = \left(0, \ldots, \hat{B}_u^{\circ\gamma}, \ldots, 0\right).$$

It is now necessary to briefly discuss the first term on the RHS. The following result is taken from Kusuoka and Stroock [33], but a comprehensive proof is provided in the next section.

Proposition 65. *Given $m \in \mathbb{N}$, there exist constants $C_m, \mu_m \in (0,\infty)$ such that for all $T > 0$*

$$\mathbb{P}\left(\inf_{a \in S^{N^{0,\emptyset}_{m-1}-1}} \int_0^T \left[\sum_{\gamma \in \mathcal{A}_{0,\emptyset}(m-1)} T^{-\frac{\|\gamma\|}{2}-\frac{1}{2}} a_\gamma \hat{B}_t^{\circ\gamma} \right]^2 dt \leq \frac{1}{n} \right) \leq C_m \exp\{-n^{\mu_m}\}. \tag{157}$$

Proof. The proof of this result requires a detour. For a detailed proof, consult the next section of the appendix. □

As a result of this strong bound, which is incidentally much stronger than that which is required for invertibility, it is very easy to deduce the following two equivalent properties:

Corollary 66. *For any $m \in \mathbb{N}$, and $p \in [1, \infty)$, there holds:*

$$\mathbb{E}\left(\inf_{a \in S^{N^{0,\emptyset}_{m-1}-1}} \int_0^T \left[\sum_{\gamma \in \mathcal{A}_{0,\emptyset}(m-1)} T^{-\frac{\|\gamma\|}{2}-\frac{1}{2}} a_\gamma \hat{B}_t^{\circ\gamma} \right]^2 dt \right)^{-p} < \infty. \tag{158}$$

And, equivalently, for all $q \in [1, \infty)$

$$\mathbb{P}\left(\inf_{a \in S^{N^{0,\emptyset}_{m-1}-1}} \int_0^t \left[\sum_{\gamma \in \mathcal{A}_{0,\emptyset}(m-1)} T^{-\frac{\|\gamma\|}{2}-\frac{1}{2}} a_\gamma \hat{B}_t^{\circ\gamma} \right]^2 dt \leq \frac{1}{n} \right) < C_{m,q} n^{-q}. \tag{159}$$

The usefulness of the above might not be immediately clear, so turn attention back to the lower bound obtained for $(\xi, M(t,x)\xi)$. The fact that any $\alpha \in \mathcal{A}(m)$ can be expressed as $\alpha = j * \gamma$ for some $1 \leq j \leq d$ and $\gamma \in \mathcal{A}_{0,\emptyset}(m-1)$ is used. This allows the effective utilisation of the structure of $a^0_{\cdot,\alpha}(t,x)$.

$$\begin{aligned}
&\left\| \sum_{\alpha \in \mathcal{A}(m)} \xi_\alpha \left[\frac{t}{y}\right]^{-\frac{\|\alpha\|}{2}} \int_0^{t/y\wedge\cdot} a^0_{\cdot,\alpha}(u,x) du \right\|_H^2 \\
&= \left\| \sum_{j=1}^d \sum_{\gamma \in \mathcal{A}_{0,\emptyset}(m-1)} \xi_{j*\gamma} \left[\frac{t}{y}\right]^{-\frac{\|\gamma\|+1}{2}} \int_0^{t/y\wedge\cdot} a^0_{\cdot,j*\gamma}(u,x) du \right\|_H^2 \\
&= \left\| \sum_{\gamma \in \mathcal{A}_{0,\emptyset}(m-1)} \left[\frac{t}{y}\right]^{-\frac{\|\gamma\|+1}{2}} \int_0^{t/y\wedge\cdot} (\xi_{j*\gamma} \hat{B}_u^{\circ\gamma})_{j=1,\ldots,d}\, du \right\|_H^2 \\
&= \left\| \int_0^{t/y\wedge\cdot} \left(\sum_{\gamma \in \mathcal{A}_{0,\emptyset}(m-1)} \left[\frac{t}{y}\right]^{-\frac{\|\gamma\|+1}{2}} \xi_{j*\gamma} \hat{B}_u^{\circ\gamma} \right)_{j=1,\ldots,d} du \right\|_H^2 \\
&= \sum_{j=1}^d \int_0^{t/y\wedge\cdot} \left[\sum_{\gamma \in \mathcal{A}_{0,\emptyset}(m-1)} \left[\frac{t}{y}\right]^{-\frac{\|\gamma\|+1}{2}} \xi_{j*\gamma} \hat{B}_u^{\circ\gamma} \right]^2 du.
\end{aligned}$$

It can also easily be shown that by taking $\inf_{\xi \in S_{N_{m-1}}}$ of both sides:

$$\inf_{\xi \in S_{N_{m-1}}} \left\| \sum_{\alpha \in \mathcal{A}(m)} \xi_\alpha \left[\frac{t}{y}\right]^{-\frac{\|\alpha\|}{2}} \int_0^{t/y \wedge .} a^0_{.,\alpha}(u,x)du \right\|_H^2$$

$$= \inf_{\xi \in S_{N_{m-1}}} \sum_{j=1}^{d} \int_0^{t/y \wedge .} \left[\sum_{\gamma \in \mathcal{A}_{0,\emptyset}(m-1)} \left[\frac{t}{y}\right]^{-\frac{\|\gamma\|+1}{2}} \xi_{j*\gamma} \hat{B}^{\circ\gamma}_{s,u} \right]^2 du$$

$$= \inf_{a \in S_{N^{0,\emptyset}_{m-1}-1}} \int_0^{t/y \wedge .} \left[\sum_{\gamma \in \mathcal{A}_{0,\emptyset}(m-1)} \left[\frac{t}{y}\right]^{-\frac{\|\gamma\|+1}{2}} a_\gamma \hat{B}^{\circ\gamma}_u \right]^2 du,$$

recalling that $N^{0,\emptyset}_{m-1} = N^{0,\emptyset}_{m-1} + 2$ This is precisely why the upper bound derived in Proposition 65 was introduced. It enables a precise control over the tail behaviour of $(\xi, M(t,x)\xi)$. The various pieces of analysis are now synthesised. In what follows, note that

$$\mathbb{P}\left(\frac{1}{2}X - Y \leq \frac{1}{n}\right) = \mathbb{P}\left(\frac{1}{2}X - Y \leq \frac{1}{n}, Y < \frac{1}{n}\right) + \mathbb{P}\left(\frac{1}{2}X - Y \leq \frac{1}{n}, Y \geq \frac{1}{n}\right)$$
$$\leq \mathbb{P}\left(Y \geq \frac{1}{n}\right) + \mathbb{P}\left(X \leq \frac{4}{n}\right).$$

This gives:

$$\mathbb{P}\left(\inf_{\xi \in S^{N_{m-1}}} (\xi, M(t,x)\xi) < \frac{1}{n}\right)$$

$$\leq \mathbb{P}\left(\inf_{a \in S_{N^{0,\emptyset}_{m-1}+1-1}} \int_0^{t/y} \left[\sum_{\gamma \in \mathcal{A}_{0,\emptyset}(m-1)} \left[\frac{t}{y}\right]^{-\frac{\|\gamma\|+1}{2}} a_\gamma \hat{B}^{\circ\gamma}_u \right]^2 du < \frac{4}{n} \right)$$

$$+ \mathbb{P}\left(\inf_{\xi \in S^{N_{m-1}}} \left[\sum_{\alpha \in \mathcal{A}(m)} \xi^2_\alpha \right] \left[\int_0^{t/y} \sum_{\alpha \in \mathcal{A}(m)} \sum_{i=1}^{d} \left[\frac{t}{y}\right]^{-\|\alpha\|} r_{i,\alpha}(u,x)^2 du \right] \geq \frac{1}{n} \right)$$

$$= \mathbb{P}\left(\inf_{a \in S_{N^{0,\emptyset}_{m-1}+1-1}} \int_0^{t/y} \left[\sum_{\gamma \in \mathcal{A}_{0,\emptyset}(m-1)} \left[\frac{t}{y}\right]^{-\frac{\|\gamma\|+1}{2}} a_\gamma \hat{B}^{\circ\gamma}_u \right]^2 du < \frac{4}{n} \right)$$

$$+ \mathbb{P}\left(\int_0^{t/y} \sum_{\alpha \in \mathcal{A}(m)} \sum_{i=1}^{d} \left[\frac{t}{y}\right]^{-\|\alpha\|} r_{i,\alpha}(u,x)^2 du \geq \frac{1}{n} \right).$$

The program is almost complete. The following is deduced from Proposition 20,

Lemma 67. *There holds, for all $p \in [1, \infty)$,*

$$\sup_{\substack{x \in \mathbb{R}^N \\ t \in (0,1]}} \mathbb{E}\left(\int_0^t \sum_{\alpha \in \mathcal{A}(m)} \sum_{i=1}^d t^{-\|\alpha\|-1} r_{i,\alpha}(u,x)^2 du\right)^p < \infty.$$

Proof. We may apply the semimartingale rate bound obtained in the proof of Proposition 20. Indeed, we observe that:

$$\xi_t := \int_0^t u(s,x) dB_s + \int_0^t v(s,x) ds,$$

$$u(s,x) \equiv 0,$$

$$v(s,x) = \sum_{\alpha \in \mathcal{A}(m)} \sum_{i=1}^d t^{-\|\alpha\|} r_{i,\alpha}(s,x)^2.$$

Observe from Proposition 49, noting $\|\alpha\| \le m$,

$$\sup_{\substack{x \in \mathbb{R}^N \\ t \in (0,T]}} \mathbb{E}\left(t^{-r_u} \mid u(t,x) \mid\right)^p < \infty, \qquad \sup_{\substack{x \in \mathbb{R}^N \\ t \in (0,T]}} \mathbb{E}\left(t^{-r_v} \mid v(t,x) \mid\right)^p < \infty,$$

where $r_v = 0$ and r_u is arbitrarily large. Hence it follows that:

$$\sup_{\substack{x \in \mathbb{R}^N \\ t \in (0,T]}} \mathbb{E}\left(t^{-r} \mid \xi_t^x \mid\right)^p < \infty,$$

where $r_\xi = r_v + 1 = 1$, as required. □

The proof can now be completed.

$$\mathbb{P}\left(\inf_{a \in S^{N^{0,\emptyset}_{m-1}-1}} \left\{\int_0^{t/y} \left[\sum_{\gamma \in \mathcal{A}_{0,\emptyset}(m-1)} \left[\frac{t}{y}\right]^{-\frac{\|\gamma\|+1}{2}} a_\gamma \hat{B}_u^{\circ\gamma}\right]^2 du\right\} < \frac{4}{n}\right)$$
$$+ \mathbb{P}\left(\int_0^{t/y} \sum_{\alpha \in \mathcal{A}(m)} \sum_{i=1}^d \left[\frac{t}{y}\right]^{-\|\alpha\|} r_{i,\alpha}(u,x)^2 du \ge \frac{1}{n}\right)$$
$$= \mathbb{P}\left(\inf_{a \in S_{N^{0,\emptyset}_{m-1}+1-1}} \left\{\int_0^{t/y} \left[\sum_{\gamma \in \mathcal{A}_{0,\emptyset}(m-1)} \left[\frac{t}{y}\right]^{-\frac{\|\gamma\|+1}{2}} a_\gamma \hat{B}_u^{\circ\gamma}\right]^2 du\right\} < \frac{4}{n}\right)$$

$$+\mathbb{P}\left(\int_0^{t/y}\sum_{\alpha\in\mathcal{A}(m)}\sum_{i=1}^{d}\left[\frac{t}{y}\right]^{-\|\alpha\|-1} r_{i,\alpha}(u,x)^2du \geq \frac{y}{nt}\right)$$
$$< C_{m,q}\left(\frac{4}{n}\right)^q + \tilde{C}_{m,q}\left(\frac{nt}{y}\right)^q \leq C_{m,q}\left(\frac{4}{n}\right)^q + \tilde{C}_{m,q}\left(\frac{n}{y}\right)^q.$$

It is important to note that the above bounds hold $\forall t \in (0,1]$ and $\forall x \in \mathbb{R}^N$. The decision to introduce $y \geq 1$ should become clear. Without it, the analysis would fail. Indeed, there is a clever choice of y such that Lemma 64 holds. Set

$$y = \frac{n^2}{4},$$

so that

$$\frac{n}{y} = \frac{4}{n}.$$

And finally, combining this with the above we obtain:

$$\mathbb{P}\left(\inf_{\xi\in S^{N_m-1}}(\xi, M(t,x)\xi) < \frac{1}{n}\right) < \tilde{\tilde{C}}_{m,p}\frac{1}{n^q},$$

as required.

In the next section regularity results about the inverse of the matrix are proved. These results shall be fundamental to the integration by parts formula.

A.2 Diffuseness of Iterated Stratonovich Integrals

It was seen in the last section that invertibility of the Malliavin covariance matrix can be achieved if Proposition 65 holds. Its statement is recalled and it is sought to prove this result using the work of Kusuoka/Stroock in [33] as a guide.

Proposition 68. *For any $m \in \mathbb{N}$, and $p \in [1,\infty)$, there holds:*

$$\mathbb{E}\left(\inf_{a\in S^{N^{0,\emptyset}_{m-1}-1}}\int_0^T\left[\sum_{\gamma\in\mathcal{A}_{0,\emptyset}(m-1)} T^{-\frac{\|\gamma\|}{2}-\frac{1}{2}}a_\gamma\hat{B}_t^{\circ\gamma}\right]^2dt\right)^{-p} = C_{m,p} < \infty. \quad (160)$$

And, equivalently, for all $q \in [1,\infty)$

$$\mathbb{P}\left(\inf_{a\in S^{N^{0,\emptyset}_{m-1}-1}}\int_0^T\left[\sum_{\gamma\in\mathcal{A}_{0,\emptyset}(m-1)} T^{-\frac{\|\gamma\|}{2}-\frac{1}{2}}a_\gamma\hat{B}_t^{\circ\gamma}\right]^2dt \leq \frac{1}{n}\right) < C_{m,q}n^{-q}. \quad (161)$$

Proof. The proof of this important result is begun through simplification of the problem. By considering the distribution of the iterated Stratonovich integrals one is able to make a change of variable to the integral. Indeed, note that:

$$\hat{B}^{\circ\gamma}_{st} \stackrel{\mathcal{D}}{=} s^{\frac{\|\gamma\|}{2}} \hat{B}^{\circ\gamma}_{t},$$

Hence it may be deduced:

$$\begin{aligned}
&\int_0^T \left[\sum_{\gamma\in\mathcal{A}_{0,\emptyset}(m-1)} T^{-\frac{\|\gamma\|}{2}-\frac{1}{2}} a_\gamma \hat{B}^{\circ\gamma}_t \right]^2 dt \\
&\qquad \stackrel{\mathcal{D}}{=} \int_0^T \left[\sum_{\gamma\in\mathcal{A}_{0,\emptyset}(m-1)} T^{-\frac{1}{2}} a_\gamma \hat{B}^{\circ\gamma}_{\frac{t}{T}} \right]^2 dt \\
&\qquad \stackrel{u=t/T}{=} \int_0^1 \left[\sum_{\gamma\in\mathcal{A}_{0,\emptyset}(m-1)} a_\gamma \hat{B}^{\circ\gamma}_u \right]^2 du.
\end{aligned}$$

Hence, the problem is reduced to showing that for each $p \geq 1$, there exists $C > 0$ s.t.

$$\mathbb{P}\left(\inf_{a\in S^{N^{0,\emptyset}_{m-1}-1}} \int_0^1 \left[\sum_{\gamma\in\mathcal{A}_{0,\emptyset}(m-1)} a_\gamma \hat{B}^{\circ\gamma}_u \right]^2 du < \frac{1}{n} \right) < Cn^{-p}, \tag{162}$$

for all $n \geq 1$.

Iterated Stratonovich integrals arise in a very natural way from the geometry of this problem. That said, one must often turn to the more established results in stochastic integration to do an accurate analysis of them. These results are almost always phrased in terms of Itô integration and the semimartingales resulting therefrom. Hence, attention is switched to iterated Itô integrals via the following proposition. The moral of the story is that, although undoubtedly different objects, iterated Itô and Stratonovich integrals are equally as diffuse.

Proposition 69. *Define* $\hat{B}^{\circ L}_t := (\hat{B}^{\circ\alpha}_t)_{\|\alpha\|\leq L}$ *and* $\hat{B}^{L}_t := (\hat{B}^{\alpha}_t)_{\|\alpha\|\leq L}$. *Then, for all* $L \in \mathbb{N}$ *there exist constant matrices* $A_L, \tilde{A}_L \in \mathbb{R}^{N_L\times N_L}$ *such that*

$$\textit{(i):}\quad \hat{B}^{\circ L}_t = A_L \hat{B}^{L}_t \qquad \textit{and} \qquad \textit{(ii):}\quad \hat{B}^{L}_t = \tilde{A}_L \hat{B}^{\circ L}_t.$$

i.e. A_L *is invertible with* $A_L^{-1} = \tilde{A}_L$.
Moreover, it follows that the existence of constants $C_m, \mu_m \in (0,\infty)$

$$\mathbb{P}\left(\inf_{\sum a_\gamma^2=1} \int_0^1 \left[\sum_{\gamma\in\mathcal{A}_{0,\emptyset}(m-1)} a_\gamma \hat{B}^{\circ\gamma}_t \right]^2 dt \leq \frac{1}{n} \right) \leq C_m \exp\{-n^{\mu_m}\},$$

is equivalent to the existence of constants $\tilde{C}_m, \tilde{\mu}_m \in (0, \infty)$ *such that*

$$\mathbb{P}\left(\inf_{\sum a_\gamma^2 = 1} \int_0^1 \left[\sum_{\gamma \in \mathcal{A}_{0,\emptyset}(m-1)} a_\gamma \hat{B}_t^\gamma \right]^2 dt \leq \frac{1}{n} \right) \leq \tilde{C}_m \exp\{-n^{\tilde{\mu}_m}\}.$$

Proof of (i) (adapted from the proof of Lemma A.12 in Kusuoka and Stroock [33]). (i) is approached by using an induction argument on L. Clearly if $L = 1$ then there is little to prove as $\hat{B}_t^{\circ L} = \hat{B}_t^L$. Hence, as $A_L = I_{d \times d} = \tilde{A}_L$. Now assume that the result holds for $L \leq k$. i.e. for all α such that $\|\alpha\| \leq k$ there holds, for some deterministic constants: $a_{\alpha,\beta}^k$, $\|\beta\| \leq k$.

$$\hat{B}_t^{\circ\alpha} = \sum_{\|\beta\| \leq k} a_{\alpha,\beta}^k \hat{B}_t^\beta.$$

It is clear one need only prove, for suitable constants $a_{\alpha,\beta}^{k+1}$, $\|\beta\| \leq k+1$ for $\|\alpha\| = k+1$

$$\hat{B}_t^{\circ\alpha} = \sum_{\|\beta\| \leq k+1} a_{\alpha,\beta}^{k+1} \hat{B}_t^\beta.$$

Let $\alpha = (\alpha', \alpha^*)$ where $\|\alpha'\| = k-1$ if $\alpha^* = 0$, and $\|\alpha'\| = k$ if $\alpha^* \in \{1, \ldots, d\}$. The cases $\alpha* = 0$ and $\alpha* \in \{1, \ldots, d\}$ are treated separately. Assume first that $\alpha* = 0$. Then

$$\begin{aligned}
\hat{B}_t^{\circ\alpha} = \int_0^t \hat{B}_s^{\circ\alpha'} ds &= \int_0^t \sum_{\|\beta\| \leq k} a_{\alpha',\beta}^k \hat{B}_s^{\circ\beta} ds \\
&= \sum_{\|\beta\| \leq k} a_{\alpha',\beta}^k \hat{B}_t^{\beta*0} \\
&= \sum_{\substack{\|\beta\| \leq k+1 \\ \beta* = 0}} a_{\alpha',\beta'}^k \hat{B}_t^{\beta*0} \\
&= \sum_{\|\beta\| \leq k+1} a_{\alpha,\beta}^{k+1} \hat{B}_t^\beta,
\end{aligned}$$

where $a_{\alpha,\beta}^{k+1} = \begin{cases} a_{\alpha',\beta'}^k & \text{if } \beta* = 0 \\ 0 & \text{if } \beta* \neq 0. \end{cases}$

Now assume $\alpha * \in \{1, \ldots, d\}$:

$$\begin{aligned}\hat{B}_t^{\circ\alpha} = \int_0^t \hat{B}_s^{\circ\alpha'} \circ dB_s^{\alpha *} &= \int_0^t \hat{B}_s^{\circ\alpha'} dB_s^{\alpha *} + \frac{1}{2}\int_0^t \hat{B}_s^{\circ\alpha''} ds\, 1_{\{\alpha * = (\alpha')*\}} \\ &= \int_0^t \sum_{\|\beta\| \le k} a_{\alpha',\beta}^k \hat{B}_s^\beta dB_s^{\alpha *} \\ &\quad + \frac{1}{2}\int_0^t \sum_{\|\beta\| \le k} a_{\alpha'',\beta}^k \hat{B}_s^\beta ds\, 1_{\{\alpha * = (\alpha')*\}} \\ &= \sum_{\substack{\|\beta\| \le k+1 \\ \beta * = \alpha *}} a_{\alpha',\beta'}^k \hat{B}_t^\beta + \frac{1_{\{\alpha * = (\alpha')*\}}}{2} \sum_{\substack{\|\beta\| \le k+1 \\ \beta * = 0}} a_{\alpha'',\beta'}^k \hat{B}_t^\beta \\ &= \sum_{\|\beta\| \le k+1} a_{\alpha,\beta}^{k+1} \hat{B}_t^\beta,\end{aligned}$$

where $a_{\alpha,\beta}^{k+1} = \begin{cases} a_{\alpha',\beta'}^k & \text{if } \alpha * = \beta *, \\ \frac{1}{2} a_{\alpha'',\beta'}^{k-1} & \text{if } \beta * = 0, \alpha * = (\alpha')*, \\ 0 & \text{otherwise.} \end{cases}$

This completes the argument. As (ii) can be proved in an analogous manner, its proof is omitted. It is now shown how $(i), (ii)$ imply the remaining equivalence result. Note that if A_L is invertible, then A_L^T is also invertible with $(A_L^T)^{-1} = (A_L^{-1})^T$. Moreover, from invertibility

$$0 < c_{\min} := \min_{|\xi|=1} \left| A_L^T \xi \right|.$$

Adopting the shorthand notation $\hat{B}_t^{\circ L}$, $\hat{B}_t^L$ employed above, there holds:

$$\begin{aligned}\inf_{|\xi|=1} \int_0^1 \left(\xi, \hat{B}_t^{\circ L}\right)^2 dt &= \inf_{|\xi|=1} \int_0^1 \left(\xi, A_L \hat{B}_t^L\right)^2 dt \\ &= \inf_{|\xi|=1} \int_0^1 \left(A_L^T \xi, \hat{B}_t^L\right)^2 dt \\ &\ge \inf_{|\nu|=1} \int_0^1 \left(\nu, \hat{B}_t^L\right)^2 dt\; c_{\min}^2.\end{aligned}$$

A similar estimate can be made from (ii). These estimates prove the remaining claim of the proposition. □

Before tackling Proposition 65 in earnest, some supplementary results about iterated Itô integrals are required.

Lemma 70. *Fix $l \in \mathbb{N}$. There exists $C_l < \infty$ and $\nu_l > 0$ such that for all $\alpha \in \mathcal{A}$ with $\|\alpha\| = l$, there holds:*

$$\mathbb{P}\left(\sup_{t\in(0,1]} \left| \hat{B}_t^{\alpha} \right| \geq n \right) \leq C_l \exp\left(-\frac{1}{2} n^{\nu_l} \right), \tag{163}$$

for all $n \geq 1$.

Proof (adapted from the proof of Lemma A.7 in Kusuoka and Stroock [33]). Fundamental use of the following martingale inequality is made. For $K_1, K_2 \geq 0$

$$\mathbb{P}\left(\sup_{t\in(0,T]} \mid M_T \mid \geq K_1,\ \langle M \rangle_T \leq K_2 \right) \leq 2\exp\left\{ -\frac{K_1^2}{2K_2} \right\}.$$

This result is proved by expressing the above martingale as time-changed Brownian motion (run at the "speed" of its quadratic variation, see Karatzas and Shreve [26, Theorem 3.4.6]), and then using the following two inequalities:

$$\mathbb{P}(\sup_{t\in(0,T]} \mid B_t \mid \geq K) \leq 2\mathbb{P}(B_T \geq K),$$

$$\int_x^{\infty} \frac{1}{\sqrt{2\pi}} e^{-u^2/2} du \leq e^{-x^2/2}, \qquad x \geq 0.$$

The latter is seen by splitting consideration into two cases: $x \in [0,1)$ and $x \geq 1$.

The relation in question can be obtained by iterative applications of this martingale inequality. Define $\nu_N \equiv 2$, and in what follows allow ν_i to be chosen optimally afterwards. First assume that $\alpha \in \{1,\ldots,d\}^N$.

$$\begin{aligned}
&\mathbb{P}\Big[\sup_{t\in(0,1]} \left| \hat{B}_t^{\alpha} \right| \geq K \Big] \\
&\leq \mathbb{P}\Big[\sup_{t\in(0,1]} \left| \hat{B}_t^{\alpha} \right| \geq K,\ \langle \hat{B}^{\alpha'} \rangle_1 < K^{\nu_N} \Big] + \mathbb{P}\Big[\langle \hat{B}^{\alpha} \rangle_1 \geq K^{\nu_N} \Big] \\
&= \mathbb{P}\Big[\sup_{t\in(0,1]} \left| \hat{B}_t^{\alpha} \right| \geq K,\ \langle \hat{B}^{\alpha'} \rangle_1 < K^{\nu_N} \Big] + \mathbb{P}\Big[\int_0^1 \left| \hat{B}_t^{\alpha'} \right|^2 dt \geq K^{\nu_N} \Big] \\
&\leq \mathbb{P}\Big[\sup_{t\in(0,1]} \left| \hat{B}_t^{\alpha} \right| \geq K,\ \langle \hat{B}^{\alpha'} \rangle_1 < K^{\nu_N} \Big] + \mathbb{P}\Big[\sup_{t\in(0,1]} \left| \hat{B}_t^{\alpha'} \right| \geq K^{\nu_N/2} \Big] \\
&\leq \sum_{i=1}^{N} \mathbb{P}\Big[\sup_{t\in(0,1]} \left| \hat{B}_t^{\alpha(N+1-i)} \right| \geq K^{\nu_i/2},\ \langle \hat{B}^{\alpha(i-1)} \rangle_1 < K^{\nu_{i-1}} \Big] \\
&\leq \sum_{i=1}^{N} 2\exp\left(-\frac{1}{2} K^{\nu_i - \nu_{i-1}} \right),
\end{aligned}$$

where $\alpha^{(i)}$ denotes that shortening of the multi-index $\alpha = (\alpha_1, \ldots, \alpha_N)$ by i. i.e. $\alpha^{(i)} = (\alpha_1, \ldots, \alpha_{N-i})$ (additionally: $\alpha^{(0)} = \alpha$).

Now choose ν_i for $i = 1, \ldots, N$ given that $\nu_N = 2$ and $\nu_0 \geq 0$. In fact, ν_0 can be chosen arbitrarily for $K \geq 1$. If it is assumed that $\nu_i - \nu_{i-1} \equiv \delta > 0$ for $i = 1, \ldots, N$, then

$$\sum_{i=1}^{N} \nu_i - \nu_{i-1} = N\delta \quad \Rightarrow \quad \delta = \frac{2}{N},$$

and $\nu_i = \frac{2i}{N}$. Assembling these facts gives:

$$\mathbb{P}\Big(\sup_{t\in(0,1]} \Big| \hat{B}_t^{\alpha} \Big| \geq K \Big) \leq 2N \exp\Big(-\frac{1}{2} K^{\frac{2}{N}} \Big), \tag{164}$$

for arbitrary $|\,\alpha\,| = N$. Assuming instead that $\|\alpha\| = l$ and noting that $|\,\alpha\,| \leq \|\alpha\|$ so that $\frac{l}{2} \leq |\,\alpha\,| \leq l$ gives the same upper bound with N replaced by l. i.e. $C_l = 2l$ and $\nu_l = 2/l$.

Now observe that if $\alpha_i = 0$ for some $i = 1, \ldots, N$ the situation is even simpler:

$$\mathbb{P}(\sup_{t\in(0,1]} \Big| \hat{B}_t^{(\alpha_1,\ldots,\alpha_i)} \Big| \geq K) \leq \mathbb{P}(\sup_{t\in(0,1]} \Big| \hat{B}_t^{(\alpha_1,\ldots,\alpha_i-1)} \Big| \geq K),$$

as $\sup_{t\in(0,T]} \Big| \int_0^t \hat{B}_s^{\alpha} ds \Big| \leq T \sup_{t\in(0,T]} \Big| \hat{B}_t^{\alpha} \Big|$. Therefore, one needs only apply the martingale inequality Card $\{i : \alpha_i \neq 0\}$ times. i.e. $(2\,|\,\alpha\,| - \|\alpha\|)$ times. Hence, for a general α,

$$\mathbb{P}\Big(\sup_{t\in(0,1]} \Big| \hat{B}_t^{\alpha} \Big| \geq K \Big) \leq 2(2\,|\,\alpha\,| - \|\alpha\|) \exp\Big(-\frac{1}{2} K^{\frac{2}{2|\alpha|-\|\alpha\|}} \Big).$$

However, for any α such that $\|\alpha\| = l$ the identified constants in (164) are still appropriate, as $\sup_{\|\alpha\|=l}(2\,|\,\alpha\,| - \|\alpha\|) = l$. □

The main consequence of the above lemma is the following:

Proposition 71. *It suffices to show the existence of C_m, μ_m such that for all $n \geq 1$, there holds*

$$\sup_{a\in S^{N_{m-1}^{0,\emptyset}-1}} \mathbb{P}\Bigg(\int_0^1 \Bigg[\sum_{\alpha\in\mathcal{A}_{0,\emptyset}(m-1)} a_\alpha \hat{B}_t^{\alpha} \Bigg]^2 dt \leq \frac{1}{n} \Bigg) \leq C_m \exp\{-n^{\mu_m}\}. \tag{165}$$

Adapted from the proof of Lemma 2.3.1 in Nualart [51]. There is some constant M_m such that for all $n \geq 1$, $S^{N_{m-1}^{0,\emptyset}-1}$ contains some finite set $\Sigma(n)$ with

$$|\,\Sigma(n)\,| \le M_m n^{2N_m} \qquad \text{and} \qquad S^{N_{m-1}^{0,\emptyset}-1} \subset \bigcup_{c\in\Sigma(n)} B_{1/\sqrt{2}n}.$$

Observe, for fixed $a^c \in S^{N_{m-1}^{0,\emptyset}-1} \cap B_{1/\sqrt{2}n}(c)$, there holds

$$\min_{c\in\Sigma(n)} \int_0^1 \left[\sum_{\alpha\in\mathcal{A}_{0,\emptyset}(m-1)} c_\alpha \hat{B}_t^\alpha\right]^2 dt$$

$$= \min_{c\in\Sigma(n)} \int_0^1 \left[\sum_{\alpha\in\mathcal{A}_{0,\emptyset}(m-1)} (c_\alpha - a_\alpha^c + a_\alpha^c)\hat{B}_t^\alpha\right]^2 dt$$

$$\le 2 \min_{c\in\Sigma(n)} \left(\int_0^1 \left[\sum_{\alpha\in\mathcal{A}_{0,\emptyset}(m-1)} (c_\alpha - a_\alpha^c)\hat{B}_t^\alpha\right]^2 dt + \int_0^1 \left[\sum_{\alpha\in\mathcal{A}_{0,\emptyset}(m-1)} a_\alpha^c \hat{B}_t^\alpha\right]^2 dt\right)$$

$$\le 2 \min_{c\in\Sigma(n)} \int_0^1 \left[\sum_{\alpha\in\mathcal{A}_{0,\emptyset}(m-1)} (c_\alpha - a_\alpha^c)\hat{B}_t^\alpha\right]^2 dt + 2 \min_{c\in\Sigma(n)} \int_0^1 \left[\sum_{\alpha\in\mathcal{A}_{0,\emptyset}(m-1)} a_\alpha^c \hat{B}_t^\alpha\right]^2 dt$$

$$\le 2 \min_{c\in\Sigma(n)} |c-a^c|^2 \int_0^1 \sum_{\alpha\in\mathcal{A}_{0,\emptyset}(m-1)} \left|\hat{B}_t^\alpha\right|^2 dt + 2 \min_{c\in\Sigma(n)} \int_0^1 \left[\sum_{\alpha\in\mathcal{A}_{0,\emptyset}(m-1)} a_\alpha^c \hat{B}_t^\alpha\right]^2 dt$$

$$\le 2\frac{1}{2n^2} \int_0^1 \sum_{\alpha\in\mathcal{A}_{0,\emptyset}(m-1)} \left|\hat{B}_t^\alpha\right|^2 dt + 2 \min_{c\in\Sigma(n)} \int_0^1 \left[\sum_{\alpha\in\mathcal{A}_{0,\emptyset}(m-1)} a_\alpha^c \hat{B}_t^\alpha\right]^2 dt.$$

Now, the above upper bound holds for any $a^c \in S^{N_{m-1}^{0,\emptyset}-1} \cap B_{1/\sqrt{2}n}(c)$, in particular, it must hold upon taking the infimum over all $a \in S^{N_{m-1}^{0,\emptyset}-1}$, as $S^{N_{m-1}^{0,\emptyset}-1} = \bigcup_{c\in\Sigma(n)} S^{N_{m-1}^{0,\emptyset}-1} \cap B_{1/2n^2}(c)$. This gives:

$$\min_{c\in\Sigma(n)} \int_0^1 \left[\sum_{\alpha\in\mathcal{A}_{0,\emptyset}(m-1)} c_\alpha \hat{B}_t^\alpha\right]^2 dt \le 2\frac{1}{2n^2} \int_0^1 \sum_{\alpha\in\mathcal{A}_{0,\emptyset}(m-1)} \left|\hat{B}_t^\alpha\right|^2 dt$$

$$+ 2 \inf_{a\in S^{N_{m-1}^{0,\emptyset}-1}} \int_0^1 \left[\sum_{\alpha\in\mathcal{A}_{0,\emptyset}(m-1)} a_\alpha \hat{B}_t^\alpha\right]^2 dt. \tag{166}$$

Furthermore, it is evident that:

$$\mathbb{P}\left[\inf_{a\in S^{N^{0,\emptyset}_{m-1}-1}}\int_0^1\left[\sum_{\alpha\in\mathcal{A}_{0,\emptyset}(m-1)}a_\alpha\hat{B}^\alpha_t\,dt\right]^2\leq\frac{1}{n}\right]$$

$$\leq\mathbb{P}\left(\min_{c\in\Sigma(n)}\int_0^1\left[\sum_{\alpha\in\mathcal{A}_{0,\emptyset}(m-1)}a^c_\alpha\hat{B}^\alpha_t\,dt\right]^2\leq\frac{3}{n}\right)$$

$$+\mathbb{P}\left(\int_0^1\sum_{\alpha\in\mathcal{A}_{0,\emptyset}(m-1)}\left|\hat{B}^\alpha_t\right|^2dt\geq n\right).$$

Using (166) to proceed, it is seen that:

$$\mathbb{P}\left[\inf_{\xi\in S^{N^{0,\emptyset}_{m-1}-1}}\int_0^1\left[\sum_{\alpha\in\mathcal{A}_{0,\emptyset}(m-1)}a_\alpha\hat{B}^\alpha_t\,dt\right]^2\leq\frac{1}{n}\right]$$

$$\leq\mathbb{P}\left[\min_{c\in\Sigma(n)}\int_0^1\left[\sum_{\alpha\in\mathcal{A}_{0,\emptyset}(m-1)}c_\alpha\hat{B}^\alpha_t\,dt\right]^2\leq\frac{3}{n}\right]+\mathbb{P}\left[\int_0^1\sum_{\alpha\in\mathcal{A}_{0,\emptyset}(m-1)}\left|\hat{B}^\alpha_t\right|^2dt\geq n\right]$$

$$\leq\sum_{c\in\Sigma(n)}\mathbb{P}\left[\int_0^1\left[\sum_{\alpha\in\mathcal{A}_{0,\emptyset}(m-1)}c_\alpha\hat{B}^\alpha_t\,dt\right]^2\leq\frac{3}{n}\right]+\mathbb{P}\left[\sup_{t\in(0,1]}\sum_{\alpha\in\mathcal{A}_{0,\emptyset}(m-1)}\left|\hat{B}^\alpha_t\right|^2\geq n\right]$$

$$\leq M_mK^{2N^{0,\emptyset}_{m-1}}\sup_{\xi\in S^{N^{0,\emptyset}_{m-1}-1}}\mathbb{P}\left[\int_0^1\left[\sum_{\alpha\in\mathcal{A}_{0,\emptyset}(m-1)}a_\alpha\hat{B}^\alpha_t\,dt\right]^2\leq\frac{3}{n}\right]$$

$$+N^{0,\emptyset}_{m-1}\max_{\alpha\in\mathcal{A}_{0,\emptyset}(m-1)}\mathbb{P}\left[\sup_{t\in(0,1]}\left|\hat{B}^\alpha_t\right|^2\geq\frac{n}{N^{0,\emptyset}_{m-1}}\right]$$

$$\leq M_mn^{2N^{0,\emptyset}_{m-1}}B_m\exp\left(-\left[\frac{n}{3}\right]^{\mu_m}\right)+N^{0,\emptyset}_{m-1}\max_{\substack{k=0,\dots,m-1\\\|\alpha\|=k}}\mathbb{P}\left[\sup_{t\in(0,1]}\left|\hat{B}^\alpha_t\right|\geq\sqrt{\frac{n}{N^{0,\emptyset}_{m-1}}}\right]$$

$$\leq M_mn^{2N^{0,\emptyset}_{m-1}}B_m\exp\left(-\left[\frac{n}{3}\right]^{\mu_m}\right)+N^{0,\emptyset}_{m-1}\max_{k=0,\dots,m-1}C_m\exp\left(-\frac{1}{2}\left[\frac{n}{N^{0,\emptyset}_{m-1}}\right]^{\frac{\nu_m}{2}}\right)$$

$$\leq A_m\exp\left(-n^{\lambda_m}\right),$$

for some (large) constant A_m and (small) $\lambda_m>0$, for all $n\geq 1$. Both (163) and (165) have been used. □

The goal is now reasonably clear. If inequality (165) can be proved, then the claim will have been justified. Before turning to this proof in earnest, another supporting result is proved. Note that the rest of the proof is, unless otherwise stated, taken from the appendix (p. 73 and onwards) of Kusuoka and Stroock [33].

Lemma 72. *Assume $a \in S^{N^{0,\emptyset}_{m-1}-1}$ such that $|\, a_\emptyset \,| < 1$.*[13] *Then there are constants $Q_m < \infty$ and $\nu_m > 0$ such that:*

$$\mathbb{P}\left(\int_0^1 \Big[\sum_{\alpha \in \mathcal{A}_{0,\emptyset}(m-1)} a_\alpha \hat{B}^\alpha_t \Big]^2 dt \le \frac{1}{n} \right) \le Q_m \exp\left\{ -\frac{1}{2} \left(\frac{\left[|\, a_\emptyset \,| - \frac{1}{\sqrt{n}} \right] \vee 0}{\sqrt{1 - a_\emptyset^2}} \right)^{2\nu_m} \right\}. \tag{167}$$

Proof. The starting point is noting that:

$$\begin{aligned}
\left(\int_0^1 \Big[\sum_{\alpha \in \mathcal{A}_{0,\emptyset}(m-1)} a_\alpha \hat{B}^\alpha_t \Big]^2 dt \right)^{\frac{1}{2}} &\ge |\, a_\emptyset \,| - \left(\int_0^1 \Big[\sum_{1 \le \|\alpha\| \le m-1} a_\alpha \hat{B}^\alpha_t \Big]^2 dt \right)^{\frac{1}{2}} \\
&\ge |\, a_\emptyset \,| - \sqrt{1 - a_\emptyset^2} \int_0^1 \Big[\sum_{1 \le \|\alpha\| \le m-1} \left| \hat{B}^\alpha_t \right|^2 dt \Big]^{\frac{1}{2}} \\
&\ge |\, a_\emptyset \,| - \sqrt{1 - a_\emptyset^2} \sup_{t \in (0,1]} \Big[\sum_{1 \le \|\alpha\| \le m-1} \left| \hat{B}^\alpha_t \right|^2 \Big]^{\frac{1}{2}}.
\end{aligned}$$

Consequently,

$$\sup_{t \in (0,1]} \sum_{1 \le \|\alpha\| \le m-1} \left| \hat{B}^\alpha_t \right|^2 \ge \left(\frac{\left[|\, a_\emptyset \,| - \left(\int_0^1 \left[\sum_{\alpha \in \mathcal{A}_{0,\emptyset}(m-1)} a_\alpha \hat{B}^\alpha_t \right]^2 dt \right)^{\frac{1}{2}} \right] \vee 0}{\sqrt{1 - a_\emptyset^2}} \right)^2.$$

In particular,

$$\begin{aligned}
&\mathbb{P}\left(\int_0^1 \Big[\sum_{\alpha \in \mathcal{A}_{0,\emptyset}(m-1)} a_\alpha \hat{B}^\alpha_t \Big]^2 dt \le \frac{1}{n} \right) \\
&\le \mathbb{P}\left(\sup_{t \in (0,1]} \sum_{1 \le \|\alpha\| \le m-1} \left| \hat{B}^\alpha_t \right|^2 \ge \left(\frac{\left(|\, a_\emptyset \,| - \frac{1}{\sqrt{n}} \right) \vee 0}{\sqrt{1 - a_\emptyset^2}} \right)^2 \right) \\
&\le Q_m \ \exp\left\{ -\frac{1}{2} \left(\frac{\left[\left(|\, a_\emptyset \,| - \frac{1}{\sqrt{n}} \right) \vee 0 \right]}{\sqrt{1 - a_\emptyset^2}} \right)^{2\nu_m} \right\},
\end{aligned}$$

for some Q_m, ν_m, where (163) has been used. □

[13]Indeed, the consideration is trivial if this condition is violated.

A semimartingale inequality from Norris [50] is now recalled, which plays an identical role to a similar martingale inequality in Kusuoka and Stroock [33].

Lemma 73. *Assume $a, y \in \mathbb{R}$. Let $\beta = (\beta)_{t\geq 0}$ be a one-dimensional previsible process, and let $\gamma = (\gamma_t := (\gamma_t^1, \ldots, \gamma_t^d))_{t\geq 0}$, $u = (u_t := (u_t^1, \ldots, u_t^d))_{t\geq 0}$ be d-dimensional previsble processes. Moreover, assume $B = (B_t)_{t\geq 0}$ is a d-dimensional Brownian motion. Define,*

$$b_t = b + \int_0^t \beta_s ds + \int_0^t \gamma_s^i dB_s^i,$$

$$Y_t = y + \int_0^t b_s ds + \int_0^t u_s^i dB_s^i.$$

Then for any $q > 8$ and some $\nu < (q-8)/9$, there is a constant $C = C(q,\nu)$ (independent of K) such that

$$\mathbb{P}\left[\int_0^1 Y_t^2 dt < \frac{1}{n}, \int_0^1 |b_t|^2 + |u_t|^2 dt \geq \frac{1}{n^{1/q}}, \sup_{t\in(0,1]} |\beta_t| \vee |\gamma_t| \vee |b_t| \vee |u_t| \leq n\right]$$
$$\leq C \exp\{-n^\nu\}. \tag{168}$$

Remark 74. Upon checking the above result in Norris [50], the keen reader would observe that the result is stated in a different fashion. Namely, the bound

$$\sup_{t\in(0,T]} |\beta_t| \vee |\gamma_t| \vee |b_t| \vee |u_t| \leq M,$$

is assumed up to some bounded stopping time T, as an extra condition. The resulting statement is then phrased in terms of some constant, which depends on M. i.e. $C = C(q,\nu,M)$ in (168). This constraint has been circumvented by letting the constant M depend also on n (indeed: $M = n$). The observation that C is then of the form $C = \hat{C}(q,\nu)n^l$ for some $l \in \mathbb{N}$, is then made. This observation is a result of tracking the constant in the proof of the lemma. This does not affect (168) as there is some larger constant $\tilde{C}$ and smaller $\tilde{\nu}$, which can be chosen such that $\hat{C}(q,\nu)n^l \exp\{-n^\nu\} \leq \tilde{C}(q,\nu)\exp\{-n^{\tilde{\nu}}\}$, for all $n \geq 1$.

The proof of the bound in Proposition 71 is done via a strong induction argument. The base case $m - 1 = 0$ is trivial. Assume therefore, that (165) holds for $0 \leq m - 1 \leq k - 1$. Let $a \in S^{N_k^{0,\emptyset}-1}$. Define, using the notation of Lemma 73, the following:

$$Y_t := \sum_{\|\alpha\|\leq k} a_\alpha \hat{B}_t^\alpha,$$

$$
\begin{aligned}
b_t &:= \sum_{\substack{1\le\|\alpha\|\le k\\ \alpha^*=0}} a_\alpha \hat{B}_t^{\alpha'},\\
u_t^i &:= \sum_{\substack{1\le\|\alpha\|\le k\\ \alpha^*=i}} a_\alpha \hat{B}_t^{\alpha'},\\
\beta_t &:= \sum_{\substack{1\le\|\alpha\|\le k\\ \alpha^*=0,(\alpha')^*=0}} a_\alpha \hat{B}_t^{\alpha''}, \quad \text{for } |\alpha|\ge 2,\\
\gamma_t^i &:= \sum_{\substack{1\le\|\alpha\|\le k\\ \alpha^*=0,(\alpha')^*=i}} a_\alpha \hat{B}_t^{\alpha''}, \quad \text{for } |\alpha|\ge 2,\\
y &:= a_\emptyset,\\
b &:= 0.
\end{aligned}
$$

With these definitions it is easy to see

$$
\begin{aligned}
b_t &= b + \int_0^t \beta_s ds + \int_0^t \gamma_s^i dB_s^i,\\
Y_t &= y + \int_0^t a_s ds + \int_0^t u_s^i dB_s^i.
\end{aligned}
$$

Using Lemma (72) consideration may be split into two separate cases. Assume first that $1 - a_\emptyset^2 \le n^{-1/2q}$, where $q \ge 1$. So that

$$
\sqrt{1-a_\emptyset^2} \le n^{-1/4q},
$$

and

$$
\begin{aligned}
&|\, a_\emptyset \,| \ge \{(1-n^{-1/2q}) \vee 0\}^{1/2}\\
\Rightarrow &\left[\,|\, a_\emptyset \,| - \frac{1}{\sqrt{n}}\right] \vee 0 \ge (1-2n^{-1/2q}) \vee 0.
\end{aligned}
$$

Then, by (167):

$$
\begin{aligned}
\mathbb{P}\Big(\int_0^1 \Big[\sum_{\alpha\in\mathcal{A}_{0,\emptyset}(k)} a_\alpha \hat{B}_t^\alpha\Big]^2 dt \le \frac{1}{n}\Big) &\le Q_k \exp\left\{-\frac{1}{2} n^{\nu_k/2q}\Big(\big[1-\frac{2}{n^{1/2q}}\big]\vee 0\Big)^{2\nu_k}\right\}\\
&\le P_k \exp\left\{-n^{\lambda_k}\right\},
\end{aligned}
$$

for some (large) constant P_k and (small) λ_k, as required. Suppose now that $1-a_\emptyset^2 \geq 1/n^{1/2q}$. Then it is clear that

$$\left\{\int_0^1 \Big[\sum_{\|\alpha\|\leq k} a_\alpha \hat{B}_t^\alpha\Big]^2 dt \leq \frac{1}{n}\right\} \subset E_1 \cup E_2 \cup E_3,$$

where

$$E_1 = \left\{\int_0^1 Y_t^2 \leq \frac{1}{n}, \int_0^1 |\, b_t\,|^2 + |\, u_t\,|^2 \, dt \geq \frac{1}{n^{1/q}},\right.$$

$$\left.\sup_{t\in(0,1]} |\,\beta_t\,| \vee |\,\gamma_t\,| \vee |\, b_t\,| \vee |\, u_t\,| \leq n\right\},$$

$$E_2 = \left\{\sup_{t\in(0,1]} |\,\beta_t\,| \vee |\,\gamma_t\,| \vee |\, b_t\,| \vee |\, u_t\,| > n\right\},$$

$$E_3 = \left\{\int_0^1 |\, b_t\,|^2 + |\, u_t\,|^2 \, dt < \frac{1}{n^{1/q}}\right\}.$$

It is now shown that $\mathbb{P}(E_i) \leq C_i \exp\{-n^{\nu_i}\}$ for $i = 1,2,3$. For $i = 1,2$, Lemma 73 and Lemma 70 imply respectively, the required bounds (i.e. independent of $a \in S^{N_{m-1}^{0,\emptyset}-1}$). The case $i = 3$ is handled using the inductive hypothesis.

Define

$$N_j := \sum_{\substack{1\leq\|\alpha\|\leq k-\|(j)\| \\ \alpha^*=j}} a_\alpha^2$$

As $\sum_{j=0}^d N_j = 1 - a_\emptyset^2 \geq 1/n^{1/2q}$, there exists $j_0 \in \{0,\ldots,d\}$ such that $N_{j_0} \geq 1/(d+1)n^{1/2q}$. Moreover, $|\, b_t\,|^2 + |\, u_t\,|^2 \geq \Big|\sum_{\substack{1\leq\|\alpha\|\leq k-\|(j_0)\| \\ \alpha^*=j_0}} a_\alpha \hat{B}_t^{\alpha'}\Big|^2$. Thus, using the inductive hypothesis,

$$\begin{aligned}
\mathbb{P}(E_3) &\leq \mathbb{P}\left(\int_0^1 \Big|\sum_{\substack{1\leq\|\alpha\|\leq k-\|(k_0)\| \\ \alpha^*=j_0}} a_\alpha \hat{B}_t^{\alpha'}\Big|^2 dt \leq \frac{1}{n^{1/q}}\right) \\
&= \mathbb{P}\left(\frac{1}{N_{j_0}}\int_0^1 \Big|\sum_{\substack{1\leq\|\alpha\|\leq k-\|(k_0)\| \\ \alpha^*=j_0}} a_\alpha \hat{B}_t^{\alpha'}\Big|^2 dt \leq \frac{1}{N_{j_0} n^{1/q}}\right) \\
&\leq C_{k-1} \exp\{-(N_{j_0} n^{1/q})^{\nu_{k-1}}\} \\
&\leq C_{k-1} \exp\{-(n^{1/2q}/(d+1))^{\nu_{k-1}}\} \\
&\leq C_k \exp\{-n^{\nu_k}\},
\end{aligned}$$

for some C_k, ν_k. In applying the inductive hypothesis, care has been taken to check that $\left(\sum_{\substack{1\le\|\alpha\|\le k-\|(k_0)\| \\ \alpha^*=k_0}} a_\alpha^2\right)/N_{k_0} = 1$. This finishes the proof. □

We now move on the prove Lemma 23 which was fundamental to establishing relationships between the elements of our integration by parts formula. This is done in two stages: in the first stage we focus on demonstrating the result for $\mathcal{K}_r$, that is, those elements which are *smooth* processes. We then supplement this for the non-smooth case with additional comments/proofs where appropriate.

Lemma 75 (Properties of Kusuoka–Stroock Smooth Processes). *The following hold*

1. *Suppose* $f \in \mathcal{K}_r(E)$, *where* $r \ge 0$. *Then, for* $i = 1, \dots, d$,

$$\int_0^\cdot f(s,x)dB_s^i \in \mathcal{K}_{r+1}(E) \quad \text{and} \quad \int_0^\cdot f(s,x)ds \in \mathcal{K}_{r+2}(E).$$

2. $a_{\alpha,\beta}, b_{\alpha,\beta} \in \mathcal{K}_{(\|\beta\|-\|\alpha\|)\vee 0}$ *where* $\alpha, \beta \in \mathcal{A}(m)$.
3. $k_\alpha \in \mathcal{K}_{\|\alpha\|}(H)$, *where* $\alpha \in \mathcal{A}(m)$.
4. $D^{(\alpha)}u := \langle Du(t,x), k_\alpha\rangle_H \in \mathcal{K}_{r+\|\alpha\|}$ *where* $u \in \mathcal{K}_r$ *and* $\alpha \in \mathcal{A}(m)$.
5. *If* $M^{-1}(t,x)$ *is the inverse matrix of* $M(t,x)$, *then* $M^{-1}_{\alpha,\beta} \in \mathcal{K}_0$, $\alpha, \beta \in \mathcal{A}(m)$.
6. *If* $f_i \in \mathcal{K}_{r_i}$ *for* $i = 1, \dots, N$, *then*

$$\prod_{i=1}^N f_i \in \mathcal{K}_{r_1+\ldots+r_N} \quad \text{and} \quad \sum_{i=1}^N f_i \in \mathcal{K}_{\min(r_1,\ldots,r_N)}.$$

Proof. **(1)** It is clear that if $f(t,.)$ is smooth and $\partial_\alpha f(.,.)$ is continuous then the same is true of $\int_0^\cdot f(s,x)dB_s^i$ for $i = 0, \dots, d$, with

$$\partial_\alpha \int_0^\cdot f(s,x)dB_s^i = \int_0^\cdot \partial_\alpha f(s,x)dB_s^i.$$

For $k \ge 1$, $p \in [1,\infty)$, $i = 1, \dots, d$, we have (note that the dependence of the norms on the Hilbert space E has been suppressed):

$$\left\|\int_0^t \partial_\alpha f(s,x)dB_s^i\right\|_{k,p}^p = \mathbb{E}\left\|\int_0^t \partial_\alpha f(s,x)dB_s^i\right\|^p + \sum_{j=1}^k \mathbb{E}\left\|D^j\int_0^t \partial_\alpha f(s,x)dB_s^i\right\|_{H^{\otimes j}}^p. \tag{169}$$

Focussing for a moment of the LHS, and assuming w.l.o.g. $p \ge 2$ (as there holds monotonicity of norms in p), we see that for $j = 0, \dots, k$, there holds

$$
\begin{aligned}
&\mathbb{E}\left\| D^j \left[\int_0^t \partial_\alpha f(s,x) dB_s^i \right] \right\|_{H^{\otimes j}}^p \\
&= \mathbb{E}\left\| \int_0^t D^j \partial_\alpha f(s,x) dB_s^i + \int_0^t D^{j-1} \partial_\alpha f(s,x) \otimes e_i ds \right\|_{H^{\otimes j} \otimes E}^p \\
&\le 2^{p-1} \left[\mathbb{E}\left\| \int_0^t D^j \partial_\alpha f(s,x) dB_s^i \right\|_{H^{\otimes j}}^p + \mathbb{E}\left\| \int_0^t D^{j-1} \partial_\alpha f(s,x) \otimes e_i ds \right\|_{H^{\otimes j}}^p \right] \\
&\le \tilde{C}_p \left[\mathbb{E} \int_0^t t^{\frac{1}{2}(p-1)} \left\| D^j \partial_\alpha f(s,x) \right\|_{H^{\otimes j}}^p + t^{p-1} \left\| D^{j-1} \partial_\alpha f(s,x) \right\|_{H^{\otimes (j-1)}}^p ds \right] \\
&\le \tilde{C}_p t^{\frac{1}{2}(p-1)} \left[\int_0^t \mathbb{E} \left\| D^j \partial_\alpha f(s,x) \right\|_{H^{\otimes j}}^p ds + \int_0^t \mathbb{E} \left\| D^{j-1} \partial_\alpha f(s,x) \right\|_{H^{\otimes (j-1)}}^p ds \right] \\
&\le \tilde{C}_p t^{\frac{1}{2}(p-1)} \int_0^t \| \partial_\alpha f(s,x) \|_{k,p}^p ds \\
&\le \tilde{C}_p t^{\frac{1}{2}(p-1)} \int_0^t s^{rp/2} \sup_{\substack{x \in \mathbb{R}^N \\ v \in (0,1]}} v^{-rp/2} \| \partial_\alpha f(v,x) \|_{k,p}^p ds \\
&\le \tilde{\tilde{C}}_p t^{\frac{1}{2}(p[r+1])},
\end{aligned}
$$

where we have used Burkholder–Davis–Gundy inequality, Jensen's inequality and Hölder's inequality for finite sums. Note that the above holds for $j = 0$ by taking D^{j-1} to be the zero map. The upper bound is independent of $x \in \mathbb{R}^N$ and by a simple rearrangement, and combining with (169), the result follows. Note that the result for $\int_0^\cdot f(s,x)ds$ is proved similarly.

(2) The fact that $a_{\alpha,\beta}(t,.)$, $b_{\alpha,\beta}(t,.)$ are smooth with partial derivatives which are jointly continuous in $(t,x) \in (0,1] \times \mathbb{R}^N$ and that $a_{\alpha,\beta}, b_{\alpha,\beta} : [0,T] \times \mathbb{R}^N \to \mathbb{D}^\infty$ follows from Proposition 18. The fact that the appropriate bound holds for $a_{\alpha,\beta}$ with rate $r = (\|\beta\| - \|\alpha\|) \wedge 0$ follows from applying the expression for $a_{\alpha,\beta}$, given in (47), and Proposition 20. The corresponding result for $b_{\alpha,\beta}$ is derived in an analogous way to $a_{\alpha,\beta}$.

(3) This follows easily from **(1)** and **(2)**.

(4) From Nualart [51][Proposition 1.3.3] we have the following:

$$
\langle Du, k_\alpha \rangle_H = u\, \delta(k_\alpha) - \delta(u\, k_\alpha)
$$

Moreover, we know that $u, k_\alpha \in \mathbb{D}^\infty$, and that $\delta : \mathbb{D}^\infty \to \mathbb{D}^\infty$[14], hence it is clear that $\langle Du, k_\alpha \rangle_H \in \mathbb{D}^\infty$. The existence of regular derivatives of all orders follows from direct differentiation. The required bounds follows easily from 6.

(5) Our first observation is that if $f \in \mathcal{K}_r(E)$, where $r \ge 0$, then $g(t,x) := t^{-s/2} f(t,x)$ satisfies $g \in \mathcal{K}_{r-s}(E)$. This is obvious, and from this basic observation we note that $M_{\alpha,\beta}(t,x) := t^{-(\|\alpha\|+\|\beta\|)/2} \langle k_\alpha(t,x), k_\beta(t,x) \rangle_H$ must

[14]cf. Proposition 1.5.4 in Nualart [51]

satisfy $M_{\alpha,\beta} \in \mathcal{K}_0$. This comes from applying the above observation, along with (**3**) and (**4**) of this Lemma. To prove the same about elements of the inverse of $M(t,x)$ we first note that smoothness (in x) and joint continuity (in (t,x)) follows from the inverse function theorem. To prove Malliavin differentiability and the corresponding bounds, we use the ideas of the proof of Nualart [51, Lemma 2.1.6]. That is, we seek to prove the following:

Lemma 76. *Let $A(.,.)$ be a square random matrix, which is invertible almost surely and such that $| \det A(t,x) |^{-1} \in L^p$ for all $p \geq 1$. Assume further that the elements of $A_{\alpha,\beta}(t,x) \in \mathbb{D}^\infty$ and satisfy:*

$$\sup_{\substack{x \in \mathbb{R}^N \\ s \in [0,t]}} \left\| A_{\alpha,\beta}(s,x) \right\|_{k,p} < \infty.$$

Then $A^{-1}_{\alpha,\beta}(t,x) \in \mathbb{D}^\infty$ and the elements satisfy:

$$\sup_{\substack{x \in \mathbb{R}^N \\ s \in [0,t]}} \left\| A^{-1}_{\alpha,\beta}(s,x) \right\|_{k,p} < \infty. \tag{170}$$

The proof of this lemma is almost identical to the proof of Nualart [51, Lemma 2.1.6]. One merely needs to take care in showing (170). This is done easily by using a Hölder-type inequality for the seminorms $\|.\|_{k,p}$ (cf. Nualart [51, Proposition 1.5.6].

Remark 77. If we hadn't chosen to mutliply and divide the elements of the matrix $\hat{M}(t,x) := (\langle k_\alpha(t,x), k_\beta(t,x)\rangle)$ by $t^{\frac{\|\alpha\|+\|\beta\|}{2}}$, when forming the matrix M, then more care would have been required to ensure that the rate of decay of the inverse (as a Kusuoka Stroock process) is independent of the dimension of the matrix. Indeed, it can be shown the inverse of the determinant of $\hat{M}$ is bounded above by a rate which **is** dimension dependent. However, this dimensionality dependence disappears when one considers the product with the adjugate matrix, which has the equal and opposite dimensionality dependence.

(**6**) It is clear that smoothness, joint continuity and Malliavin differentiability are inherited from the constituent functions. The second property remains to be shown. Consider $\prod_{i=1}^N f_i$. It may be shown that for the kth Malliavin derivative the following Leibniz-type rule holds:

$$D^k \prod_{i=1}^N f_i = \sum_{i_1+\ldots+i_N=k} \binom{k}{i_1,\ldots,i_N} D^{i_1} f_1 \otimes \ldots \otimes D^{i_N} f_N.$$

Now noting that, if $i_1 + \ldots + i_N = k$, we have

$$\left\| D^{i_1} f_1 \otimes \ldots \otimes D^{i_N} f_N \right\|_{H^{\otimes k}} = \prod_{j=1}^N \left\| D^{i_j} f_j \right\|_{H^{\otimes i_j}},$$

so that

$$\begin{aligned}
&\left\|\prod_{i=1}^{N} f_i(t,x)\right\|_{k,p}^{p} \\
&= \mathbb{E}\left|\prod_{i=1}^{N} f_i(t,x)\right|^{p} + \sum_{j=1}^{k}\left\|D^j\prod_{i=1}^{N} f_i(t,x)\right\|_{H^{\otimes j}}^{p} \\
&= \mathbb{E}\left|\prod_{i=1}^{N} f_i(t,x)\right|^{p} + \sum_{j=1}^{k}\left\|\sum_{i_1+\ldots+i_N=j}\binom{j}{i_1,\ldots,i_N}\bigotimes_{m=1}^{N} D^{i_m} f_m(t,x)\right\|_{H^{\otimes j}}^{p} \\
&\leq \mathbb{E}\left|\prod_{i=1}^{N} f_i(t,x)\right|^{p} + \sum{}' C(p,j)\left\|\bigotimes_{m=1}^{N} D^{i_m} f_m(t,x)\right\|_{H^{\otimes j}}^{p} \\
&\leq \prod_{i=1}^{N}\|f_i(t,x)\|_{L^{p_i}(\Omega)}^{p} + \sum{}' C(p,j)\prod_{m=1}^{N}\|D^{i_m} f_m(t,x)\|_{H^{\otimes i_m}}^{p},
\end{aligned}$$

where $p^{-1} = p_1^{-1} + \ldots p_N^{-1}$, applying Hölder's Generalised Inequality. Whence, letting $r = \sum_{i=1}^{N} r(i)$ we see that

$$\begin{aligned}
\sup_{t\in(0,1],x\in\mathbb{R}^N} t^{-r/2}\left\|\prod_{i=1}^{N} f_i(t,x)\right\|_{k,p}^{p} &\leq \prod_{i=1}^{N}\sup_{\substack{t\in(0,1],\\ x\in\mathbb{R}^N}} t^{-\frac{r_i}{2}}\|f_i(t,x)\|_{L^{p_i}(\Omega)}^{p} \\
&\quad + \sum{}' C(p,j)\prod_{m=1}^{N}\sup_{\substack{t\in(0,1],\\ x\in\mathbb{R}^N}} t^{-\frac{r_i}{2}}\left\|D^{i_m} f_m(t,x)\right\|_{H^{\otimes i_m}}^{p} \\
&< \infty.
\end{aligned}$$

To see that $\sum_{i=1}^{N} f_i \in \mathcal{K}_{\min(r_1,\ldots,r_N)}$. We note that $\mathcal{K}_r \subset \mathcal{K}_s$ for $r \leq s$. Hence, it should clear that the sum is contained in that $\mathcal{K}_r$ in which **all** of its terms are contained. Namely, $\mathcal{K}_{\min(r_1,\ldots,r_N)}$. A full proof is omitted. □

We now extend the result to coincide with the stated one

Lemma 23 (Properties of Kusuoka–Stroock processes).

1. *Suppose* $f \in \mathcal{K}_r^{loc}(E,n)$, *where* $r \geq 0$. *Then, for* $i = 1,\ldots,d$,

$$\int_0^{\cdot} f(s,x)dB_s^i \in \mathcal{K}_{r+1}^{loc}(E,n) \quad \text{and} \quad \int_0^{\cdot} f(s,x)ds \in \mathcal{K}_{r+2}^{loc}(E,n).$$

2. $a_{\alpha,\beta}, b_{\alpha,\beta} \in \mathcal{K}_{(\|\beta\|-\|\alpha\|)\vee 0}^{loc}(k-m)$ *where* $\alpha,\beta \in \mathcal{A}(m)$.
3. $k_\alpha \in \mathcal{K}_{\|\alpha\|}^{loc}(H,k-m)$, *where* $\alpha \in \mathcal{A}(m)$.

4. *$D^{(\alpha)}u := \langle Du(t,x), k_\alpha\rangle_H \in \mathcal{K}^{loc}_{r+\|\alpha\|}(n \wedge [k-m])$ where $u \in \mathcal{K}^{loc}_r(n)$ and $\alpha \in \mathcal{A}(m)$.*
5. *If $M^{-1}(t,x)$ is the inverse matrix of $M(t,x)$, then $M^{-1}_{\alpha,\beta} \in \mathcal{K}^{loc}_0(k-m)$, $\alpha, \beta \in \mathcal{A}(m)$.*
6. *If $f_i \in \mathcal{K}^{loc}_{r_i}(n_i)$ for $i = 1, \ldots, N$, then*

$$\prod_{i=1}^{N} f_i \in \mathcal{K}^{loc}_{r_1+\ldots+r_N}(\min_i n_i) \quad \text{and} \quad \sum_{i=1}^{N} f_i \in \mathcal{K}^{loc}_{\min_i r_i}(\min_i n_i).$$

Moreover, if we assume the vector fields $V_0, \ldots, V_d$ are also uniformly bounded, then (2)–(5) hold with $\mathcal{K}^{loc}$ replaced by $\mathcal{K}$.

Proof. The proof of this lemma is very similar to the corresponding lemma in the second chapter. Notes are made on where the proof differs, rather than providing a full and extensive reproof, to avoid repetition.

Proof of 1. It is clear that if $f(t,.)$ n-times differentiable and $\partial_\alpha f(.,.)$ is continuous then the same is true of $\int_s^\cdot f(u,x)dB^i_u$ for $i = 0, \ldots, d$, with

$$\partial_\alpha \int_s^\cdot f(u,x)dB^i_u = \int_s^\cdot \partial_\alpha f(u,x)dB^i_u.$$

The remainder of the proof is analogous.

Proof of 2. The fact that $a_{\alpha,\beta}(t,.), b_{\alpha,\beta}(t,.)$ are k-times differentiable with partial derivatives of order $|\gamma|$, which are jointly continuous in (t,x), and which are in $\mathbb{D}^{k-|\gamma|,p}$ for all $p \geq 1$ follows from Proposition 18. The appropriate bounds can be seen to hold by observing the expression for $a_{\alpha,\beta}$ and applying Proposition 20. The corresponding result for $b_{\alpha,\beta}$ is derived in an analogous way.
Proof of 3. This follows easily from (**1**) and (**2**).
Proof of 4. From Nualart [51][Proposition 1.3.3] we have the following:

$$\langle Du, k_\alpha\rangle_H = u\,\delta(k_\alpha) - \delta(u\,k_\alpha)$$

Moreover, we know that for each $p \geq 1$, there holds $u \in \mathbb{D}^{n,p}$, $k_\alpha \in \mathbb{D}^{(k-m-1),p}$, and that $\delta : \mathbb{D}^{k,p} \to \mathbb{D}^{k-1,p}$ (see, e.g. Proposition 1.5.4 in Nualart [51]), hence it is clear that $\langle Du, k_\alpha\rangle_H \in \mathbb{D}^{n\wedge(k-m-1),q}$ for any $q \geq 1$. The existence of regular derivatives of orders less that $n \wedge (k-m-1)$ follows from direct differentiation, and the required bounds follow from 6.
Proof of 5. The k-times differentiability of the inverse (in x) and joint continuity (in (t,x)) is a result of the inverse function theorem. The Malliavin differentiability of the matrix inverse can be deduced by extending Lemma 76, for square matrices with elements of general Malliavin differentiability.
Proof of 6. It is clear and straightforward to demonstrate that the differentiability and joint continuity are inherited from the constituent functions. The level of differentiability is a result of the product rule for differentiation. The second

property of a K-S-process can be shown in a similar way, making sure to take care of the finite level of differentiability.

A.3 Convergence of the Cubature Method in the Absence of the V_0-Condition

The proof of the convergence of the cubature methods hinges on the control of the L_2-norms of the iterated integrals $I_{f_{\alpha,\varphi}}(t)$, $\alpha = (i_1, \ldots, i_r) \in \mathcal{A}$ in terms of the supremum norm of the gradient bounds of $f_{\alpha,\varphi} := V_{i_1} \ldots V_{i_r}\varphi$ and $V_i f_{\alpha,\varphi}$, $i = 1, \ldots, d$ with the function φ being replaced by $P_t\varphi$. In particular we need to be able to control $V_0 P_t \varphi$ (and higher derivatives involving $V_0 P_t \varphi$). However, under the UFG condition, gradient bounds are available only for derivatives in the directions $V_{[\alpha]}$, $\alpha \in \mathcal{A}$ which explicitly excludes V_0 (see Sect. 2.3 for the definition of $\mathcal{A}$ and Corollaries 31, 32 and respectively 38 for the corresponding bounds). We need to find a way to "hide" V_0. We succeed to do this by employing the Stratonovitch expansion not of $P_t\varphi$, but of $t \to P_{T-t}\varphi(X_t)$, $t \in [0, T]$. Assume $g \in C_b^\infty([0, T] \times \mathbb{R}^N)$. Then, by applying Itô's lemma for Stratonovich integrals, we see that

$$g(T-t, X_t^x) = g(T, x) + \sum_{i=0}^{d} \int_0^t \tilde{V}_i g(T-s, X_s^x) \circ dB_s^i, \tag{171}$$

where $\tilde{V}_i$, $i = 0, \ldots, d$ are the vector fields on $[0, T] \times \mathbb{R}^N$ defined as:

$$\tilde{V}_0 := V_0 - \partial_t, \quad \tilde{V}_i := V_i, \quad i = 1, \ldots, d.$$

Equation (171) may be iterated to obtain the following expansion for $g(T-t, X_t^x)$:

$$g(T-t, X_t^x) = \sum_{\{\alpha,\ \|\alpha\| \leq m\}} (\tilde{V}_\alpha g)(T, x) \hat{B}_t^{\circ\alpha} + R_m(t, x, g), \quad m = 2, 3, \ldots, \tag{172}$$

where $\tilde{V}_\alpha = \tilde{V}_{i_1} \ldots \tilde{V}_{i_r}$ for $\alpha = (i_1, \ldots, i_r)$ and

$$R_m(t, x, g) = \sum_{\substack{\|\alpha\| = m+1 \\ \|\alpha\| = m+2, \alpha = 0 * \beta, \|\beta\| = m}} \tilde{I}_{\tilde{V}_\alpha g}(t) \tag{173}$$

In (173), $\tilde{I}_{\tilde{V}_\alpha g}(t)$, $\alpha = (i_1, \ldots, i_r)$ is defined as

$$\tilde{I}_{\tilde{V}_\alpha g}(t) := \int_0^t \int_0^{s_0} \cdots \left(\int_0^{s_{r-2}} \tilde{V}_\alpha g(T - s_{r-1}, X_{s_{r-1}}^x) \circ dW_{s_{r-1}}^{i_1} \right) \circ \cdots \circ dW_{s_1}^{i_{r-1}} \circ dW_{s_0}^{i_r}.$$

This Taylor expansion will in due course be applied to the diffusion semigroup $P_t f$. Note, in particular, that the cubature measure of order l, where $l \geq m$, agrees with the Wiener measure on the iterated Stratonovich integrals of (172). Therefore the convergence rate will be given by their difference on the "remainder term" R_m. Indeed, it is a simple exercise to show that:

$$\sqrt{\mathbb{E}[R_m(t,x,g)^2]} \leq C \sum_{\substack{\|\alpha\|=m+1 \\ \|\alpha\|=m+2, \alpha=0*\beta, \|\beta\|=m}} t^{\|\alpha\|/2} \|\tilde{V}_\alpha g\|, \tag{174}$$

where

$$\|\tilde{V}_\alpha g\| = \sup_{s\in[0,t], x\in\mathbb{R}^N} |\tilde{V}_\alpha g(T-s,x)|.$$

The expectation $\mathbb{E}[R_m(t,x,g)^2]$ in (174) can be exchanged with the expectation with respect to the cubature measure $\mathbb{Q}_t$, that is, $\mathbb{E}_{\mathbb{Q}_t}[R_m(t,x,g)^2]$ with the result still holding. The following inequality is therefore immediate:

$$\left|\mathbb{E}\left[g(T-t,X_t^x)\right] - \mathbb{E}_{\mathbb{Q}_t}\left[g(T-t,X_t^x)\right]\right| \leq C \sum_{j=m+1}^{m+2} t^{j/2} \sup_{\|\alpha\|=j} \|\tilde{V}_\alpha g\|. \tag{175}$$

The above is an upper bound for the error of a finite measure based on a single application of the cubature formula. Iterated applications of the cubature over the partition $\mathcal{D}$ will give us the correct rate. The Markovian property of the cubature method and the semigroup property of the diffusion allow us to deduce the required uppers bounds based on (175). Again, we emphasize that the difference between what is done here and the earlier proof is that the control on $V_0 P_t \varphi$ is no longer necessary. Instead we need a control along the vector field $\tilde{V}_0 = \partial_t - V_0$ which is available as $P_t f$ is smooth along the vector fields $V_1, \ldots, V_d$ and also for each $(t,x) \in (0,T] \times \mathbb{R}^N$

$$(\partial_t - V_0) P_t f(x) = \sum_{i=1}^{d} V_i^2 P_t f(x) = \frac{1}{t}\mathbb{E}\left[f(X_t^x)\Phi_1(t,x)\right] \tag{176}$$

for a suitably chosen Kusuoka function $\Phi_1(t,x)$ (see Corollary 28). This result may be iterated to prove a corollary similar to Corollary 32.

Corollary 78. *Let $f \in \mathcal{C}_b^\infty(\mathbb{R}^N,\mathbb{R})$. If the vector fields $V_0,\ldots,V_d$ are uniformly bounded then, under the UFG condition, there exists a constant $C_\alpha < \infty$ such that:*

$$\|\tilde{V}_\alpha P_t f\|_\infty \leq \frac{C_\alpha}{t^{(\|\alpha\|-1)/2}} \|\nabla f\|_\infty. \tag{177}$$

Proof. The proof hinges on the observation that $\tilde{V}_\alpha P_t f$ satisfies the following convenient identity

$$\tilde{V}_\alpha P_t f = \sum_{i=1}^{\|\alpha\|} \sum_{\substack{\beta_1,\ldots,\beta_i \in \mathcal{A}, \\ \|\beta_1\|+\ldots+\|\beta_i\|=\|\alpha\|}} c_{\alpha,\beta_1,\ldots,\beta_i} V_{[\beta_1]} \ldots V_{[\beta_i]} P_t f, \tag{178}$$

where $c_{\alpha,\beta_1,\ldots,\beta_i} \in \mathbb{R}$. This is proved by induction over the "length" m of the multi-index α, $m = \|\alpha\|$. The case $\|\alpha\| = 1$ is trivial and $\|\alpha\| = 2$ follows from the first identity in (176). We outline next the inductive step. If $\alpha = (i_1, \ldots, i_r)$, $\|\alpha\| = m + 1$, $m > 1$ and $i_1 \neq 0$, then by the inductive hypothesis

$$\tilde{V}_\alpha P_t f = \sum_{i=1}^{\|\alpha\|-1} \sum_{\substack{\beta_1,\ldots,\beta_i \in \mathcal{A}, \\ \|\beta_1\|+\ldots+\|\beta_i\|=\|\alpha\|-1}} c_{(\alpha_2,\ldots\alpha_r),\beta_1,\ldots,\beta_i} V_{[i_1]} V_{[\beta_1]} \ldots V_{[\beta_i]} P_t f,$$

as, by definition $V_{[i_1]} = V_{i_1}$. If $i_1 = 0$, note that

$$\begin{aligned}(\partial_t - V_0) V_{[\beta_1]} \ldots V_{[\beta_i]} &= [(\partial_t - V_0), V_{[\beta_1]}] V_{[\beta_2]} \ldots V_{[\beta_i]} + V_{[\beta_1]} (\partial_t - V_0) V_{[\beta_2]} \ldots V_{[\beta_i]} \\ &= V_{[(\beta_1,0)]} V_{[\beta_2]} \ldots V_{[\beta_i]} + V_{[\beta_1]} (\partial_t - V_0) V_{[\beta_2]} \ldots V_{[\beta_i]},\end{aligned}$$

since, as ∂_t commutes with $V_{[\beta_1]}$, we have

$$[(\partial_t - V_0), V_{[\beta_1]}] = -[V_0, V_{[\beta_1]}] = V_{[(\beta_1,0)]}$$

By applying the same procedure to the second term and iterating, we obtain eventually that

$$\begin{aligned}&(\partial_t - V_0) V_{[\beta_1]} \ldots V_{[\beta_i]} P_t f \\ &\quad = V_{[(\beta_1,0)]} V_{[\beta_2]} \ldots V_{[\beta_i]} P_t f + \ldots + V_{[\beta_1]} V_{[\beta_2]} \ldots V_{[(\beta_i,0)]} P_t f \\ &\quad\quad + \sum_{j=1}^{d} V_{[\beta_1]} \ldots V_{[\beta_i]} V_{[j]} V_{[j]} P_t f(x).\end{aligned}$$

The last identity together with the induction hypothesis gives us (178) also for the case $i_1 = 0$. From (178) we deduce (177) by using Corrolary 32. □

It is important to note that derivatives along $\tilde{V}_0 := \partial_t - V_0$ add 1 to the rate as a power of t. Let Q_t be the Markov operator defined in (89) corresponding to the m-perfect family of stochastic processes, $\bar{X}(x) = \{\bar{X}_t(x)\}_{t\in[0,\infty)}$ for $x \in \mathbb{R}^d$, constructed by the cubature method as described in Example 41. The following result simply tells us that Lemma 44 holds true also in the absence of the V_0 condition.

Lemma 79. *Under the UFG condition the exists a constant $C = C_T > 0$ independent of $s, t \in [0, T]$ such that*

$$\|P_t(P_s\varphi) - Q_t(P_s\varphi)\|_\infty \le C \|\nabla\varphi\|_\infty \sum_{j=m+1}^{m+2} \frac{t^{\frac{j}{2}}}{s^{\frac{j-1}{2}}}, \tag{179}$$

where $\varphi \in C_b^1(\mathbb{R}^N, \mathbb{R})$.

Proof. Immediate from (175) and Corollary 78. □

Following Lemma 79, it is now immediate that the same rates of convergence such as those described in Sect. 3.4 are valid for the approximation given by the cubature method in the absence of the cubature measure. Let $T, \gamma > 0$ and $\pi_n = \{t_j = (\frac{j}{n})^\gamma T\}_{j=0}^n$ be a partition of the interval $[0, T]$ where $n \in \mathbb{N}$ is such that $\{h_j = t_j - t_{j-1}\}_{j=1}^n \subseteq (0, 1]$. Just as in the Sect. 3.4, let us define the function,

$$\Upsilon^1(n) = \begin{cases} n^{-\frac{1}{2}\min(\gamma,(m-1))} & \text{if } \gamma \neq m-1 \\ n^{-(m-1)/2}\ln n & \text{for } \gamma = m-1 \end{cases}$$

and let $\mathcal{E}^{\gamma,n}(\varphi)$ be the cubature error In the following,

$$\mathcal{E}^{\gamma,n}(\varphi) := \left\| P_T\varphi - Q_{h_n}^m Q_{h_{n-1}}^m \cdots Q_{h_1}^m \varphi \right\|_\infty$$

for $\gamma \in \mathbb{R}$, $n \in \mathbb{N}$. The proof of the following theorem is identical with that of Theorem 46 and Corollary 47.[15]

Theorem 80. *Under the UFG condition, there exists a constant $C = C(\gamma, T) > 0$ such that, for any $\varphi \in C_b^1(\mathbb{R}^N, \mathbb{R})$,*

$$\mathcal{E}^{\gamma,n}(\varphi) \le C\,\Upsilon^1(n)\,\|\nabla\varphi\|_\infty + \left\| P_{h_1}\varphi - Q_{h_1}^m\varphi \right\|_\infty \tag{180}$$

In particular, if $\gamma \ge m-1$ there exists a constant $C' = C'(\gamma, T) > 0$ then,

$$\mathcal{E}^{\gamma,n}(\varphi) \le \frac{C'}{n^{\frac{m-1}{2}}} \|\nabla\varphi\|_\infty .$$

Acknowledgements We should like to thank the editors of the volume, Vicky Henderson and Ronnie Sircar for their understanding and patience whilst handling our submission. One of the authors (D.C.) would like to thank Nizar Touzi for the invitation to give a series of lectures on the subject in February–April 2010 at the Institute Henri Poincaré. This work grew out of the notes of the IHP lectures. The material in the second section is based on the corresponding chapter in the PhD thesis of C. Nee [44]. The material in Sects. 3.1–3.3 is based on the paper [13].

[15]Theorem 80 can be extended to cover the rate of convergence for test functions $\varphi \in C_b^p(\mathbb{R}^N, \mathbb{R})$, in the same manner as the corresponding results in Theorem 46 and Corollary 47.

The work of D. Crisan and K. Manolarakis was partially supported by the EPSRC Grant No: EP/H0005500/1. C. Nee's work was supported by an EPSRC Doctoral Training Award.

Special thanks are due to Salvator Ortiz–Latorre who read large portions of the first draft and suggested many corrections and improvements. The authors would also like to thank the anonymous referee for the careful reading of the manuscript and the constructive remarks.

References

1. G.B. Arous, Flots et series de Taylor stochastiques. Probab. Theory Relat. Fields **81**, 29–77 (1989)
2. V. Bally, D. Talay, The law of the Euler scheme for stochastic differential equations I. Convergence rate of the distribution function. Probab. Theory Relat. Fields **104**, 43–60 (1996)
3. B. Bouchard, N. Touzi, Discrete time approximation and Monte Carlo simulation for backward stochastic differential equations. Stoch. Process. Their Appl. **111**, 175–206 (2004)
4. B. Bouchard, X. Warin, Valuation of American options—new algorithm to improve on existing methods. Working paper, 2010
5. K. Burrage, P.M. Burrage, High strong order methods for non-commutative stochastic ordinary differential equation systems and the Magnus formula. Physica D **133**(1–4), 34–48 (1999)
6. K. Burrage, P.M. Burrage, T. Tian, Numerical methods for strong solutions of stochastic differential equations: An overview. Proc. R. Soc. Lond. A **460**, 373–402 (2004)
7. F. Castell, Asymptotic expansion of stochastic flows. Probab. Theory Relat. Fields **96**, 225–239 (1993)
8. F. Castell, J. Gaines, An efficient approximation method for stochastic differential equations by means of exponential lie series. Math. Comput. Simul. **38**(1), 13–19 (1995)
9. P. Cheridito, M. Soner, N. Touzi, N. Victoir, Second-order backward stochastic differential equations and fully non linear parabolic PDEs. Commun. Pure Appl. Math. **60**(7), 1081–1110 (2007)
10. M. Clark, An efficient approximation for a class of stochastic differential equations, in *Advances in Filtering and Optimal Stochastic Control*, vol. 42. Lecture Notes in Control and Information Sciences (Springer, Berlin, 1982), pp. 69–78
11. M. Clark, Asymptotically optimal quadrature formulae for stochastic integrals, in *Proceedings of the 23rd Conference on Decision and Control*, Las Vegas, Nevada (IEEE, 1984), pp. 712–715
12. D. Crisan, F. Delarue, Sharp derivative bounds for solutions of degenerate semi-linear partial differential equations. J. Funct. Anal. **263**(10), 3024–3101 (2012)
13. D. Crisan, S. Ghazali, *On the Convergence Rate of a General Class of SDEs*, (World Scientific, New Jersey, 2007), pp. 221–248
14. D. Crisan, T. Lyons, Minimal entropy approximations and optimal algorithms. Monte Carlo Methods Appl. **8**(4), 343–355 (2002)
15. D. Crisan, K. Manolarakis, Numerical solution for a BSDE using the Cubature method. Application to non linear pricing. SIAM J. Financ. Math. **3**, 534–571 (2012)
16. N. El Karoui, C. Kapoudjan, E. Pardoux, M. Quenez, Reflected solutions of backward SDEs and related obstacle problems. Ann. Probab. **25**(2), 702–737 (1997)
17. N. El Karoui, S. Peng, M. Quenez, Backward stochastic differential equations in finance. Math. Financ. **7**(1), 1–71 (1997)
18. T. Fawcett, Problems in Stochastic Analysis: Connections between Rough Paths and Non-Commutative Harmonic Analysis. Ph.D. thesis, University of Oxford, 2003
19. S. Ghazali, The Global Error in Weak Approximations of Stochastic Differential Equations. Ph.D. thesis, Imperial College London, 2006
20. E. Gobet, C. Labart, Error expansion for the discretization of backward stochastic differential equations. Stoch. Process. Their Appl. **117**(7), 803–829 (2007)

21. E. Gobet, J.P. Lemor, X. Warin, A regression based Monte Carlo method to solve backward stochastic differential equations. Ann. Appl. Probab. **15**(3), 2172–2002 (2005)
22. L.G. Gyurkó, T.J. Lyons, Efficient and practical implementations of cubature on Wiener space, in *Stochastic Analysis 2010* (Springer, Heidelberg, 2011), pp. 73–111
23. S.L. Heston, A closed-form solution for options with stochastic volatility with applications to bond and currency options. Rev. Financ. Stud. **6**(2), 327–343 (1993)
24. L. Hörmander, Hypoelliptic second order differential equations. Acta Math. **119**, 147–171 (1967)
25. D. Jerison, Sánchez-Calle, Estimates for the heat kernel for a sum of squares of vector fields. Indiana Univ. Math. **35**(4), 835–854 (1986)
26. I. Karatzas, S. Shreve, *Brownian Motion and Stochastic Calculus* (Springer, New York, 1991)
27. P. Kloeden, E. Platen, *Numerical Solutions of Stochastic Differential Equations* (Springer, Berlin, 1999)
28. H. Kunita, Stochastic differential equations and stochastic flows of diffeomorphisms. Lect. Notes Math. **1097**, 143–303 (1984)
29. S. Kusuoka, Approximation of expectations of diffusion processes based on Lie algebra and Malliavin calculus. Adv. Math. Econ. (Springer, Tokyo) **6**, 69–83 (2004)
30. S. Kusuoka, Malliavin Calculus revisited. J. Math. Sci. Univ. Tokyo **10**, 261–277 (2003)
31. S. Kusuoka, S. Ninomiya, A new simulation method of diffusion processes applied to finance, in *Stochastic Processes and Applications to Mathematical Finance* (World Scientific, River Edge, 2004), pp. 233–253
32. S. Kusuoka, D. Stroock, Applications of the Malliavin calculus. I. Stochastic analysis (Katata/Kyoto, 1982). in *North-Holland Math. Library*, vol. 32 (North-Holland, Amsterdam, 1984), pp. 271–306
33. S. Kusuoka, D, Stroock, Applications of the Malliavin calculus - II. J. Faculty Sci. Univ. Tokyo **1**, 1–76 (1985)
34. S. Kusuoka, D. Stroock, Applications of the Malliavin calculus III. J. Faculty Sci. Univ. Tokyo **34**(1A), 391–422 (1987)
35. C. Litterer, T. Lyons, High order recombination and an application to cubature on Wiener space. Ann. Appl. Probab. **22**(4), 1301–1327 (2012)
36. T. Lyons, N. Victoir, Cubature on Wiener space. Proc. R. Soc. Lond. **468**, 169–198 (2004)
37. J. Ma, J. Zhang, Representation theorems for backward stochastic differential equations. Ann. Appl. Probab. **12**(4), 1390–1418 (2002)
38. J. Ma, P. Protter, J. Yong, Solving forward–backward SDEs expicitly—a four step scheme. Probab. Theory Relat. fields **122**(2), 163–190 (1994)
39. W. Magnus, On the exponential solution of differential equations for a linear operator. Commun. Pure Appl. Math. **7**, 649–673 (1954)
40. P. Malliavin, Stochastic calculus of variations and hypoelliptic operators, in *Proceedings of the International Symposium on Stochastic Differential Equations*, Kyoto, 1976
41. A. Millet, M. Sanz-Sole, A simple proof of the support theorem for diffusion processes, Séminaire de probabilitiés, XXVIII **1583**, 36–48 (1994)
42. G. Milstein, A method of second-order accuracy integration of stochastic differential equations. Theory Probab. Appl. **23**, 396–401 (1976)
43. G. Milstein, Weak approximation of solutions of systems of stochastic differential equations. Theory Probab. Appl. **30**, 750–766 (1985)
44. C. Nee, Sharp Gradient Bounds for the Diffusion Semigroup. Ph.D. thesis, Imperial College London, 2011
45. N. Newton, An asymptotically efficient difference formula for solving stochastic differential equations. Stochastics **19**(3), 175–206 (1986)
46. N. Newton, An efficient approximation for stochastic differential equations on the partition of symmetrical first passage times. Stoch. Stoch. Rep. **29**(2), 227–258 (1990)
47. S. Ninomiya, A partial sampling method applied to the Kusuoka approximation. Mt. Carlo Methods Appl. **9**(1), 27–38, (2003)

48. M. Ninomiya, S. Ninomiya, A new higher-order weak approximation scheme for stochastic differential equations and the Runge-Kutta method. Financ. Stoch. **13**(3), 415–443 (2009)
49. S. Ninomyia, N. Victoir, Weak approximation scheme of stochastic differential equations and applications to derivatives pricing. Appl. Math. Financ. **15**(2), 107–121 (2008)
50. J.R. Norris, Simplified Malliavin calculus. Séminaire de probabilités **20**, 101–130 (1986)
51. D. Nualart, in *The Malliavin Calculus and Related Topics*, 2nd edn. Probability and Its Applications, New York (Springer, Berlin, 2006)
52. E. Platen, W. Wagner, On a Taylor formula for a class of Itô processes. Probab. Math. Stat. **3**, 37–51 (1982)
53. M. Soner, N. Touzi, Well posedness of second order backward SDEs. Stoch. Process. Their Appli. **121**, 265–287 (2011)
54. D.W. Stroock, S.R.S. Varadhan, Diffusion processes with continuous coefficients. Commun. Pure Appl. Math **22**, 345–400 (1969)
55. D.W. Stroock, S.R.S. Varadhan, On the support of diffusion processes with applications to the strong maximum principle, in *Proceedings of the Sixth Berkeley Symposium on Mathematical Statistics and Probability*, vol. 3. (1970), pp. 333–359, Probability Theory
56. D.W. Stroock, S.R.S. Varadhan, On degenerate elliptic-parabolic operators of second order and their associated diffusions. Commun. Pure. Appl. Math. **25**, 651–713 (1972)
57. D. Talay, Efficient numerical schemes for the approximation of expectations of functionals of the solution of an SDE and applications. Filter. Control Random Process. **61**, 294–313 (1984)
58. D. Talay, Discrétisation d'une e.d.s. et calcul approché d'ésperances de fontionelles de la solution,. Math. Model. Numer. Anal. **20**, 141–179 (1986)
59. D. Talay, L. Tubaro, Expansion of the global error for numerical schemes solving stochastic differential equations. Stoch. Anal. Appl. **8**(4), 483–509 (1990)
60. J. Zhang, A numerical scheme for BSDEs. Ann. Appl. Probab. **14**(1), 459–488 (2004)

LECTURE NOTES IN MATHEMATICS

Springer

Edited by J.-M. Morel, B. Teissier; P.K. Maini

Editorial Policy (for Multi-Author Publications: Summer Schools / Intensive Courses)

1. Lecture Notes aim to report new developments in all areas of mathematics and their applications - quickly, informally and at a high level. Mathematical texts analysing new developments in modelling and numerical simulation are welcome. Manuscripts should be reasonably selfcontained and rounded off. Thus they may, and often will, present not only results of the author but also related work by other people. They should provide sufficient motivation, examples and applications. There should also be an introduction making the text comprehensible to a wider audience. This clearly distinguishes Lecture Notes from journal articles or technical reports which normally are very concise. Articles intended for a journal but too long to be accepted by most journals, usually do not have this "lecture notes" character.

2. In general SUMMER SCHOOLS and other similar INTENSIVE COURSES are held to present mathematical topics that are close to the frontiers of recent research to an audience at the beginning or intermediate graduate level, who may want to continue with this area of work, for a thesis or later. This makes demands on the didactic aspects of the presentation. Because the subjects of such schools are advanced, there often exists no textbook, and so ideally, the publication resulting from such a school could be a first approximation to such a textbook. Usually several authors are involved in the writing, so it is not always simple to obtain a unified approach to the presentation.
   For prospective publication in LNM, the resulting manuscript should not be just a collection of course notes, each of which has been developed by an individual author with little or no coordination with the others, and with little or no common concept. The subject matter should dictate the structure of the book, and the authorship of each part or chapter should take secondary importance. Of course the choice of authors is crucial to the quality of the material at the school and in the book, and the intention here is not to belittle their impact, but simply to say that the book should be planned to be written by these authors jointly, and not just assembled as a result of what these authors happen to submit.
   This represents considerable preparatory work (as it is imperative to ensure that the authors know these criteria before they invest work on a manuscript), and also considerable editing work afterwards, to get the book into final shape. Still it is the form that holds the most promise of a successful book that will be used by its intended audience, rather than yet another volume of proceedings for the library shelf.

3. Manuscripts should be submitted either online at www.editorialmanager.com/lnm/ to Springer's mathematics editorial, or to one of the series editors. Volume editors are expected to arrange for the refereeing, to the usual scientific standards, of the individual contributions. If the resulting reports can be forwarded to us (series editors or Springer) this is very helpful. If no reports are forwarded or if other questions remain unclear in respect of homogeneity etc, the series editors may wish to consult external referees for an overall evaluation of the volume. A final decision to publish can be made only on the basis of the complete manuscript; however a preliminary decision can be based on a pre-final or incomplete manuscript. The strict minimum amount of material that will be considered should include a detailed outline describing the planned contents of each chapter.
   Volume editors and authors should be aware that incomplete or insufficiently close to final manuscripts almost always result in longer evaluation times. They should also be aware that parallel submission of their manuscript to another publisher while under consideration for LNM will in general lead to immediate rejection.

4. Manuscripts should in general be submitted in English. Final manuscripts should contain at least 100 pages of mathematical text and should always include
   - a general table of contents;
   - an informative introduction, with adequate motivation and perhaps some historical remarks: it should be accessible to a reader not intimately familiar with the topic treated;
   - a global subject index: as a rule this is genuinely helpful for the reader.

   Lecture Notes volumes are, as a rule, printed digitally from the authors' files. We strongly recommend that all contributions in a volume be written in the same LaTeX version, preferably LaTeX2e. To ensure best results, authors are asked to use the LaTeX2e style files available from Springer's web-server at
   ftp://ftp.springer.de/pub/tex/latex/svmonot1/ (for monographs) and
   ftp://ftp.springer.de/pub/tex/latex/svmultt1/ (for summer schools/tutorials).
   Additional technical instructions, if necessary, are available on request from:
   lnm@springer.com.

5. Careful preparation of the manuscripts will help keep production time short besides ensuring satisfactory appearance of the finished book in print and online. After acceptance of the manuscript authors will be asked to prepare the final LaTeX source files and also the corresponding dvi-, pdf- or zipped ps-file. The LaTeX source files are essential for producing the full-text online version of the book. For the existing online volumes of LNM see:
   http://www.springerlink.com/openurl.asp?genre=journal&issn=0075-8434.
   The actual production of a Lecture Notes volume takes approximately 12 weeks.

6. Volume editors receive a total of 50 free copies of their volume to be shared with the authors, but no royalties. They and the authors are entitled to a discount of 33.3 % on the price of Springer books purchased for their personal use, if ordering directly from Springer.

7. Commitment to publish is made by letter of intent rather than by signing a formal contract. Springer-Verlag secures the copyright for each volume. Authors are free to reuse material contained in their LNM volumes in later publications: a brief written (or e-mail) request for formal permission is sufficient.

Addresses:

Professor J.-M. Morel, CMLA,
École Normale Supérieure de Cachan,
61 Avenue du Président Wilson, 94235 Cachan Cedex, France
E-mail: morel@cmla.ens-cachan.fr

Professor B. Teissier, Institut Mathématique de Jussieu,
UMR 7586 du CNRS, Équipe "Géométrie et Dynamique",
175 rue du Chevaleret,
75013 Paris, France
E-mail: teissier@math.jussieu.fr

For the "Mathematical Biosciences Subseries" of LNM:

Professor P. K. Maini, Center for Mathematical Biology,
Mathematical Institute, 24-29 St Giles,
Oxford OX1 3LP, UK
E-mail : maini@maths.ox.ac.uk

Springer, Mathematics Editorial I,
Tiergartenstr. 17,
69121 Heidelberg, Germany,
Tel.: +49 (6221) 4876-8259
Fax: +49 (6221) 4876-8259
E-mail: lnm@springer.com